SOCIAL LEARNING IN ANIMALS:
THE ROOTS OF CULTURE

SOCIAL LEARNING
IN ANIMALS:
THE ROOTS
OF CULTURE

Edited by

Cecilia M. Heyes

Department of Psychology
University College London
London, United Kingdom

Bennett G. Galef, Jr.

Department of Psychology
McMaster University
Hamilton, Ontario, Canada

Academic Press
San Diego   New York   Boston   London
Sydney   Tokyo   Toronto

Front cover photograph: Pygmy chimps, © Zoological Society of San Diego.

This book is printed on acid-free paper. ∞

Academic Press, Inc.
A Division of Harcourt Brace & Company
525 B Street, Suite 1900, San Diego, California 92101-4495

United Kingdom Edition published by
Academic Press Limited
24-28 Oval Road, London NW1 7DX

Library of Congress Cataloging-in-Publication Data

Social learning in animals : the roots of culture / edited by Cecilia
    M. Heyes, Bennett G. Galef, Jr.
            p.   cm.
    Includes bibliographical references and index.
    ISBN 0-12-273965-5 (alk. paper)
    1. Social behavior in animals. 2. Learning in animals.
I. Heyes, Cecilia M.   II. Galef, Bennett G.
QL775.S645   1996
591.51--dc20   95-52124
                                               CIP

PRINTED IN THE UNITED STATES OF AMERICA
96   97   98   99   00   01   BC   9   8   7   6   5   4   3   2   1

Contents

CHAPTER 2

Cultural Transmission of Feeding Behavior in the Black Rat (Rattus rattus)   17

JOSEPH TERKEL

CHAPTER 3

Social Enhancement of Food Preferences in Norway Rats: A Brief Review   49

BENNETT G. GALEF, JR.

CHAPTER 4

Social Learning in Monkeys: Primate "Primacy" Reconsidered 65

DOROTHY M. FRAGASZY AND ELISABETTA VISALBERGHI

CHAPTER 5

Copying and Mate Choice 85

LEE A. DUGATKIN

PART 2

Imitation

CHAPTER 10

CHAPTER 11

CHAPTER 14

Studies of Imitation in Chimpanzees and Children   291
ANDREW WHITEN AND DEBORAH CUSTANCE

CHAPTER 15

CHAPTER 16

Contributors

Numbers in parentheses indicate the pages on which the authors' contributions begin.

ROBERT BOYD (129) Department of Anthropology, University of California, Los Angeles, California 90024.

DEBORAH CUSTANCE (291) Scottish Primate Research Group, School of Psychology, University of St. Andrews, Fife KY16 9JU, Scotland.

LEE A. DUGATKIN (85) Department of Biology, University of Louisville, Louisville, Kentucky 40292.

DOROTHY FRAGASZY (65) Department of Psychology, University of Georgia, Athens, Georgia 30602.

BENNETT G. GALEF, JR. (3, 49) Department of Psychology, McMaster University, Hamilton, Ontario, Canada L8S 4K1.

LUC-ALAIN GIRALDEAU (107) Department of Biology, Concordia University, Montréal, Québec, Canada H3A 1M8.

CECILIA M. HEYES (211, 371) Department of Psychology, University College London, London WC1E 6BT, United Kingdom.

MICHAEL A. HUFFMAN (267) Department of Zoology, Kyoto University, Kyoto, Japan.

ANDREW P. KING (155) Department of Psychology, Indiana University, Bloomington, Indiana 47408.

KEVIN N. LALAND (129) Sub-Department of Animal Behaviour, University of Cambridge, Madingley, Cambridge CB3 8AA, United Kingdom.

Louis Lefebvre (107) Department of Biology, McGill University, Montréal, Québec, Canada H3G 1B1.

Andrew N. Meltzoff (347) Department of Psychology, University of Washington, Seattle, Washington 98195-7920.

Bruce R. Moore (245) Department of Psychology, Dalhousie University, Halifax, Nova Scotia, Canada B3H 4J1.

Robert R. Provine (179) Department of Psychology, University of Maryland Baltimore County, Baltimore, Maryland 21228.

Peter J. Richerson (129) Division of Environment Studies, Center for Population Biology, University of California, Davis, California 95616.

Joseph Terkel (17) Department of Zoology, Tel Aviv University, Tel Aviv 69978 Israel.

Michael Tomasello (319) Department of Psychology and Yerkes Primate Center, Emory University, Atlanta, Georgia 30322.

Elisabetta Visalberghi (65) Istituto di Psicologia, Consiglio Nazionale delle Ricerche, Rome, Italy.

Meredith J. West (155) Department of Psychology, Indiana University, Bloomington, Indiana 47408.

Andrew Whiten (291) Scottish Primate Research Group, School of Psychology, University of St. Andrews, Fife KY16 9JU, Scotland.

Thomas R. Zentall (221) Department of Psychology, University of Kentucky, Lexington, Kentucky 40506.

Preface

In August 1994, a group of 46 behavioral scientists from nine countries met at Madingley Hall, a few miles west of Cambridge, England, to discuss recent progress in studies of social learning and imitation. The workshop, generously supported by the Human Frontier Science Program (Grant WS-98/93 to C. Heyes), was organized by Cecilia Heyes (University College London) and Bennett Galef (McMaster University) in response to the rapidly growing interest of both experimental psychologists and behavioral biologists in studies of imitation and other forms of social influence affecting the acquisition and expression of behavior.

The present book is very different from the Madingley workshop. Only a few of the working participants have contributed chapters to this book. Also, several eminent researchers who did not attend were invited to participate.

Our reason for organizing a multiauthored text in animal social learning is straightforward. During the past 10 years, there has been an impressive increase in the number of researchers around the world who either have become interested in the effects of social interactions on learning or have discovered the potential contribution of studies of imitation learning to our understanding of the cognitive world of nonlinguistic animals. Because of the simultaneous development of the field from several foci, both geographical and intellectual, the study of social learning has become particularly scattered. Some researchers, particularly primatologists working in the tradition of developmental or comparative psychology, have focused on the question of whether animals are able to imitate. Others, usually biologists with a background in behavioral ecology, have been more concerned with the impact of social learning on the development of adaptive behavioral repertoires by animals living in natural settings. Communication among researchers working in different intellectual traditions, with different species and in different countries, has not kept pace with developments in the area.

This book is intended to facilitate communication by providing researchers from different parts of the globe, working in different intellectual traditions and studying species that range from guppies to human adults, with the opportunity to examine views about the current state of knowledge and appropriate directions for future research in the area. These chapters provide a snapshot of the state of the field for the extensive community of experimental, developmental, cognitive, and comparative psychologists, primatologists, ornithologists, behavioral ecologists, and anthropologists.

The book reflects the breadth of approaches that is currently being used in studies of social learning and imitation. It provides an introduction to the area for the neophyte as well as a starting point for the more knowledgeable seeking to update their information about the field. The book's various chapters provide both succinct reviews of recent findings and statements of diverse theoretical perspectives on controversial issues.

Those interested in developments in the area during the last decade would, perhaps, benefit from comparing the table of contents of the present volume with that of an earlier edited work on the same general topic (Zentall & Galef, 1988). Some subjects, such as the study of "true" imitation learning, be it in rats, parrots, or primates, have flourished during the past decade. Others, for example, research on social learning of arbitrary instrumental responses, are less salient. Still others, such as studies of the role of social learning in sexual selection, have only recently appeared on the scene. Such innovation, together with the substantial progress in understanding that has occurred during the last decade, makes the present volume particularly timely.

Those of us active in the study of social learning and imitation are truly excited about the progress that has been made in the last few years. We hope that the present collection provides readers with an understanding of the causes of that excitement, as well as a feeling for all that remains to be done to explore fully the role of social interaction in the acquisition of behavior and the implications of an ability to imitate for our understanding of animal cognition.

We thank all contributors to the Madingley workshop for making it a lively and productive meeting; Anneliese Bohn and Barry Everitt of the Human Frontier Science Program for their assistance in planning the program; and Mark Gardner, Kevin Laland, Chris Mitchell, Liz Ray, and Kate Bradford for their help in ensuring that the workshop ran smoothly.

The editors decided the order in which their names would appear on the title page of this volume through a wager. The number of thousands of shares traded on the New York Stock Exchange on November 18, 1994 (as printed in *The Times* of London on November 19) was even.

CECILIA M. HEYES AND BENNETT G. GALEF, JR.

PART 1

Social Learning

1

Introduction

Bennett G. Galef, Jr.

Department of Psychology
McMaster University
Hamilton, Ontario
Canada L8S 4K1

SOCIAL LEARNING AND IMITATION

Those who study social learning in animals do so because of an interest in one of two quite different issues. Psychologists working in the area usually want to know whether nonhuman animals can imitate behaviors they observe. It is assumed that any situation in which animals "from an act witnessed learn to do an act" (Thorndike, 1911) provide an opportunity both to investigate the cognitive abilities of animals and to compare the mental processes of humans with those of other animals.

Researchers whose work on social learning is part of a broader interest in animal behavior or behavioral ecology more often study social learning to understand the role of social interactions in the development of patterns of behavior that enhance the fitness of free-living animals. Consequently, those with a background in biology tend to be interested in the role of social learning in the lives of animals regardless of whether a particular instance of social learning results from imitation or from some presumably less sophisticated social-learning process (Galef, 1988b).

This divergence in approach to the study of social learning in animals, reflected in the organization of this book, is historical as well as contemporary. There are two, distinct, century-old traditions in the study of animal social learning: the first focused on the implications of animal imitation for understanding the relationship between the cognitive capacities of humans and other animals, the second concerned with ways in which social learning might contribute to survival and reproductive success in natural circumstances.

Like so much else in the history of the life sciences, both traditions have their roots in the work of Charles Darwin.

Imitation

The first systematic studies of imitation in animals were motivated by a controversy between Charles Darwin and Alfred Russell Wallace as to whether human mind had evolved from animal mind by purely natural processes. Darwin believed that the mind of man had, as his protégé George Romanes (1884) stated, "slowly evolved from lower types of psychical existence." Wallace, on the other hand, argued quite logically from a set of evolutionary postulates different from those espoused by Darwin, that natural selection could not, in principle, have produced human mind.

Darwin believed that sexual selection and Lamarkian evolution (pangenesis) acted together with natural selection to produce adaptation, while Wallace believed that natural selection was the only natural process involved in evolution. Wallace also believed that natural selection could produce only those features of animals necessary for their survival, while Darwin saw natural selection as based on competition between naturally occurring variants and, therefore, capable of generating adaptations that increased fitness beyond the mere minimum necessary for survival.

Wallace had spent many years traveling in South America and Malaysia. His observations of the way of life in "primitive" societies led him to conclude that the mental abilities and moral sensibilities of humans living in a state of nature far exceeded their requirements for survival (Wallace, 1870, 1886). Wallace argued from first principles that such unnecessary elaboration could not be a product of natural selection. And, because in Wallace's view natural selection was the sole secular process active in evolution, if natural selection could not explain the intelligence and morality to be observed in traditional human societies, then these characteristics of primitive man must be attributed to some supernatural cause.

In the latter part of the 19th century the disagreement between Darwin and Wallace about the origin of human mind and morals was an important philosophical and psychological issue. It motivated George Romanes, Darwin's immediate heir in behavioral matters, to search (with Darwin's active assistance) for evidence of mental and emotional faculties that were shared by animals and humans. Romanes argued that if humans and animals could be shown to share faculties of mind, then the probability of separate origins of human and animal intelligence would be greatly reduced.

The Victorians considered humans (at least adult, European, male humans) to be rational beings capable of self-conscious thought; they viewed animals as creatures whose behavior was guided almost entirely by mindless instinct. Demonstration of learning by imitation in both humans and animals was seen as providing evidence of a continuity of mind that would not be predicted on Wallace's theory of a supernatural origin of human intelligence. Indeed, imitation was a particularly important test case for Romanes's and Darwin's position because imitation was thought by Darwin and his contemporaries to be a faculty of mind that reached its highest level of expression not in rational, adult male Europeans, but in infants, savages, women, and the feeble minded (Darwin, 1871; Romanes, 1884). If human mind had evolved from animal mind, then those who viewed phylogeny as one Great Chain of Being, as did most of Darwin's contemporaries (Galef, 1988a), would expect humans of lower type to exhibit continuity in mental facilities with species at the higher end of the animal continuum.

The search for evidence of true imitation learning in animals, still underway more than 100 years after Darwin instigated it, has proved surprisingly long and hard. Indeed, reasonably convincing experimental demonstrations of true learning by imitation in animals (described in the second section of this volume) have become available only during the present decade.

Social Learning

The roots of animal behaviorists' work on social learning can be traced to a second of Darwin's interests, quite distinct from his desire to secure evidence of a continuity of animal and human mind.

Darwin spent a summer morning, sometime during the 1870s, watching bumble bees feed by cutting small holes at the base of flowers of the kidney bean plant, and then sucking out nectar through the holes they had made. The following day, Darwin observed that large numbers of honey bees were feeding from the holes cut by the bumble bees. Darwin wrote in his journal:

> I think that the hive bees [honey bees] either saw the humble bees [bumble bees] cutting the holes and understood what they were doing and immediately profited by their labour; or that they merely imitated the humble-bees after they cut the holes and when sucking at them. Yet I feel sure that if anyone who had not known the previous history had seen every single hive-bee, without a moment's hesitation flying with the utmost celerity and precision from the underside of one flower to another, and then

rapidly sucking the nectar, he would have declared that it was a beautiful case of instinct.

(Romanes, 1884)

In this brief paragraph, Darwin anticipates the second major approach of 20th-century scientists to the study of social learning. Here, Darwin speaks of a possible need to invoke social learning of some kind to explain propagation of an adaptive pattern of behavior through a population of free-living animals.[1] Although Darwin and others of his generation did not pursue studies of the role of social learning in the acquisition of adaptive behavior with nearly the energy with which they undertook investigations of imitation as a source of insight into the relationship of animal and human mind, Darwin was clearly aware of the possibility that social learning could play a role in the spread of adaptive instrumental behaviors through natural populations.

Those working today in this second Darwinian approach to the study of social learning, like Darwin in the passage quoted above, often cannot know how behavior is transmitted from one individual to another when social learning is first observed in its natural setting. However, unlike Darwin and his contemporaries, 20th-century scientists have undertaken extensive observations and experiments under controlled conditions to determine precisely how social interactions might contribute to the spread or maintenance of behaviors that play an important role in the survival and reproduction of free-living animals.

Perhaps Darwin was wise to leave unexplored the mechanisms supporting the social learning that he observed in his hive bees, because, as so clearly demonstrated in the remaining chapters of the first half of this volume, the task of discovering just how the behavior of one animal influences that of its fellows has usually proven unexpectedly long and complex.

---

1. In the interest of historical accuracy, it should be noted that, as suggested in the preceding quotation, Darwin and his contemporaries most often used the term imitation to refer to blind, mindless copying. Hence, Darwin's use of the phrase "merely imitated" and similar usages elsewhere in the work of Victorian naturalists (Romanes, 1882; Wallace, 1870).

Since the end of the 19th century (Thorndike, 1898), most students of animal behavior have used the term imitation to refer to a cognitively complex form of social learning requiring some form of cross-modal matching that, in turn, is believed to rest on relatively sophisticated cognitive processes. Consequently, in contemporary terminology, Darwin's "merely imitating" honey bees, would be described as engaging in some nonimitative form of social learning, perhaps local enhancement (Thorpe, 1956).

Synthesis

One might argue that, given the marked differences in the objectives of studies of social learning and studies of imitation, there is little reason to treat the two areas of research as part of a single research enterprise. There is, however, still extensive fruitful interchange (as well as some less healthy bickering) among those segments of the research community studying imitation and those studying social learning processes more generally.

Evidence that animals may be able to imitate provides a constant reminder to those analyzing instances of social learning in behaviorist terms, that behaviorist analyses may not invariably suffice to explain the transmission of behavior from one individual to another. It is always possible that animals acquire novel motor skills by observing and then imitating the behavior of others.

Conversely, by describing relatively simple processes that can support transmission of behavior from one individual to another, those studying mechanisms of social learning constrain the range of phenomena that are accepted as truly imitative by students of imitation. Such constraint helps those interested in imitation to avoid focusing their attention on instances of behavioral transmission that rest on behavioral mechanisms less sophisticated than those that imitation is believed to require. What are these alternative behavioral social learning mechanisms that might be confused by the unwary with true imitation learning? Well that is a long story.

In recent years, there have been at least four major attempts to bring some order to the chaotic historical nomenclature of nonimitative social learning (Galef, 1988b; Heyes, 1994; Mitchell, 1989; Whiten & Ham, 1992). Although each is useful to the beginner seeking entree into the now-voluminous literature on animal social learning, none has proven of much use to researchers when analyzing the behavioral mechanisms that support a newly discovered, socially learned behavior. It is, therefore, not surprising to find that the authors of the eight chapters in the first section of the book (whose work is a fair sample of approaches in use by those interested in the contribution of social learning in the development of adaptive behavioral repertoires in animals) ignore almost totally the elaborate taxonomies developed during the last decade. This failure to attend to attempts at categorizing nonimitative social learning processes is probably a blessing to the field in that it keeps researchers focused, as they should be, on analyses of proximal causes of instances of social learning and avoids futile controversies over less interesting taxonomic issues.

STUDIES OF SOCIAL LEARNING

The core of the field of social learning lies, of course, not in endless (and possibly pointless) controversies over how the various instances of social learning in animals should be categorized, but in analyses of the ways in which acquisition of behavior by one animal can be influenced by social interaction with others of its species. The last two decades have seen a rapid proliferation of such analyses contributing immeasurably to our understanding of both the causes and functions of social learning in animals. The eight chapters that constitute the remainder of the first section of this book provide a representative sample of this recent work, indicating how much has been accomplished in a relatively brief span of years.

Roof Rats and Pine Cones

The first substantive chapter provides a summary of the work of Joseph Terkel and his associates at the University of Tel Aviv who have been studying the transmission and maintenance of a unique pattern of feeding behavior observed in populations of black rats (*Rattus rattus*) that inhabit the pine forests of Israel. These rats are unique in that they survive on a diet consisting entirely of seeds extracted from pine cones. The rats of the pine forests of Israel are of interest to students of animal social learning because only young rats raised by dams that strip seeds from cones learn to strip seeds in a fashion allowing more energy to be acquired from pine seeds than is expended in extracting them.

Terkel's work is as close an approximation as we have of the modern extension of Darwin's interest in social learning processes. Like Darwin, Terkel begins with the observation of an unusual pattern of behavior that has spread through a free-living population of animals. However, unlike Darwin, Terkel is not satisfied with speculation as to the processes supporting propagation of the socially influenced behavior he observed. Instead, he undertakes a truly elegant series of experiments elucidating the necessary conditions for occurrence of that social learning.

Terkel's work is of particular interest because it is the sole instance in which social learning has been demonstrated to permit members of a species to occupy an ecological niche which would otherwise be closed to them. Of course, when animals enter a new niche, they are exposed to new selective pressures. Consequently, Terkel's work hints at the exciting possibility that social learning may play a direct role in niche creation and, consequently, in evolutionary process itself.

Social Induction of Food Preference

The second chapter describes some of the quarter century of research that Galef and his collaborators have undertaken to understand the role of social learning in the development of adaptive patterns of food selection by Norway rats. When living in natural circumstances, Norway rats, like other dietary generalists, are faced with a wealth of ingestible items: some of nutritional value, some worthless, and some actually harmful to eat. A weanling rat learning to construct a nutritionally adequate diet while faced with the steady depletion and impending exhaustion of its internal reserves, faces a formidable challenge.

Galef and his associates have described redundant social learning processes, each acting to facilitate naive young rats' identification and selective ingestion of valuable foods. Galef begins, as did Terkel, with field observations of behaviors that appear to be socially transmitted. He then brings those behaviors into the laboratory, analyzes the social learning processes involved in their propagation through a population, and then attempts to understand how the behavioral mechanisms he uncovers operate in natural circumstances.

In the case described here, rats are shown to extract olfactory information from conspecifics that have recently eaten a food. This socially acquired information allows its extractors to identify the foods that other rats have eaten. It also causes the recipients of the information to exhibit substantially enhanced preferences for those foods they discover that other rats have eaten. The results of Galef's laboratory studies suggest that rats in nature can use socially obtained information not only to select conspecifics to follow to food and foods to eat, but also to aid in identification of foods to avoid eating.

Social Learning about Food by Capuchin Monkeys

The work of Fragaszy and Visalberghi, focusing on the transfer of patterns of food selection from adult to young capuchin monkeys (*Cebus apella*), develops further the theme of social influences on food choice. The results of Fragaszy and Visalberghi's studies reveal that the feeding behavior of capuchins, like that of the rodents that Galef and Terkel have studied, is modified in important ways by simple social interactions. Social influence depends on tolerance of adults for juveniles, food scrounging by juveniles, and cofeeding of juveniles with adults, not on active interference by adults in the food choices of young. Consequently, Fragaszy and Visalberghi see no need to use higher-order "cognitive" processes to explain social influences on food choice either in the capuchins that they have so extensively examined, or in the free-living savanna baboons (*Papio cynocephalus*) Barbara King has studied in Kenya and whose social learning about foods seems very similar to that seen in capuchins.

In general, Fragaszy and Visalberghi reject the tendency they see in the published literature to overinterpret and anthropomorphize the behavior of primates. And they make the important and surely valid point that, like interspecific differences in ecology, interspecific differences in psychology, in the social dynamics or emotional expressiveness of individual species, may have contributed to the evolution of species' differences in reliance on social learning.

Mate Choice by Guppies

Lee Dugatkin's chapter on the role of social learning in mate selection by guppies extends the study of animal social learning to incorporate another of Darwin's interests, sexual selection. Although it has been known since Darwin first introduced the concept of sexual selection into evolutionary biology that females select males with whom to mate on the basis of males' physical characteristics, it is only quite recently that we have empirical evidence of a role of social interaction in female mate choice.

Dugatkin discovered that female guppies prefer to associate with a male after they have seen that male courting another female. As is so often the case in studies of social learning, a long series of control experiments were necessary to demonstrate that females were, in fact, influenced by the behavior of other females when choosing males with whom to affiliate, but the necessary work has been done and has provided rich rewards.

Dugatkin's work promises insight into a range of intriguing phenomena from the skewed reproductive success of males in lek-mating species to the function of male sailfin mollies's (*Poecilia latipinna*) infertile matings with female Amazon mollies (*P. formosa*) (Schlupp, Marler, & Ryan, 1994). In each case, the tendency of females to prefer to affiliate with males that they have observed interacting with other females provides a potential explanation of the behavior.

The Mathematical Basis of Comparative Analysis of Social Learning

Students of social learning have started to move beyond analyses of the mechanisms supporting social learning to study of the ecological circumstances under which social learning is most likely to occur. Such studies, both theoretical and empirical, of the ecology of social learning hold great promise that is just beginning to be realized.

In his autobiography, Darwin (1950) deeply regretted not having proceeded far enough in his own study of mathematics to gain some understanding of the basic principles of the subject, for, says Darwin, "men thus endowed seem to have an extra sense." If Darwin were with us today and were aware of the power of mathematical analyses to elucidate questions in ecology and evolution his regrets over his lack of mathematical training would surely be increased many fold.

Lefebvre and Giraldeau discuss mathematical techniques suitable for analyzing behavioral data to determine whether gregarious or opportunistic species exhibit adaptive specializations that are specific to social learning. They argue convincingly that the simple finding that, for example, members of a gregarious species learn from conspecifics more rapidly than do members of a closely related solitary species is not in itself sufficient to establish that the gregarious species is specialized for social learning. It is always possible that members of the gregarious species learn faster than do members of the solitary species when learning individually as well as when learning socially. Consequently, Lefebvre and Giraldeau argue, the appropriate test for existence of specialized social learning is not a simple comparison between two or more species on a social-learning task. To provide compelling evidence of ecological influence on social learning (and to bring such work to the standard developed in other areas of comparative biology) species must be compared in performance on both social-learning tasks and control tasks that involve individual learning.

Lefebvre and Giraldeau first describe linear regression and analysis of variance techniques that permit the comparisons they advocate, then they proceed to apply those techniques to data collected by Sasvari (1979, 1985) on social learning in opportunistic and specialists passerines, and by Lefebvre on social learning in social and solitary doves. While their analyses provide no evidence that social learning is an adaptive specialization either for group living or for opportunism (species that performed well on social-learning tasks also performed well on individual-learning control tasks), the techniques Lefebvre and Giraldeau describe should form the basis for all future comparative work on the ecology of social learning.

Models of Vertical and Horizontal Transmission

Laland, Richerson, and Boyd discuss the relationship between current mathematical models of social learning and empirical studies of the same phenomenon. As Laland et al. point out, while most mathematical models have assumed that social learning in animals can result in stable traditions similar to those seen in human culture, most field and laboratory studies reveal that socially transmitted information is of transient value and of limited longevity in animal populations.

In chapter seven, Laland et al. provide a model of the evolution of "horizontal"-social transmission (i.e., social transmission within a generation). This model, while consistent with Boyd and Richerson's (1985, 1988) earlier models of the evolution of vertical transmission (transmission between generations), is based on assumptions and reaches conclusions of greater relevance to empirical work on animal social learning than previous mathematical treatments in the area. In the model Laland et al. develop here, and as one might expect from the empirical data, when the probability of individual foraging success is low, social learning is likely to evolve in response to rapidly changing environments.

The extensive interactions between theoreticians and experimentalists that have occurred in the last few years appear to have begun to bear fruit; the new model speaks directly to the concerns of experimentalists. Laland et al. propose that formal theory should "identify key parameters and raise questions that inspire empirical research." Their latest model of animal traditions surely achieves the first of these goals and promises to achieve the second.

Bird Song

In Darwin's day, "naturalists [were] much divided with respect to the object of singing of birds" (Darwin, 1871); some argued that the songs of birds were intended to attract females, others that bird song was used in rivalry between males and was not produced for the "sake of charming mates." Today, we know that bird song is involved both in attracting conspecific females and in repelling conspecific males. We also know that the songs that adult males sing have been shaped by interaction of young males with both males and females of their species. Understanding the role of social interaction in shaping the development of bird song is, in large measure, the result of the work of Meredith West and Andrew King, authors of the penultimate chapter in the first section of this volume.

West and King's contribution provides a comprehensive overview of the complex forces shaping the development of song in male cowbirds and determining the effectiveness of male song in soliciting copulations from conspecific females and eliciting aggressive responses from conspecific males. The story is complicated by the fact that the observable effects of social-rearing environment on adult song, adult social behavior, and adult reproduction vary as a function of the environment in which those effects are measured. The greater the range of environments in which the behavior of subjects is observed, the more likely one is to detect the subtle yet powerful effects of early social experience on adult behavior.

In West and King's analysis, song production by an adult bird is not the end point of investigations of effects of social influences on development. Rather, "crystallized" song is but the starting point for analysis of effects of song quality on adult social and sexual behavior. The multiplicity of dependent variables that West and King employ, together with the variety of experimental settings in which they observe effects of rearing condition on adult behavior, provide a complex yet compelling picture of the multiple roles that social learning can play in development of effective vocal communications and adaptive patterns of adult social behavior in free-living cowbirds.

As noted in footnote 1, Darwin and his contemporaries used the term *imitation* to refer to a blind, mindless copying of behavior. They interpreted reports of imitative behavior as indicating a limited cognitive contribution to the acquisition of behavior. At the end of their chapter, West and King reach a conclusion reminiscent of that reached a century ago by the early British naturalists. In West and King's words, the copying of sounds may "be as reflexive and cognitively uncomplicated as the ability to breathe." Producing vocal outputs appropriate for the local social and ecological situation may be a far more complex and cognitively demanding task than is mere reproduction of sounds. Perhaps, at least in so far as imitation of vocalizations by birds is concerned, the Victorians had it right.

Yawning, Laughing, and Smiling

The final chapter in the first section of the book describes a second research program concerned with the kind of involuntary copying that captured the attention of Darwin and his contemporaries. In this case, the subject of study is the species one might imagine to be least likely to exhibit mindless imitation—that pinnacle of intellectual capacity—human beings.

Robert Provine has spent 10 years looking in great detail at yawning, laughing, and smiling—an easily studied subset of everyday human (and primate) behaviors that have the unusual property of eliciting similar behavior in those that observe them. In a series of ingenious studies involving everything from yawning through one's nose to listening to laughter played backwards, Provine seeks to describe the critical features that release such contagious behaviors, to define the conditions under which they occur, and to cast light on their potential functions. As Provine makes clear, the contagious feature of these species-typical behaviors makes them excellent candidates for study of the ontogeny and neurophysiology of complex motor patterns.

For the purposes of this book, Provine's work on contagious behaviors in humans provides a natural bridge between 19th-century work on imitation, contemporary work on social learning processes other than imitation (of which contagious behavior is a prime example), and contemporary studies of the cognitively complex imitative processes that are the focus of the second half of this volume.

REFERENCES

Boyd, R., & Richerson, P. J. (1985). *Culture and the evolutionary process.* Chicago: University of Chicago Press.

Boyd, R., & Richerson, P. J. (1988). An evolutionary model of social learning: The effects of spatial and temporal variation. In T. Zentall & B. G. Galef, Jr. (Eds.), *Social learning: Psychological and biological perspectives* (pp. 29–48). Hillsdale, NJ: Erlbaum.

Darwin, C. (1871). *The descent of man and selection in relation to sex.* (Vol. 2). London: J. Murray.

Darwin, C. (1950). *The autobiography of Charles Darwin.* Sir Francis Darwin (Ed.), New York: Collier.

Galef, B. G., Jr. (1988a). Evolution and learning before Thorndike: A forgotten epoch in the history of behavioral research. In R. C. Bolles & M. Beecher (Eds.), *Evolution and learning.* Hillsdale, NJ: Erlbaum.

Galef, B. G., Jr. (1988b). Imitation in animals: History, definition and interpretation of data from the psychological laboratory. In T. R. Zentall & B. G. Galef, Jr. (Eds.), *Social learning: Psychological and biological perspectives* (pp. 3–28). Hillsdale, NJ: Erlbaum.

Heyes, C. M. (1994). Social learning in animals: Categories and mechanisms. *Biological Reviews, 69,* 207–231, 1994.

Mitchell, R. W. (1989). A comparative developmental approach to understanding imitation. *Perspectives in Ethology, 7,* 183–215.

Romanes, G. J. (1882). *Animal intelligence.* London: Kegan Paul, Trench.

Romanes, G. J. (1884). *Mental evolution in animals.* New York: AMS Press.

Sasvari, L. (1979). Observational learning in Great, Blue & Marsh Tits. *Animal Behaviour, 27,* 767–771.

Sasvari, L. (1985). Keypeck conditioning with reinforcements in two different locations in thrush, tit and sparrow species. *Behavioural Proceseses, 11,* 245–252.

Schlupp, I., Marler, C., & Ryan, M. J. (1994). Benefit to mail sailfin mollies of mating with heterospecific females. *Science, 263,* 373–374.

Thorndike, E. L. (1898). Animal intelligence: An experimental study of the associative process in animals. *Psychological Review Monograph,* 2(Suppl. 4), 1–109.

Thorndike, E. L. (1911). *Animal intelligence.* New York: MacMillan.

Thorpe, W. H. (1956). *Learning and instinct in animals.* London: Methuen.

Wallace, A. R. (1886). Sir Charles Lyell on geological climates and the origin of species. *Quarterly Review,* **126,** 359–394.

Wallace, A. R. (1970). *Contributions to the theory of natural selection.* New York: AMS Press.

Whiten, A., & Ham, R. (1992). On the nature and evolution of imitation in the animal kingdom: Reappraisal of a century of research. *Advances in the Study of Behavior,* **21,** 239–283.

2

Cultural Transmission of Feeding Behavior in the Black Rat (Rattus rattus)

JOSEPH TERKEL

Department of Zoology
Tel Aviv University
Tel Aviv 69978, Israel

INTRODUCTION

Toward the end of the 19th-century and during the beginning of the 20th-century, most of the natural oak and pine forests in Israel were cut down. Over the last 50 years they have been replaced with large tracts of Jerusalem pine.

About 14 years ago, during a field trip to the pine forests in the north of the country with his pupils, Ran Aisner, a high school biology teacher, observed piles of bare pine cone shafts that had accumulated beneath certain pine trees (Aisner, 1981). Ran brought some of these cones to me for my opinion on the phenomenon.

When we inspected these pine cone shafts more closely, it became clear that the cones had been detached from the branches of the trees on which they grew, by gnawing, and that the animal that had done so had systematically stripped them of all their scales and removed the seeds that lay concealed beneath. If pine cone shafts in a similar condition had been found in other parts of the world, there is no doubt that squirrels would have been considered the most likely agents of the cone destruction (Smith & Balda, 1979). However, as there are no squirrels in Israel, this possibility was not considered in the present case.

Although the crossbill (*Loxia curvirostra*) eats pine seeds, it uses a different method of feeding than does the squirrel. Moreover, the number of crossbills in Israel is relatively small compared to the large quantity of discarded pine cones that had been found; so this possibility was also an unrealistic one. At this point we had absolutely no clue to the identity of the animal feeding on pine cones in the forests of Israel.

During the first stage of our investigations, in order to determine whether the pine seed eating animal in Israel was diurnal or nocturnal, we concentrated our efforts on capturing it. Accordingly, Ran placed plastic sheets below those trees where we had found large piles of discarded cone shafts. We found that most of the activity, determined by counting the number of cones and scales that accumulated on the plastic sheet, occurred during the hours of darkness. Since Ran also found rodent-like feces on the plastic sheets, it became quite clear that we were dealing with a nocturnal rodent. At this point he placed traps on the branches of the pine trees and shortly afterwards caught the first black rats (*Rattus rattus*).

We brought these wild black rats to the animal house at Tel Aviv University, gave them a short period of habituation, and then provided them with pine cones. Because of the rats' extreme timidity in our presence, we were unable at first to observe them actually manipulate the cones. However, the bare cone shafts that we found in the cages each morning, left us in no doubt whatsoever that these rats had indeed stripped the scales from the cones, eaten the seeds, and discarded the bare shafts, which were identical to those found in the pine forests of Israel. This was our first evidence, albeit indirect, that black rats in the pine forests of Israel open pine cones and obtain nourishment from the seeds (Aisner, 1981).

Since all our attempts at direct observation of the rats feeding on the pine cones in the forest had failed, Ran constructed a cylindrically shaped metal "cage," about 2.5 m high and 1.5 m in diameter, with a large Perspex window, enabling convenient observation and photography. The size of the container permitted the introduction of a large section of pine-tree trunk with branches, complete with cones, allowing the rats to move freely on the branches and to exhibit their natural feeding behavior.

After the rats became accustomed to captive conditions, it was possible to see them leave their nest boxes each evening to feed on the pine cones. They began by climbing cautiously over the pine branches, stopping occasionally to examine some of the cones (see Fig. 1). Having finally chosen a cone, a rat would detach it from the branch—a feat requiring considerable skill to avoid dropping the cone.

The rat would then carry the cone to a favored feeding branch where it would

Fig. 1. A black rat on a branch of a Jerusalem pine, starting to detach cone from branch. From Aisner & Terkel (1992).

systematically strip off the scales in a stereotyped fashion. Starting at the base of the cone (Fig. 2A), the rat systematically stripped the hard scales one by one, following their spiral order around the shaft (Figs. 2B and 2C).

The scale itself is actually removed by the rat prising the free, upper distal edge from the cone's surface with its teeth and then gnawing off the proximal end (Fig. 3). This efficient technique avoids wasting energy by gnawing more of each scale than is necessary. After the rat has completed its stripping, only the bare shaft is left (Fig. 2D), which the rat then drops to the ground beneath the feeding site. In the pine forest itself, bare cone shafts accumulate beneath such feeding sites (Aisner, 1981; Aisner & Terkel, 1985, 1992a,b).

Some populations of black rats appear to have retained a persistent food preference for pine seeds for several years, despite the availability of edible fruit nearby. We therefore asked how such a food preference, and the skill to exploit it, had developed. This approach has formed the subject of our investigations for the past few years.

Fig. 2. Pine cones in different stages of being stripped. (A) Intact cone; arrow indicates where rat would begin stripping. (B) Cone with several rows of scales removed. (C) Half stripped cone. (D) Bare cone shaft after all scales have been removed. Modified from Aisner & Terkel (1992).

GENERAL OBSERVATIONS AND AIMS

We were quite surprised to trap black rats in pine trees, since there had been no previous indication that these rodents inhabited pine forests in Israel. On the other hand, black rats are known to dwell in the roofs of buildings and to obtain food from a wide variety of sources, including storehouses and food production factories in widely differing geographical areas (Canby, 1977; Lore & Flannelly, 1977).

The black rat is an opportunistic feeder (Rozin, 1976), and it is this flexible feeding behavior, as well as highly adaptable behavioral mechanisms, that has led to the rats' worldwide distribution and successful invasion of new habitats.

Our field observations were carried out in two forests near Kibbutz Ramot Menashe, an agricultural settlement in central Israel. Each forest covered about 15 ha, and was located at least 2 km from any human population. The trees had been

Fig. 3. A black rat actively stripping a pine cone. Note the use of the rat's lower incisors in prising up the scales.

planted 20–50 years ago, at a density of 300 per ha, and are today 15–20 m high. Because of their high density, the pines are poorly developed and only those at the forest edges receive sufficient light and rainfall to allow them to develop a generous root area and heavy cone crop. We found most of the black rat activity taking place in these well-developed trees.

The rats nest in the pine trees—a niche occupied by tree squirrels, wood mice, and other animals in other parts of the world. The rats skillfully interweave thin pine twigs to construct their nests, which vary in size from 25 × 30 × 35 cm to 40 × 50 × 100 cm. The nests, located 5–10 m above ground, are almost completely enclosed, with only one opening, 4–5 cm in diameter. They are attached by twin twigs to branches of the tree at some distance from the trunk, possibly to evade predators. Nests contain food remains, droppings, intact and stripped cones, and small, thin twigs (Aisner & Terkel, 1991, 1992b).

In Israel, rats are the sole exploiters of this arboreal habitat and, thus, have no competitors for the pine seeds that provide their main source of nourishment. They obtain their water by licking dew from the pine needles. In the laboratory, when we fed our trapped rats solely on pine seeds (and water), they remained healthy, gave birth, and successfully raised their litters to weaning. As far as we know, the species *Rattus* has not previously been included in the list of animals known to exploit pine seeds for food (Smith & Balda, 1979).

We never succeeded in trapping rats when traps were placed on the forest floor, which is consistent with our belief that they seldom leave the pine trees for the ground. The rats obtain the seeds directly from the cones, which are attached to the upper branches of the trees and, except following accidental breakage of a branch, such as occurs during a winter storm, are never found on the ground. The pine cones are of a type that responds to extreme heat (such as fire) by lifting their scales and releasing the seeds. During exceptionally hot summer days when this process occurs, the seeds are unavailable to the rats because after release they are blown to the ground by the wind.

Considering the limited variety of food sources available to the rats in the relatively sterile pine forest, we hypothesized that the ability of the rats to strip the scales from cones in the efficient spiral manner, and to obtain the highly nourishing seeds, may be the clue to their successful survival in the pine forests.

In the first part of this report I shall describe a series of experiments performed by Ran Aisner, designed to determine: (a) whether naive adult rats could acquire the pine cone opening technique either by trial and error learning or by observing rats with experience in stripping cones; (b) whether the flavor of milk from mothers feeding on pine seeds could bias the preference of nursing pups toward opening cones; (c) whether pups born and raised by mothers with no experience with pine cones could learn to open cones as successfully as did pups born to mothers experienced in stripping cones; (d) whether pine cone opening and seed extracting behavior is an inherited trait transmitted genetically from one generation to the next, or whether it is acquired by learning through a process of social facilitation; and finally, (e) whether naive adult rats could acquire this skill when provided with the clue of cones with exposed seeds.

EXPERIMENTS

In our study we used rats of both sexes from each of two environments: rats trapped in the pine forest and already experienced in stripping pine cones, which we termed "strippers"; and rats trapped in urban habitats such as dwellings or agricultural storehouses, with no previous experience with pine cones, which we termed "naive."

Trial-and-Error Learning

In our first experiment we examined the ability of naive, adult black rats to learn to open pine cones through a process of trial-and-error learning (Aisner & Terkel, 1992). The naive rats were housed in cages either individually or in pairs and were supplied with pine cones, either detached or still on a branch. We restricted their food (rat chow) to 85% of normal intake to induce a state of constant slight hunger, while monitoring the animals' physical condition at all times. Since naive rats showed little interest in pine cones unless they were extremely hungry, once weekly we deprived them of rat chow for 48 h, during which time they were provided only with fresh pine cones at the same stage of ripeness as those on which adult rats normally feed in nature. By watching our subjects on a regular schedule and examining the pine cones in their cages each day, we determined whether they had learned to open cones. Due to the rats' hunger, most overcame their natural timidity in the presence of observers and began examining the cones immediately. We determined the stripping ability of those rats that did not manipulate the cones during observation periods by examining the state of the cones. Because of the structure of the cones, a simple examination sufficed to show us whether a rat had acquired the opening skill.

When we terminated the experiment after three months, we found that none of the naive animals had learned to open the cones efficiently by trial and error learning (see Table 1).

Under the experimental conditions some of the naive rats gnawed at the cones

TABLE 1

Does Trial-and-Error Learning Result in Pine Cone Opening Behavior in Naive Adult Black Rats?[a]

Conditions	No. of animals tested	No. of animals acquiring technique
Solitary or in pairs	32	0
Supplied with intact cones and rat chow		

[a]Modified from Aisner and Terkel (1992).

Fig. 4. Four pine cones which have been gnawed in random fashion by naive black rats.

in a random fashion (see Fig. 4) and ate some seeds. However, the way in which they obtained the seeds was inefficient and they would have died of starvation if we had not supplied them with supplementary food.

Observation of Adult Rats by Other Adult Rats

In our second experiment we examined whether naive adult rats could acquire the pine-cone-opening skill through observation of experienced stripper rats (Aisner & Terkel, 1992a). Each naive adult rat was paired with a stripper rat of the same sex, with pine cones available in the cage throughout the entire experiment so that the naive rats could observe the strippers opening the cones and feeding on the seeds. The naive rats had access to the pine cones at all times. To maintain their physical condition, they were removed from the cage once daily and supplied with 85% of normal intake of rat chow. We used the same procedure to determine whether a rat had learned to open the pine cones as we had used in the first experiment. We found that even after three months of exposure to strippers, the ability of naive adult rats to open cones was not enhanced. None of the 15 naive animals learned to open cones efficiently. They simply continued with their random gnawing (see Table 2).

Although naive adult rats were unable to acquire the pine-cone-opening technique either by trial and error or by observing an experienced rat in the lab, in

---

TABLE 2

Does Observational Learning Result in Pine Cone Opening Behavior in Naive Adult Black Rats? [a]

Conditions	No. of animals tested	No. of animals acquiring technique
Housed with "strippers" Supplied with intact cones and rat chow	15	0

[a] Modified from Aisner and Terkel (1992).

nature, or in the pine forests, pine cone opening is perpetuated. We wondered whether stripper mothers might be transmitting the pine seed flavor in their diet through their milk to the young (Galef & Henderson, 1972; Galef & Clark, 1972; Le Magnen & Tallon, 1968). The mothers might, in this way, provide a flavor cue that would shape the preference of the pups for pine seeds at weaning, thus enhancing the pups' ability to learn the stripping technique.

Effects of Feeding Mothers Pine Seeds

In this experiment we examined the ability of pups, born to naive mothers fed on pine seeds, to learn to open the cones (Aisner & Terkel, 1992a). We fed seven lactating dams solely on pine seeds that we extracted manually from cones. After weaning, we provided their pups with pine cones which, with a supplement of rat chow, comprised their main source of nourishment, and we then determined the pups' ability to obtain seeds from the cones. None of the pups succeeded in acquiring the stripping technique, and they were unable to open the cones and obtain nourishment from the seeds (see Table 3). The pups became progressively weaker and would have died had we not provided them with supplementary food.

Living with Mothers That Strip Cones

At this point in our study, we still did not know how naive rats acquired the pine-cone-opening technique. Since in nature the ability to strip pine cones appears in one generation of wild rats after another, we next examined whether pups born and raised by stripper mothers were able to learn to open cones (Aisner & Terkel,

TABLE 3

Does Flavor of Mother's Milk Influence Acquisition of Pine Cone Opening Behavior in Black Rat Pups?[a]

Conditions	No. of pups	No. of pups acquiring technique
Pups nursing from naive mothers fed only pine seeds during lactation. Young tested with pine cones	77	0

[a] Modified from Aisner and Terkel (1992).

992a). Five dams experienced in opening the cones were caged individually and llowed to raise their own young while feeding on seeds, which they extracted from ones provided ad libitum throughout the duration of the three month experiment. he growing pups remained in their mother's proximity while she opened cones id fed on the seeds.

When we tested them at three months of age, 31 out of 33 pups born to ipper mothers were able to open pine cones efficiently and obtain the seeds (see ble 4). Only two of the pups were unable to strip the scales, but continued instead gnaw randomly at the cones (see Fig. 4).

Ran Aisner and I next attempted to verify that the stripping ability is trans-red from the mother to her offspring by a process of social, rather than genetic,

TABLE 4

Do "Stripper" Mothers Transmit Pine Cone Opening Technique to Their Offspring?[a]

Conditions	No. of pups	No. of pups acquiring technique
Pups exposed to own "stripper" mothers opening cones and to intact cones	33	31

[a] Modified from Aisner and Terkel (1992).

transmission (Aisner & Terkel, 1992a). We cross-fostered pups born to naive mothers on stripper mothers and vice versa. The experimental design was as follows: we removed eight pups from two naive mothers on Day 5 postpartum and exchanged them with seven pups from two stripper mothers that had given birth at the same time. The two naive mothers suckled a total of six of their own pups and seven foster pups, and the two stripper mothers raised six of their own pups and eight foster pups. Because the naive rats were unable to open pine cones, we provided them with rat chow in addition to the pine cones provided throughout the experiment. The stripper mothers had a continuous supply of intact cones which they stripped, feeding exclusively on the seeds.

We tested the ability of the growing pups to open pine cones. At three months of age, none of the pups born to and raised by naive mothers, nor those born to stripper mothers and reared by naive mothers, were able to strip cones and obtain the seeds. However, all pups born to stripper mothers and left with them, as well as all those pups born to naive mothers and reared by stripper mothers, learned to open cones and obtain the seeds (see Table 5).

This reciprocal fostering experiment clearly indicates that the ability to strip pine cones is acquired through the social proximity of the pups to a dam experienced in stripping.

TABLE 5

Does Cross-Fostering of Pups to "Stripper" or Naive Mothers Influence Pine Cone Opening Behavior of Pups? [a]

Conditions	No. of pups	No. of pups acquiring technique
Naive mother raising own and pups from "stripper" mother. Pups exposed to intact cones	13	0
"Stripper" mother raising own and foster pups from naive mother. Pups exposed to mother opening cones and to intact cones	14	14

[a] Modified from Aisner and Terkel (1992).

Exposure to Partially Opened Cones

The intact, closed cone does not provide the pups with any clue to its potential as a food source, while those cones which experienced dams are in the process of stripping do provide such clues. The nutritious seeds are revealed as the pups gather around the dam's mouth while she is stripping cones. We wondered whether exposing pups to partially opened cones would affect their motivation, and later their ability, to learn to open closed cones (Aisner & Terkel, 1992a). We therefore heated pine cones in an oven to 50°C, causing the scales to lift, thus exposing the seeds beneath. When naive rats were presented with such cones they were able to obtain the exposed seeds without difficulty.

In the present experiment we provided three naive rats, from midpregnancy, with pine cones with exposed seeds. As their pups developed and started to eat solid food they also fed on the seeds from these cones. Although the growing pups had removed exposed seeds from partially opened cones, such early exposure to the pine cones' nutritional potential did not enhance their learning to spirally strip the cones (see Table 6).

Having found that naive adult rats, which were provided with clues to the nutritional potential of the pine cone's contents, did not learn to strip the cones, while pups kept with stripper dams did acquire the technique, we decided next to concentrate on the potential clues available to pups from the cones that their mothers were stripping. In the following experiment, we focused on the structure of the cones with respect to the pup's ability to learn to strip them.

The spiral structure of the pine cone determines that the scales of the first row

TABLE 6

Does Experiencing Cones with Exposed Seeds Affect Cone Opening Behavior in Weanling Pups?[a]

Conditions	No. of pups	No. of pups acquiring technique
Naive mother raising own young. Pups experience cones with exposed seeds and tested with closed pine cones	17	0

[a]Modified from Aisner and Terkel (1992).

at the base (the widest part) partially cover the scales on the next row. Therefore, in order for the rats to extract the seeds efficiently, they must strip the scales systematically, beginning with the first row at the base of the cone, completing the entire spiral, and then continuing with the second row, and so on. During this experiment, Professor Jeff Galef visited our laboratory and suggested that we provide already started cones as a clue for the rat pups.

Daphna Yanai and I (Yanai & Terkel, 1991) next determined whether cones that had been partially opened in the spiral manner (i.e., in which the first few rows had been stripped, similar to cone B in Fig. 2) would provide the pups with an important clue in the process of their learning the efficient stripping technique. To determine to what extent the stimulus of a partially opened cone promotes learning to open cones, we provided naive adult rats with pine cones from which the first few rows of scales had been removed by a stripper rat, or cones from which we had already removed the first few rows of scales. In both cases, the rats received cones for which the direction in which to continue opening had already been given (see Fig. 5).

In a pilot study we found that when we provided naive adult rats with cones from which the first four rows of scales had been removed, most of the rats were able to continue to open the rest of the cone and feed on the seeds. However, when we subsequently provided such rats with intact cones, they were unable to strip the cones and only gnawed at them randomly. We therefore decided to "assist" the rats by presenting them with a series of graded clues: We provided them with cones with a consecutively decreasing number of previously stripped rows of scales, from

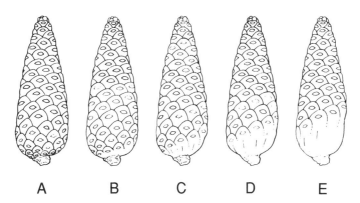

A   **B**   **C**   **D**   **E**

Fig. 5. Pine cones in different stages of opening, with the number of rows of previously stripped scales decreasing consecutively from right to left. Modified from Yanai (1989).

four rows to zero (intact cones). We proceeded as follows: The rats were exposed to cones with four rows of scales removed and monitored to determine whether they could continue to strip the rest of the cone in the spiral manner. Those rats that succeeded were then given cones with three rows of scales removed, and so on. If the rats were unable to continue opening a new grade of cones, following a period of one week, we scored them as able to strip cones only from the previous grade (Yanai, 1989).

We divided the experiment into two parts: First we presented 51 naive adult rats, housed individually, with cones from which four rows of scales had been stripped. As in our previous experiments, we restricted their food to 85% of normal intake in order to induce a state of constant slight hunger. We examined the cones daily, and a rat was considered to be a stripper only if it was observed to open cones systematically, until only the bare shaft remained. The experiment lasted four weeks. A rat that did not succeed in opening the cones during that period was considered as not having acquired the stripping technique.

Thirty-five of 51 rats that started the experiment learned to continue to open pine cones and feed on the seeds when provided with spirally opened cones (see Table 7). The remainder of the rats (31.4%) either failed to relate to the cones in any way as a potential source of food, or simply gnawed on them in a random fashion.

In the second part of this experiment we provided 20 rats, which had previously learned to continue to strip cones from which four rows of scales had been removed, with cones with a steadily decreasing number of stripped rows of scales. This part of the experiment also lasted four weeks.

Ninety percent of the rats that had a demonstrated ability to continue to strip

TABLE 7

Does Exposure of Naive Adult Black Rats to Partially Stripped Cones Enhance Their Ability to Continue Stripping the Cones?[a]

Conditions	No. of animals tested	No. of animals opening cones after 4 weeks
Naive adults. Solitary, supplied with cones stripped of 4 rows of scales and rat chow	51	35

[a]From Yanai (1989).

cones from which we had removed four rows of scales, learned to open intact cones when presented with cones with a steadily decreasing number of previously stripped rows of scales (see Table 8).

All 20 rats in this experiment were able to strip cones with 4, 3, 2, and 1 row of scales removed. Two rats were unable to proceed to the final stage (i.e. opening intact cones).

"Shaving" Scales

Until now we have discussed only the "spiral" method by which the black rat detaches the cone from the branch and then removes the scales, beginning at the base and continuing around to the apex, following the spiral design of the overlapping scales. However, in certain instances in nature, rats strip pine cones while they are still attached to the branch. Despite our lack of conclusive evidence, it appears to us that in cases in which a rat finds itself in unfamiliar territory, or is faced with an uncertain situation (such as the invasion of new territory), it prefers not to detach a pine cone (a time-consuming activity), but to open it while it is still *in situ*. On certain isolated and prominent trees, we observed pine cones from which all scales had been removed and only the bare shafts remained attached to the tree. On branches where feeding had taken place *in situ* we also observed pine cones from which the scales had been removed only along the axis of one half of the cone. We

TABLE 8

Does Exposure of Naive Adult Black Rats to Pine Cones with a Consecutively Decreasing Number of Previous Stripped Rows of Scales Enhance Their Ability to Strip Cones?[a]

Conditions	No. of animals tested	No. of animals opening cones after 4 weeks
Adults experienced in stripping partially open cones. Solitary, supplied with cones with decreasing number of previously stripped rows of scales and rat chow	20	18

[a]From Yanai (1989).

encountered this situation in nature only when cones were attached to a branch abnormally, with the shaft of the cone parallel, rather than perpendicular, to the branch to which it was attached.

When a rat begins to strip such a cone, it removes scales only from the far side of the cone from the branch, for the simple reason that it cannot reach the scales pressed close to the branch itself. In this case, after the rat has completed its feeding activity, the partially opened cone remains attached to the branch, with the scales removed from one side only, appearing to all intents and purposes to have been shaved along the entire length of the side. Due to this appearance of the pine cone following such treatment by the rats, we termed this method of opening cones "shaving."

"Stripping" and "Shaving"

Throughout the entire period that we have been studying the phenomenon of pine-cone-opening behavior by black rats in the field, we have never encountered pine cones that were detached by rats from the branch and then shaved open. All cones that were detached by the rats in nature were stripped in the spiral manner (Fig. 6).

In the laboratory, however, we noticed that rat pups learn to open cones using both methods: spiral and shaving. It is possible that the age at which the pups first become exposed to cones influences the method which they employ to open them, and that as adults, they persist in the technique they first learned. Zohar (1987), in my laboratory, suggested in his dissertation that young pups lack either the motivation (since there are no seeds at the base of the cone beneath the first rows of scales) or the physical strength necessary to strip cones spirally around their complete circumference without reinforcement for their efforts. They therefore shave the cones because in this way they begin to obtain seeds almost immediately. A rat that has completed shaving the scales from one side of a cone then rotates it and shaves the scales from the other side.

When shaving cones, the rats encounter only a few scales that do not overlap, and in order to reach the seeds under these scales they must gnaw away relatively more scale parts than they would have to if they had used the spiral method of cone opening.

The fact that in nature we have found no evidence of the shaving method (apart from the cases mentioned above of opening cones still attached to branches), suggested to us a difference in the relative efficiency of the two methods of opening cones.

Fig. 6. Cones stripped while still attached to the branch. The rat has removed scales from one side of the cone only, using the "shaving" method. From Aisner (1984).

Exposing Pups to Half-Opened Cones

The importance of started cones in providing a clue to the rats as to how to continue to open cones, was further confirmed in an experiment conducted by Yanai (1989) in which she presented naive rats with cones from which four rows of scales had been removed only halfway around the circumference of the base. We termed such cones "started to 180°," as opposed to cones from which rows of scales had been removed around the entire circumference (360°), as in the previous experiment.

Twenty-four individually housed naive rats were provided daily with two pine cones started to 180° and only two pellets of rat chow. Their permanent slight hunger thus increased their motivation to search for extra food. Once a week they received rat chow ad libitum for 36 h. Each morning, before we gave them the new cones, the cones from the previous day were removed and examined to determine the rat's progress in learning to open them. The experiment lasted four weeks, as most of the rats that learned to open cones began to do so within five days of the beginning of the experiment and had fully acquired the opening technique within three weeks.

We found that 22 of the 24 animals that received cones started at 180° continued to open the cones from base to apex only along one side; i.e., they employed the shaving technique (see Table 9).

In comparison to naive rats that received cones started around the entire circumference of the base, and rats that learned to continue stripping the cones spirally, the rats that were provided with cones stared only along half the circumference (180°), first completed shaving one side of the cone before continuing with the other side.

TABLE 9

Does Exposure of Naive Adult Black Rats to Partially Open Cones to either 360° or 180° Enhance Their Ability to Continue Stripping the Cones?[a]

Conditions	No. of animals tested	No. of animals opening cones	Style of opening
Naive adults. Solitary, supplied with cones stripped to 360°	51	35 (68.6%)	Sprial
Naive adults. Solitary, supplied with cones stripped to 180°	24	22 (91.7%)	Shaving

[a]From Yanai (1989).

Exposing Pups to Started Cones

Having established that naive adult rats can learn to strip pine cones when provided with appropriate clues (i.e., cones already started at the base), we now examined the role of started cones alone, in the pups' acquisition of cone-opening techniques (Zohar & Terkel, 1992). In nature, the likelihood that a pup will encounter a partially opened cone in the vicinity of its mother, is much greater than the likelihood of an adult finding a partially opened cone. This experiment, therefore, simulates a condition under which the feeding technique could be acquired and passed on to the next generation. In addition, the partially opened cones with which we provided the pups in this experiment, created a much more natural situation than the step-by-step procedure of gradually decreasing the number of previously stripped rows of scales.

We housed five naive, lactating dams, each with five pups, individually in 30 × 50 × 35 cm glass terraria with wire-mesh lids. Each terrarium was divided in half by a wire-mesh partition with an opening (2 × 2 cm) large enough to allow the pups, but not the mother, to cross from one section to the other. We provided rat chow powder for three h daily and water ad libitum. The mother and litter were housed in one side of the terrarium and we placed a fresh batch of eight cones each day in the other half. The cones were provided at four different stages of opening: 2 closed; 2 stripped of scales only at the proximal end; 2 half stripped of scales; and 2 bare shafts, stripped of all scales but still containing a few uncollected seeds. The cones had been stripped by stripper rats from our colony, so it is possible that the pups could have been attracted to them by the residual odor of conspecifics (Galef, 1982). The pups' hunger, induced by restricted rat chow, motivated them to seek additional food sources and they demonstrated interest in the pine cones introduced into the other half of the cage.

When the pups reached 80 days of age, we removed the mother from the cage and tested the pups to determine their ability to strip cones. We found that 24% of the pups had acquired the stripping technique, without having observed such feeding behavior exhibited by a model (see Table 10).

The Energetics of Stripping and Shaving

Together with Yanai and Ar (Yanai, Ar, & Terkel, 1991), we compared the efficiency of the two cone-opening methods by measuring both the energy invested in opening and the duration of the opening process.

TABLE 10

Does Exposing Pups to Pine Cones in Various Stages of Opening Enhance Their Ability to Learn to Open Cones? [a]

Conditions	No. of pups	No. of pups acquiring techniques
Naive mothers raising own young. Mothers fed on rat chow; pups exposed to pine cones in various stages of opening	25	6

Modified from Zohar and Terkel (1992).

To assess the energy required to open a pine cone we used an apparatus based on a metabolic chamber to measure the oxygen consumption of rats while they opened cones (see Fig. 7).

The 2-l volume metabolic chamber was located at the center of the apparatus and a constant temperature of 28 ± 1°C was maintained by the flow of heated

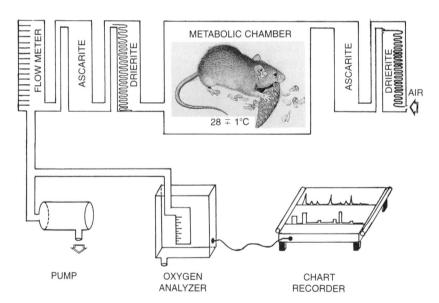

Fig. 7. Oxygen consumption measuring system. From Yanai (1989).

water through the double wall of the chamber. Dry air at room temperature was pumped into the chamber at a constant rate measured by a flowmeter. The amount of oxygen in the air leaving the metabolic chamber was measured and compared with the amount of oxygen entering the system using an oxygen analyzer. We measured oxygen consumption of the rats during rest and activity periods from curves obtained on a chart recorder (see Fig. 8). Oxygen consumption was measured in units of ml O_2/g body wt/h, and then converted from ml oxygen to Joules, the standard unit of energy. The calculation was made by deducting the oxygen consumption of a rat during a rest period from its oxygen consumption during activity. The total net oxygen consumption of a rat when opening a complete cone was calculated by multiplying the increase in oxygen consumption during opening by the time taken to open the cone.

In order to accustom rats to opening cones under experimental conditions, each animal spent several h a day for several weeks in the metabolic chamber until it began to open cones. Oxygen consumption was then assessed for each rat under three conditions: (a) at rest; (b) while opening a cone; and (c) following cone opening. To compare duration of cone opening among different rats, all animals

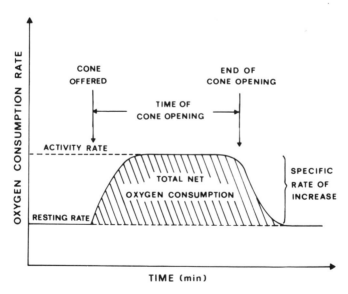

Fig. 8. Time course record of oxygen consumption during cone opening. From Yanai (1989).

were provided with cones of a standard size. Measurements were made on both shaving and stripping pine-cone-opening techniques.

As summarized in Table 11, the spiral method of opening pine cones was more efficient than the shaving method in two ways: (1) The average time required to spirally strip a standard pine cone was only one-third of that required to shave a similar cone; and (2) The rate of increase in oxygen consumption in rats spirally stripping cones was significantly lower than the increase in oxygen consumption when they shaved cones. These two variables (duration of opening and specific rate of increase in oxygen consumption), provided us with a measure of the total difference in energy cost between the two methods of opening. The spiral method of opening cones is both more energy efficient and faster than the shaving method. A rat that employs the spiral method to open the cone, therefore, both opens it more quickly and expends less energy per unit of time in doing so.

The significance of the difference in efficiency between the two methods of opening cones became clear when we calculated the relative benefit of opening the cone, that is the difference between the amount of energy provided by the pine cone and that expended in opening it. We found that a rat opening a cone using the spiral method expends 9.5% of the energy that it obtains from the cone, while a rat

TABLE 11

Duration of Cone Opening, Increase in Oxygen Consumption during Opening, and Net Oxygen Consumption for Spiral and Shaving Opening Techniques (No. of Animals in Parentheses)[a]

Style of opening	Duration per cone (min) increase (ml/(g × hr))	Net oxygen consumption	
		Specific rate (ml)	Total per cone
Spiral method	46.2 ± 7.3[d] (8)	1.35 ± 0.31[b] (4)	201.65 ± 44.18[c] (4)
Shaving method	118.9 ± 40.9 (8)	1.49 ± 0.30 (4)	476.89 ± 53.10 (4)

[a]From Yanai, Ar and Terkel (1991).
[b]$p < 0.05$, T test for repeated measurements.
[c]$p < 0.01$, T test for repeated measurements.
[d]$p < 0.01$, T test for independent measurements.

using the shaving mode expends 22.6% of the cone's available energy. If we compare the two methods, therefore, we find that spiral opening is approximately 2.5 times more efficient than shaving. A rat employing the spiral method expends only 42% of the energy expended by a rat employing the shaving method (see Table 12).

Although the shaving method of cone opening is less efficient than the spiral one, the rat nonetheless profits by it, albeit to a lesser extent than a rat that employs the more "professional" mode of spiral opening. We assume that, in the wild, rats cannot afford to use a wasteful technique for obtaining food, and that they therefore all use the spiral method whenever possible.

As described earlier, the pine cone is constructed of layers of overlapping scales arranged in a spiral structure around a central shaft. Two seeds lie concealed beneath each scale other than the first four rows of scales at the base of the cone. Because of the absence of seeds in this part of the cone, we term it the "sterile section." Thus, when a rat opens cones spirally, beginning from the sterile section, it does not gain any immediate reward. Only after stripping the first four rows of scales does the rat reach the scales that do conceal seeds, and only at this stage does it begin to obtain nutrition from the cone.

Aisner found that despite the sterile section of the cone being only one eighth of its total length, rats nonetheless invest 25% of the time they spend opening an entire cone opening this small section. The tight structure of the scales at the base of the cone requires the rat to gnaw away most of the scale parts rather than simply lift them, as can be done with the scales higher up on the cone.

In six measurements taken by Yanai in the oxygen consumption study, while monitoring the rat's behavior during cone opening, she was able to discern at which

---

TABLE 12

Energy Invested by a Rat in Opening a Pine Cone by Spiral and Shaving Techniques in Relation to the Cone's Energetic Value[a]

Style of opening	Net energy exploded/cone (j) (M ± SD)	Available energy/cone (j) (M ± SD)	Relative cost of opening
Spiral	3960 ± 751	41442 ± 1430[b]	9.55
Shaving	9366 ± 903		22.6
			Ratio 0.42

[a]From Yanai, Ar and Terkel (1991).
[b]After Aisner (1984).

point the rat completed opening the sterile section and went on to strip the rest of the cone. The time at which the rat moved from the sterile section to the seed-containing part of the cone was noted on the chart recorder, enabling us to calculate and compare the rate of oxygen consumption during opening the sterile section with the rate of oxygen consumption while gnawing the rest of the cone. We found that the increase in oxygen consumption during opening of the sterile section (1.441 $\pm$ 0.297 ml O_2/g/h) was significantly higher than that while opening the rest of the cone (1.253 $\pm$ 0.330 ml O_2/g/h).

This finding was particularly interesting because, in nature, one can find pine-cone shafts that have been stripped by black rats, but with the layers of scales of the sterile section still in place (see Fig. 9).

Rats that opened cones and left the sterile section in place did not begin to open them at the base of the shaft, but bypassed the sterile section and began by opening only those scales concealing seeds, leaving behind the shaft with the "nonbeneficial" first few rows of scales intact. Rats that do not open the sterile section, therefore, conserve energy and save 25% of the time taken to open an entire cone. A simple calculation shows that a rat that does not open the sterile section saves 27.4% of the energy expended by a rat that opens the entire cone.

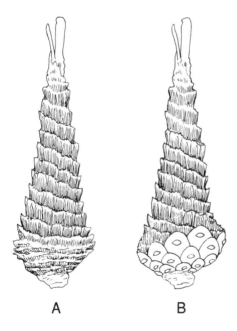

A   **B**

Fig. 9. (A) Cone with all its scales re-moved by a black rat. (B) Cone with scales removed only from above the ster-ile section. Modified from Yanai (1989).

SUMMARY AND CONCLUSIONS

The study of socially learned feeding behaviors has generally taken only one of two distinct approaches: field observation of behaviors believed to be socially transmitted (Lefebvre & Palameta, 1988), or laboratory experiments that analyze the mechanisms involved in social learning. While field observations of learned food-seeking and food-handling behavior can be both intriguing and spectacular, many available observations are anecdotal or speculative. On the other hand, most laboratory studies have dealt with behavioral tasks and objects that may not be related to the ecology or behavior of animals in the field (e.g., Gardner & Engel, 1971; Bullock & Neuringer, 1977), although there are exceptions to this tendency (e.g., Krebs, Mac-Roberts, & Cullen, 1972; Sherry & Galef, 1984). The approach that we have taken in our studies of social influences affecting food acquisition by black rats is important, because it combines field observations and laboratory experiments, allowing us to isolate and test factors that might affect acquisition of a natural feeding behavior.

When it became clear to us that naive adult black rats could not learn to open pine cones, we began to examine the acquisition of this behavior by young rats. Our finding that any pups raised by an experienced stripper dam learned to strip cones efficiently, indicated to us that the pine-cone-opening technique is not transferred genetically, from one generation to the next. It appears, rather, to involve social transmission. Opening nuts is a simpler task, and squirrels can learn to open nuts by trial and error (Eibl-Eibelsfeld, 1956) and then improve their technique through observational learning (Wiegl & Hanson, 1980).

Although Galef and Henderson (1972) showed that rat pups' diet preference at weaning can be influenced through the flavor of their mother's milk, in the case of our black rats the pups' acquisition of the cone-stripping technique and their feeding on pine seeds was independent of whether the mother had fed on pine seeds during lactation.

Social mediation of food preference can also occur through the physical presence of adults at a feeding site, encouraging social eating (Galef & Clark, 1972). In fact, rat pups may be attracted to a particular site merely by the presence of adult odors (Galef, 1977). In a similar fashion, the presence of a stripper mother may facilitate her pups' learning in several ways. Ewer (1971) observed that when rat pups begin to leave their nest and eat solid food, new behaviors appear, such as licking and sniffing the mother's mouth. We noticed that while a black rat mother is actively opening cones, stripping scales and feeding on the exposed seeds, her developing young gather around her mouth and attempt, with greater or less success, to obtain some seeds. Later in their development, we saw pups attempt to snatch cones from their mother while she was in the process of stripping them. When they succeeded, they kept the partially stripped cones and continued the stripping process on their own (Fig. 10). Kemble (1984) described similar behaviors that facilitated the learning of cricket predation by the pups of northern grasshopper mice (*Onychomys leucogaster*).

The importance of partially stripped cones in the process of learning to open cones was shown in both adult and young rats. We found that a consecutive decrease in the number of previously removed rows of scales enables the rat to develop the technique required to open an intact cone efficiently. While no naive rats learned to open cones when we presented them only with intact ones, about 70% of the naive rats successfully continued to open and completely strip the cones when given the right stimulus—a partially opened cone. Moreover, 90% of the naive adult rats that learned to strip cones from which the first four rows of scales had been removed, also learned to open intact cones after we had presented them with cones with a consecutively decreasing number of previously stripped rows of scales.

Although gradually exposing the rats to cones with a decreasing number of stripped rows of scales proved to be an appropriate way to "teach" them to strip cones, this is not a situation that occurs in nature. In the experiment in which we introduced pups to pine cones in various stages of opening simultaneously, but without going through the step-by-step procedure, only 24% of them acquired the stripping technique. In contrast, the addition of a social influence, a stripping mother, significantly increased the number of pups that learned to strip the cones to

Fig. 10. Rat pup feeding next to its mother on a partially opened cone which it has stolen from her.

almost all of them. It would appear, therefore, that the young rats cannot afford to delay their learning process beyond the period of time during which they still rely on the mother, since this is the only opportunity for them to acquire this complex feeding technique.

During the more than 10 years that we have been conducting this research and surveying pine forests, we have not come across any partially opened cones, other than a few specimens found on the ground that had fallen from the branch above while being stripped by a rat. Our hypothesis that the rat pups learn to open the cones through "stealing" partially opened ones from the mother, thus appears to be substantiated, because without stealing such cones the pups would never have the opportunity to be exposed to started cones and thus would never learn to open them. We therefore consider it possible that although the actual presence of the mother is not a precondition for learning to open cones, the partially opened cones obtained by the pups from their mother, which are a direct result of the mother's ability to open the cones, are a key factor in the learning process, permitting cultural transmission of this feeding behavior from one generation to the next.

Fisher and Hinde (1949) originally reported the spread of the phenomenon of milk-bottle opening by tits, and later interpreted it as the result of naive birds interacting with those experienced in milk-bottle opening (Hinde & Fisher, 1972). Although they emphasized the involvement of both social and nonsocial factors in the spread of this behavior, they did not determine the mechanism by which it was acquired. Sherry and Galef (1984, 1990) demonstrated that providing birds with previously opened bottles was sufficient to establish the bottle-opening behavior, and they concluded that such behaviors can be transmitted without involving social components.

The distinction between the above example and the results of our own studies, lies in the significant difference we found between pups that successfully acquire the stripping technique when exposed to the social component, in comparison with pups exposed only to nonsocial environmental changes.

The rats' ability to extract seeds from closed pine cones, and the transmission of this crucial behavior from generation to generation, are alimentary adaptations with important ecological implications. Because of the acid nature of the soil, and density of planting in Israeli pine forests, except for pine trees, they are an almost sterile habitat. For the young black rats born into this environment to survive, it is crucial that they learn to open the pine cones efficiently. Learned adaptive behavior, transferred culturally rather than genetically, has the advantage of spreading rapidly through a population, enabling its members to invade and exploit new habitats.

This is indeed the case with pine cone opening, a behavior that has allowed black rats to invade a new habitat, with its high carrying capacity.

In studies of transmission of acquired behavior, the question is regularly raised as to how the phenomenon originally started. Although we do not at present have an answer to this enigma, I believe that in this concluding part of the paper it is worth mentioning the following observations.

In many places in Israel, where there are pine groves and forests, cypress trees (*Cupressus sempervirens*) have been planted as windbreaks. We noticed that black rats inhabit these windbreaks and feed on cypress seeds (Aisner & Terkel, 1991).

Due to the basic structure of the cypress cone and simple arrangement of the seeds inside it (see Fig. 11), the technique for obtaining cypress cone seeds is relatively simple. We wondered whether rats already experienced in opening cypress cones could acquire the pine-cone-opening techniques easily. We therefore trapped rats from both cypress and pine groves and observed their motor patterns and ability to obtain nourishment from the two types of cones.

Our observation that prior experience with one variety of cone did not improve the ability of the rats to open the other, unfamiliar variety and that the behavior pattern employed in obtaining the familiar food is extremely specific, appeared to us to be most significant. Although both subpopulations of rats feed on the cones of conifers, there is a vast difference in the motor patterns and food-handling behavior used to open each type of cone.

Although we have not yet achieved all of our goals in unravelling the pine-cone-feeding phenomenon, we now have some understanding of the ontogeny of a feeding behavior that enables black rats to survive in the pine forests. Transfer of

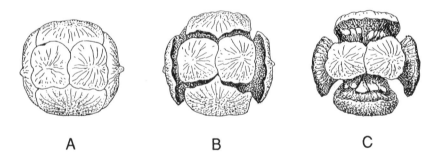

A   **B**   **C**

Fig. 11. Cypress cones at various stages of ripeness. (A) least ripe (closed), (B) partially open, suitable source of seeds, (C) overripe, seeds already dispersing.

this behavior from generation to generation is crucial for the survival of this species in an otherwise unexploited ecological niche.

The general picture that has emerged of cultural transmission in the case of cone opening is similar to other instances of transmission of feeding behavior from one generation to the next in other mammals. The significant feature in the present studies, making them different from others described in the literature, lies in our combining a complete, parametric laboratory analysis with the field observations, to provide an understanding of the processes contributing to the perpetuation of a behavior observed in a natural environment.

ACKNOWLEDGMENTS

I am grateful to Ms. Naomi Paz for preparing and editing the manuscript and to Dr. Amelia Terkel for reading and improving earlier versions; to Mr. Amikam Shoob for the photography and Mr. Walter Ferguson for the drawings. I sincerely thank my graduate students, Daphna Yanai, Ran Aisner, and Ofer Zohar, without whose dedication this work could not have been accomplished. The research was partially supported by the National Geographic Society grant #3684-85.

REFERENCES

Aisner, R. (1981). . . . of rats and pine cones. *Israel-Land and Nature,* **6,** 94–97.

Aisner, R., & Terkel, J. (1985). Habitat exploitation through cultural transmission: Pine cone feeding behaviour in black rats (*Rattus rattus*). *Proceedings of the 19th International Ethological Conference.* Toulouse: France.

Aisner, R., & Terkel, J. (1991). Sympatric black rat (*Rattus rattus*) populations. Different food handling techniques. *Mammalia,* **55,** 307–310.

Aisner, R., & Terkel, J. (1992a). Cultural transmission of pine cone opening behaviour in the black rat, *Rattus rattus. Animal Behaviour,* **44,** 327–336.

Aisner, R., & Terkel, J. (1992b). Pine cone opening behaviour in the black rat (*Rattus rattus*): ontogeny and mechanisms. In M. le Berre & L. le Guelte (Eds.), *The rodent and its environment* Paris: Chabaud (in press).

Boesch, C. (1991). Teaching among wild chimpanzees. *Animal Behaviour,* **41,** 530–532.

Bullock, D., & Neuringer, A. (1977). Social learning by following: An analysis. *Journal of Experimental Analysis of Behavior,* **25,** 103–117.

Canby, T. Y. (1977). The rat—lapdog of the devil. *National Geographic Magazine,* **152,** 60–87.

Eibl-Eibelsfeld, I. (1956). Uber Ontogenetische Entwiklung der Technik des Nusseoffnens von Eichhornche (*Sciurus vulgaris* L.). *Zeitschrift fur Saugertierkunde,* **21,** 132–134.

Ewer, R. F. (1971). The biology and behavior of a free-living population of black rats (*Rattus rattus*). *Animal Behaviour Monographs,* **4,** 127–174.

Fisher, J., & Hinde, R. A. (1949). The opening of milk bottles by birds. *British Birds,* **42,** 347–357.

Galef, B. G., Jr., & Henderson, P. W. (1972). Mother's milk: A determinant of feeding preferences of weaning pups. *Journal of Comparative Physiology and Psychology,* **78,** 213–219.

Galef, B. G., Jr., (1977). Mechanisms for the social transmission of acquired food preferences from adult to weaning rats. In L. M. Barker et al. (Eds.), *Mechanisms in food selection* (pp. 123–150). Austin, TX: Baylor University Press.

Galef, B. G., Jr., (1982). Studies of social learning in Norway rat: A brief review. *Developmental Psychobiology,* **15,** 279–295.

Galef, B. G., Jr., & Clark, M. M. (1972). Mother's milk and adult presence: Two factors determining initial dietary selection by weanling rats. *Journal of Comparative Physiology and Psychology,* **78,** 220–225.

Gardner, E. L., & Engel, D. R. (1971). Imitational and social facilitatory aspects of observational learning in the laboratory rat. *Psychonometrics Science,* **25,** 5–6.

Kemble, D. E. (1984). Effects of preweaning predatory or consummatory experience and litter size on cricket predation in northern grasshopper mice (*Onychomys leucogaster*). *Aggressive Behavior,* **10,** 55–58.

Krebs, J. R., MacRoberts, M. H., & Cullen, J. M. (1972). Flocking and feeding in the Great Tit (*Parus major*): An experimental study. *Ibis,* **114,** 507–530.

Le Magnen, J., & Tallon, S. (1968). Preference alimentaire du jeune rat induite par l'allaitement maternel. *Societe de Biologie,* Seance du 24 Fevrier 1968.

Lefebvre, L., and Palameta, B. (1988). Mechanisms, ecology and population diffusion of socially-learned, food-finding behavior in feral pigeons. In T. R. Zentall and B. G. Galef, Jr., (Eds.), *Social learning; Psychological and biological perspectives.* (pp. 141–164). Hillsdale, NJ: Erlbaum.

Lore, R., and Flannelly, K. (1977). Rat societies. *Scientific American,* **56,** 106–116.

Rozin, P. (1976). The selection of food by rats, humans and other animals. In J. Rosenblatt & R. A. Hinde, (Eds.), *Advances in the study of behavior* (Vol. 6), New York: Academic Press.

Sherry, D. F., & Galef, B. G. Jr. (1984). Cultural transmission without imitation: Milk bottle opening by birds. *Animal Behavior,* **32,** 937–938.

Sherry, D. F., & Galef, B. G., Jr. (1990). Social learning without imitation: More about milk bottle opening by birds. *Animal Behavior,* **40,** 987–989.

Smith, C. C., & Balda, R. P. (1979). Competition among insects, birds and mammals for conifer seeds. *American Zoologist,* **19,** 1065–1083.

Weigle, P. D., & Hanson, E. V. (1980). Observational learning and the feeding behaviour of the red squirrel (*Tamiasciurus hudsonicus*): The ontogeny of optimization. *Ecology,* **61,** 213–218.

Yanai, D. (1989). A comparison of different modes of pine cone opening (*Pinus halepensis*) by black rats (*Rattus rattus*). Unpublished master's thesis, Zoology Department, Tel Aviv University, Tel Aviv, Israel.

Yanai, D., & Terkel, J. (1991). Aspects of the learning ability of wild and laboratory rats encountering partially opened pine cones. *Annual Meeting of the Zoological Society of Israel,* 40–42.

Yanai, D., Ar, A., & Terkel, J. (1991). Energetic efficiency of two methods of pine cone opening behaviour by the black rat. *Sixth International Colloquium on Ecology and Taxonomy of Small African Mammals.* Mitzpe Ramon, Israel.

Zohar, O. (1987). The influence of social and environmental factors on pine cone opening learning behaviour of black rats. Unpublished master's thesis, Zoology Department, Tel Aviv University, Tel Aviv, Israel.

Zohar, O., & Terkel, J. (1992). Acquisition of pine cone stripping behaviour in black rats (*Rattus rattus*). *International Journal of Comparative Psychology,* **5,** 1–6.

3

Social Enhancement of Food Preferences in Norway Rats: A Brief Review[1]

Bennett G. Galef, Jr.

Department of Psychology
McMaster University
Hamilton, Ontario
Canada L8S 4K1

For more than 25 years, my students, coworkers, and I have been engaged in experiments designed both to analyze the behavioral processes that permit one Norway rat (*Rattus norvegicus*) to influence another's selection of foods and to determine how such social influences might facilitate the development of adaptive feeding repertoires by free-living rats.

The relative ease with which social influences on selection of food by rats can be studied in the laboratory has made social transmission of food preferences in rats a particularly fruitful model system in which to study social-learning processes at all stages in the life cycle (For reviews see Galef, 1977, 1985b, 1988, 1994): odor-bearing chemicals in a rat dam's food enter her bloodstream and cross placental membranes to infiltrate the circulation of any fetuses she is carrying. Consequently, late in gestation, fetal rats can detect the scents of at least some of the foods that their dam has eaten and will respond positively to those foods shortly after birth (Hepper, 1988). A few days after parturition, when infant rats are still totally dependent on their dam for nutriment, they receive information through their

---

1. Some portions of this article have appeared previously in Galef, B. G., Jr. (1994). Olfactory communications about foods among rats: A review of recent findings. In B. G. Galef, Jr., M. Mainardi, and P. Valsecchi (Eds.), *Behavioral Aspects of Feeding* (pp. 83–102). Chur, Switzerland: Harwood Academic Publishers.

mother's milk about flavors of foods that she is ingesting (Bronstein, Levine, & Marcus, 1975; Galef & Sherry, 1973; Galef & Henderson, 1972; Martin & Alberts, 1979). Still later in ontogeny, when weaning rats leave the safety of their natal burrow to seek their first meals of solid food in the open, they use adults of their colony as guides, foraging either at sites where adults are eating (Galef, 1971, 1981; Galef & Clark, 1971a,b) or at locations that adults have previously marked with residual olfactory cues (Galef & Beck, 1985; Galef & Heiber, 1976; Laland & Plotkin, 1991, 1993). In adolescence (and into adulthood), when rats frequently forage relatively independently, their food choices can be influenced by social interactions that occur at the home burrow at some distance from feeding sites (Galef & Wigmore, 1983; Posadas-Andrews & Roper, 1983). The scent of foods recently eaten, carried on the fur, vibrissae, and breath of a successful forager, can profoundly influence the food choices of other rats with whom the forager interacts.

Here, I first describe relatively briefly our previously reviewed (Galef, 1988), early work describing changes in rats' food preferences following social interactions at a distance from a feeding site with conspecifics that have recently eaten. I then describe in greater detail work on the phenomenon completed since the previous review was written.

OVERVIEW (1982–1986)

The Phenomenon

In 1982, Steven Wigmore and I, pursuing a lead generously provided by Barbara Strupp (See Strupp & Levitsky, 1984; Galef, 1991c), demonstrated that after a naive rat (an observer) interacted with a recently fed conspecific (a demonstrator), the observer exhibited a substantial enhancement of its preference for whatever food its demonstrator had eaten (Galef & Wigmore, 1983). Simultaneously (and independently) Posadas-Andrews and Roper (1983) discovered the same phenomenon.

Such social influence on diet selection by rats proved surprisingly robust (Galef, Kennett, & Wigmore, 1984); it is seen in a variety of situations, in rats of all postweaning ages, both sexes and several different strains (Galef et al., 1984; Richard, Grover, & Davis, 1987), as well as in house mice (*Mus domesticus*) (Valsecchi & Galef, 1989).

Social influence on the food choices of rats is also an unexpectedly powerful influence on diet selection, sometimes as powerful as learned aversions and congenital flavor preferences (Galef, 1986, 1989a): Observer rats taught a profound aversion to some food (as a result of experiencing gastrointestinal distress immediately after eating it) often completely abandoned their aversion after interacting with a demonstrator rat that had eaten the food to which the observers had learned the aversion (Galef, 1986, Galef et al., 1990). For weeks after interacting with demonstrator rats fed a diet flavored with cayenne pepper, observer rats offered a choice between a base diet and an unpalatable modification of that base diet flavored with cayenne pepper preferred the pepper-flavored diet, while control observer rats that had not interacted with demonstrators fed pepper-flavored diet strongly preferred the unflavored version of the base diet (Galef, 1989a).

Limitations on Rats' Communications about Foods

My coworkers and I were surprised to find that observer rats that interacted with unconscious demonstrators or with demonstrators that were experiencing an acute, experimentally induced, gastrointestinal distress exhibited preferences for, rather than aversions to, whatever foods their ill demonstrators had eaten. Indeed, several years of research, both in our laboratory and elsewhere (Galef, Wigmore, & Kennett, 1983; Galef, McQuoid, & Whiskin, 1990; Grover et al., 1988), have produced no evidence consistent with the view that food preferences induced in observer rats by demonstrators are influenced by the state of health of those demonstrators. Such repeated failure to find any sensitivity of observers to the well being or illness of demonstrators (coupled with the ease of finding enhanced preference for foods that demonstrators have eaten) suggests that the function of social transmission of food preferences in rats is to help them to identify potential foods, rather than to aid directly in their identification of potential poisons (Galef, 1985a).

The Analysis

Results of several of our experiments were consistent with the view that observer rats used olfactory cues emitted by demonstrators to identify foods that demonstrators had eaten: Rats developed a preference for a food fed to a demonstrator if separated from that demonstrator by a screen partition, but not if separated from a demonstrator by a transparent Plexiglas partition (Galef & Wigmore, 1983). Observers whose sense of smell had been blocked (by application of zinc sulfate solution to the nasal mucosa) failed to acquire enhanced preferences for foods their demonstrators had eaten, while intact control rats reliably exhibited such preferences (Galef & Wigmore, 1983).

Direct observation of the conditions under which observer rats acquired the food choices of their respective demonstrators indicated that for a demonstrator rat to influence the subsequent food preference of its observer, the observer had to bring its nose close to its demonstrator's mouth while they interacted (Galef & Stein, 1985). Presumably, approach to the mouth of a demonstrator rat was necessary for an observer rat to experience the scent of the food that the demonstrator had eaten.

Further experiments revealed that for an observer rat's food preferences to be affected by interaction with a demonstrator, the observer had to experience more than simple exposure to the smell of foods that their respective demonstrators had eaten, (Galef, Kennett, & Stein, 1985); observer rats that either smelled or ate a food did not develop an enhanced preference for it, while observer rats that smelled a food brushed onto the head of an anesthetized demonstrator rat did develop such a preference (Galef & Stein, 1985; Galef, Kenneth, & Stein, 1985).

I interpreted such findings as suggesting that the olfactory cues altering the food preferences of observer rats have two components: (1) a diet-identifying component (the smell associated with a food) and, (2) a contextual component (an odor produced by rats) that, acting together, were responsible for alterations in observers' food choices. We found that the diet-identifying component of the olfactory signal necessary to alter observer rats' food choices could be provided either by small amounts of food clinging to the fur and vibrissae of demonstrator rats or by the odor, escaping from the digestive system, of portions of food that had been introduced directly into the stomachs of demonstrator rats (Galef et al., 1985; Galef & Stein, 1985).

Yet other experiments showed that the contextual component of the olfactory signal necessary to modify the diet choices of observers was emitted from the anterior of anesthetized rats, but not from either the posterior of anesthetized rats or the anterior of rats recently sacrificed by anesthetic overdose (Galef et al., 1985; Galef & Stein, 1985).

RECENT DEVELOPMENTS (1986–1994)

Causal Analyses

Contextual Cues

The observation that effective contextual cues were localized at the anterior of live rats (Galef et al., 1985; Galef & Stein, 1985) led us to hypothesize that such cues might be contained in rat breath. Mass spectrographic analysis of rat breath revealed the presence, in significant quantities, of both carbonyl sulfide and carbon disulfide (CS_2) (Galef, Mason, Preti, & Bean, 1988). A subsequent test of the ability of CS_2 to provide a context within which exposure of a rat to a food odor would enhance its later preference for foods bearing that odor were successful. Both naive observer rats that we exposed to an anesthetized demonstrator rat that had eaten cinnamon-flavored diet and naive rats that we exposed to a piece of cotton batting that we had both powdered with cinnamon-flavored diet and moistened with a few drops of a dilute CS_2 solution exhibited significant (and roughly equivalent) enhancement of their subsequent preferences for cinnamon-flavored food. On the other hand, observer rats exposed to a piece of cotton batting that we both powdered with cinnamon-flavored diet and moistened with a few drops of distilled water exhibited no subsequent preference for cinnamon-flavored diet (Galef et al., 1988).

We have also asked whether experience of an odor on a conspecific produces a general enhancement of preference for that odor or enhances response to the odor only when it is associated with food. In a series of experiments (Galef, Iliffe, & Whiskin, 1994), we first exposed observer rats to demonstrator rats scented with either cinnamon or cocoa and then offered their observers choices between cinnamon- or cocoa-scented foods, cinnamon- or cocoa-scented nest materials and cinnamon- or cocoa-scented nest sites. Although, as expected, observer rats preferred food scented with the flavor they had experienced in association with a demonstrator rat, the same observers did not prefer either nest materials or nest sites bearing those same scents.

Of course, failures to find effects must always be interpreted with caution. Still, our data are not consistent with the view that the susceptibility of rats to social influences on their food preferences reflects a general enhancement of their preferences for odors experienced in association with conspecifics. Rather, experiencing an odor in association with a conspecific seems specifically to enhance rats' preferences for foods bearing that odor.

Development of Response to Demonstrators

It seemed likely that the experiences of young rats as they interacted with their dam and siblings would prove to be important in either the development or maintenance of rats' susceptibility to social influences on their food preferences. However, when we reared rat pups in total social isolation (Hall, 1975) from Day 2 or 3 postpartum to weaning and then tested them for susceptibility to social influence on food preference, we found that the effects of demonstrator rats on the food preferences of isolation-reared pups were as great as were their effects on the food preferences of pups reared by their dam and with siblings (Galef & Smith, 1994). Social enhancement of food preference developed and was maintained without interaction with conspecifics.

Extrapolation beyond the Laboratory

In all the experiments described above, observer rats interacted with demonstrators that had relatively simple recent histories of food intake; each demonstrator rat with which an observer rat interacted ate only a single food in the 24 h before interacting with its observer and each observer interacted with only a single demonstrator before it was tested for diet preference. It seems reasonable to suppose that in the world outside the laboratory: (1) free-living Norway rats often eat several different foods before interacting with colony mates, and (2) each rat interacts frequently with several of its fellows. Consequently, in natural circumstances, a rat interacting with conspecifics is likely to be exposed to an extended series of complex, food-related messages.

My students and I have carried out a number of experiments in which we fed demonstrator rats fairly complex diets before we allowed them to interact with observer rats. We then looked to see whether the observers could extract usable information from the complex, food-related olfactory signals their demonstrators provided (Galef, 1991b; Galef, Attenborough, & Whiskin, 1990; Galef & Whiskin, 1992, 1995). Rather than recount the entire history of our exploration of communications concerning food complexes, I shall describe here only two of the more elaborate situations we have examined to date.

Whiskin and I (Galef & Whiskin, 1992) first fed rats powdered rat chow to which we had added one of two combinations of four spices: either Combination A (cinnamon, anise, thyme, and cloves) or Combination B (cocoa, marjoram, cumin, and rosemary). Next we allowed individual rats that had eaten Combination A to interact for 1/2 h with individual rats that had eaten Combination B. Finally, we offered each rat a choice, for 22 h, between one of four pairs of flavored diets: (1) cinnamon-flavored vs cocoa-flavored diet, (2) anise-flavored vs marjoram-flavored diet, (3) thyme-flavored vs cumin-flavored diet, or (4) clove-flavored vs rosemary-flavored diet. (Note that one flavor in each of the four pairs of diets offered to subjects was a constituent of Combination A, the other was a constituent of Combination B.) Additional subjects that we had assigned to a control group each ate either Combination A or Combination B, but did not interact with a conspecific before we offered them a choice between one of the same four diet pairs we offered to subjects that did interact with conspecifics before testing.

During testing, the food preferences of control subjects that had eaten Combination A did not differ from the food preferences of control subjects that had eaten Combination B. However, when tested for food preference, subjects that had both eaten Combination A and interacted with a partner that had eaten Combination B, ate more of those diets flavored with a spice present in Combination B than did subjects that had eaten Combination B and interacted with a partner that had eaten Combination A. In sum, subjects developed preferences for flavors that their partners had eaten, even when those partners had eaten quite complex diets.

Other experiments revealed that observer rats can respond not only to complex single messages, but also to a succession of simple messages received from a series of demonstrators (Galef, Attenborough, & Whiskin, 1990). On each of nine occasions spread over 23 days, we allowed observer rats assigned to an experimental group to interact for 1/2 h with a demonstrator rat that had just eaten a diet unfamiliar to its observer. Each observer in a control group interacted at the same time with a demonstrator rat that we fed the same diet on which both demonstrators and observers had been maintained throughout life.

Figure 1 shows the days on which demonstrators and observers interacted, the diets fed to demonstrators in the experimental group before they interacted with their observers, and the food choices given to all subjects for 23.5 h on each day of the experiment. As can be seen in Fig. 1, on each day of the experiment when subjects interacted with demonstrators, subjects in the experimental group exhibited a significant enhancement of their preferences for the foods that their demonstrators had eaten.

Functions of Social Learning about Food

It is one thing to know that olfactory messages passing from demonstrator rats to their observers can alter the observers' later food preferences. It is quite another to understand how such socially induced changes in food choice might facilitate development of adaptive feeding repertoires in rodents living outside the laboratory. In the present section, I review several experiments the results of which suggest that olfactory communications about foods help Norway rats to decide: (1) what foods to eat, (2) what potential foods to learn to avoid eating, and (3) where to go to find food.

Learning What to Eat

Although individual rats can sometimes learn to select a single nutritionally adequate food embedded in an array of nutrient-deficient foods (Galef & Beck, 1990; Richter, Holt, & Baralare, 1938; Rozin, 1969), it is relatively easy to create

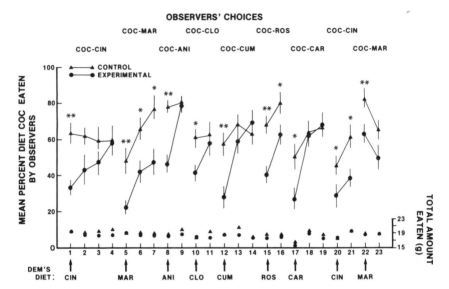

Fig. 1. Mean percentage of Diet Coc eaten and total amount eaten by observer rats in experimental and control groups. Pairs of diets shown at the top of the figure indicate the choices offered to observers on each day. Diets shown at the bottom of the figure indicate foods presented to demonstrators on days indicated by the vertical arrows. Flags indicate ± 1 SEM. * = $p < .05$, ** = $p < .01$. CIN = cinnamon-flavored diet; MAR = marjoram-flavored diet, ANI = anise-flavored diet, CLO = clove-flavored diet; CUM = cumin-flavored diet; ROS = rosemary-flavored diet; CAR = cardamom-flavored diet. Reprinted from Galef, Attenborough, & Whiskin (1990) by permission of the American Psychological Association.

situations in which naive rats have great difficulty in focusing their feeding on the sole nutritionally adequate food present in an array of foods (Beck & Galef, 1989; Galef, 1991a; Galef, Beck, & Whiskin, 1991).

Beck and I (Beck & Galef, 1989) placed individual rats in enclosures containing three protein-deficient foods (Diet, Cin, Coc, and Thy) and a single relatively unpalatable food (Diet Nut) that contained adequate protein (and all other nutrients) to support normal growth and development. Each subject was either placed alone in a cage or shared its enclosure with one or more conspecific demonstrators that we had trained to eat the protein-rich Diet Nut and to avoid eating the three, protein-deficient alternative diets present in the enclosure. Observer rats that shared their enclosures with trained demonstrators were able to grow rapidly, while rats maintained in isolation failed to thrive in the experimental situation (Beck & Galef, 1989).

To determine how trained demonstrator rats were affecting the feeding behavior of naive subjects in the test situation, we used enclosures arranged in the three ways illustrated in Fig. 2. As can be seen in the figure, we varied the diet eaten by demonstrators (protein-rich Diet Nut or protein-deficient Diet Cin) and the location of the demonstrator's food cup relative to those of their observers (whether demonstrator's ate from a cup adjacent to the cup in the observer's side of the enclosure that contained protein-rich Diet Nut or protein-deficient Diet Cin). Only observers whose demonstrators ate Diet Nut grew rapidly and the rate of growth of observers was not affected by whether demonstrators ate near the food cup containing Diet Nut or elsewhere (Fig. 3). Clearly, information obtained by observers about the food that their respective demonstrators were eating was used by naive rats to locate the most valuable of several available foods.

We have also found that the degree to which the food choices of observer rats are affected by the food choices of their demonstrators is determined by the state of health of the observers themselves. Food choices of protein-deficient observer rats were significantly more profoundly influenced by foods eaten by demonstrator rats than were food choices of protein-replete observer rats (Galef et al., 1991).

Learning What Not to Eat

The ability to use conspecifics as sources of information about which foods are safe to eat might be particularly helpful to individuals that have eaten several different unfamiliar foods (as might a weaning rat) before becoming ill. It seems reasonable to assume that it would be adaptive for a rat to act as though foods that

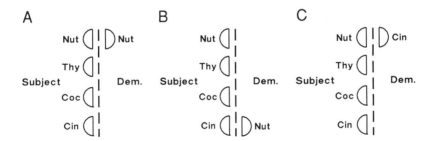

Fig. 2. Overhead schematic of the positions of food cups presented to subjects and their respective demonstrators in the Same-Food, Same-Place, Same-Food, Different-Place and Different-Food, Same-Place conditions. Nut = nutmeg-flavored diet (high in protein); Cin = cinnamon-flavored diet; Thy = thyme-flavored diet; Coc = cocoa-flavored diet (all three low in protein); Dem. = demonstrator. Reprinted from Beck and Galef (1989) by permission of the American Psychological Association.

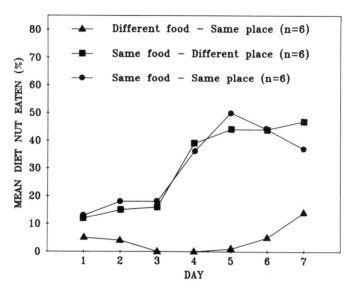

Fig. 3. Mean amount of high-protein nutmeg-flavored diet ingested by subjects as a percentage of their total intake. Reprinted from Beck and Galef (1989) by permission of the American Psychological Association.

other rats are eating are less likely to be toxic than are foods that other rats are not eating.

To allow observer rats to learn what diet another rat was eating, I fed demonstrator rats either cocoa-flavored diet (Diet Coc) or cinnamon-flavored diet (Diet Cin) and then allowed each demonstrator to interact with an observer. I next fed observer rats Diet Cin and Diet Coc in succession, then poisoned the observers with lithium-chloride. To determine whether observers had learned an aversion to Diet Cin or Diet Coc, I offered them a choice between Diet Cin and Diet Coc (Galef, 1987). Observer rats were far more likely to form an aversion to whichever food their respective demonstrator had not eaten than to the food that their respective demonstrators had eaten (See also Galef, 1989b).

Learning Where to Look for Food

Galef, Mischinger, and Malenfant (1987) found that rats will spontaneously follow trained conspecific leaders through a maze and that they will follow potential leaders that have eaten a "safe" food significantly more frequently than they will follow potential leaders that have eaten an "unsafe" food. We trained follower rats to enter the same goal box as their respective leaders. Once the followers were performing reliably in the maze, we offered each of them, in their home cage, a sample of a novel diet and poisoned them. We subsequently gave followers the opportunity to follow leaders that had eaten either the followers' standard maintenance diet or the food to which the followers had learned an aversion. We found that followers entered the same goal box as their leaders on 90% of the trials when the leaders had eaten maintenance diet on only 58% of trials on which leaders had eaten the diet to which followers had learned an aversion.

Using Studies of Causation to Elucidate Function

Because one might expect natural selection to have shaped processes supporting social learning about foods, discovery of limits on the situations in which such social learning occurs should provide some insight into the functions that social induction of food preference might serve in natural circumstances. We hypothesized that if social induction of food preference is a behavioral process evolved to permit rats to expand their feeding repertoires without incurring some of the risks inherent in sampling previously untasted potential foods, then one might expect social exposure to unfamiliar foods to be more effective in altering the food preferences of rats than social exposure to familiar foods.

In a series of experiments investigating the social induction of rats' preferences for familiar and unfamiliar foods, we found repeatedly that observer rats learned a preference for the unfamiliar rather than the familiar diet that their demonstrators were eating (Galef, 1993). Such findings are consistent with the notion that social induction of diet preference serves rats as a means of reducing the cost of adding totally new foods to their feeding repertoires. The results of subsequent experiments showed that individual experience of a food interfered with social induction of a preference for that food for only a few days after the food was eaten (Galef & Whiskin, 1994). These latter results suggest that the combined effects of individual and social learning about foods should result in free-living rats exhibiting preferences both for totally unfamiliar foods that it learns that conspecifics are eating and for foods that it learns that conspecifics are eating that it has not eaten recently.

CONCLUSIONS

Studies of social influence on food preference in rats have resulted in the discovery of a previously unsuspected major determinant of diet choice. Such studies have also provided a useful model system in which to explore the many ways in which social interactions can modulate behavioral development (Galef, 1991a; Hill, 1995). Our findings, together with those of others working in the area, have revealed multiple systems that permit naive rats to extract information from their more experienced fellows thus increasing the probability that the food choices of the naive will be beneficial.

As many other chapters in this volume make clear, neither rats nor foraging animals are unique in using socially acquired information to guide their behavior. Although the details of the processes supporting social influence on the food choices of rats may be of interest only to experts in feeding behavior, the general message that a complex of nonimitative social-learning processes can play a central role in development of locally adaptive patterns of behavior should be of importance to all with an interest in the causes and functions of the behavior of animals. Whether one's interests are in the mate choices of guppies (See Dugatkin, this volume), predator recognition by rhesus macaques (Mineka & Cook, 1988), or the spread of novel means of food extraction by European titmice (Fisher & Hinde, 1949; Sherry & Galef, 1984) investigations of social-learning processes are likely to provide insight into how locally adaptive patterns of behavior are acquired.

REFERENCES

Barnett, S. A. (1958). Experiments on "neophobia" in wild and laboratory rats. *British Journal of Psychology,* **49,** 195–201.

Beck, M., & Galef, B. G., Jr. (1989). Social influences on the selection of a protein-sufficient diet by Norway rats (*Rattus norvegicus*). *Journal of Comparative Psychology,* **103,** 132–139.

Bronstein, P. M., Levine, J. M., & Marcus, M. (1975). A rat's first bite: The nongenetic, cross-generational transfer of information. *Journal of Comparative and Physiological Psychology,* **89,** 295–298.

Fisher, J., & Hinde, R. A. (1949). The opening of milk bottles by birds. *British Birds,* **42,** 347–357.

Galef, B. G., Jr. (1971). Social effects in the weaning of domestic rat pups. *Journal of Comparative and Physiological Psychology,* **75,** 358–362.

Galef, B. G., Jr. (1977). Mechanisms for the social transmission of food preferences from adult to weanling rats. In L. M. Barker, M. Best, & M. Domjan (Eds.), *Learning mechanisms in food selection* (pp. 123–150). Waco, TX: Baylor University Press.

Galef, B. G., Jr. (1981). The development of olfactory control of feeding site selection in rat pups. *Journal of Comparative and Physiological Psychology,* **95,** 615–622.

Galef, B. G., Jr. (1985a). Direct and indirect behavioral processes for the social transmission of food avoidance. In P. Bronstein and N. S. Braveman (Eds.), *Experimental assessments and clinical applications of conditioned food aversions* (pp. 203–215). New York: New York Academy of Sciences.

Galef, B. G., Jr. (1985b). Social learning in wild Norway rats. In T. D. Johnston & A. T. Pietrewicz (Eds.), *Issues in the ecology of learning* (pp. 143–165). Hillsdale, NJ: Erlbaum.

Galef, B. G., Jr. (1986). Social interaction modifies learned aversions, sodium appetite, and both palatability and handling-time induced dietary preference in rats (*Rattus norvegicus*). *Journal of Comparative Psychology,* **100,** 432–439.

Galef, B. G., Jr. (1987). Social influences on the identification of toxic foods by Norway rats. *Animal Learning and Behavior,* **15,** 327–332.

Galef, B. G., Jr. (1988). Communication of information concerning distant diets in a social, central-place foraging species: *Rattus norvegicus.* In T. R. Zentall & B. G. Galef, Jr. (Eds.), *Social learning: Psychological and biological perspectives* (pp. 119–139). Hillsdale, NJ: Erlbaum.

Galef, B. G., Jr. (1989a). Enduring social enhancement of rats' preferences for the palatable and the piquant. *Appetite,* **13,** 81–92.

Galef, B. G., Jr. (1989b). Socially mediated attenuation of taste-aversion learning in Norway rats: Preventing development of "food phobias." *Animal Learning and Behavior,* **17,** 468–474.

Galef, B. G., Jr. (1991a). Innovations in the study of social learning in animals: A developmental perspective. In H. N. Shair, G. A. Barr, & M. A. Hofer (Eds.), *Methodological and conceptual issues in developmental psychobiology* (pp. 114–125). Oxford: Oxford University Press.

Galef, B. G., Jr. (1991b). A contrarian view of the wisdom of the body as it relates to food selection. *Psychological Review,* **98,** 218–223.

Galef, B. G., Jr. (1991c). Information centres of Norway rats: Sites for information exchange and information parasitism. *Animal Behaviour,* **41,** 295-301.

Galef, B. G., Jr. (1993). Functions of social learning about food: A causal analysis of effects of diet novelty on diet preference. *Animal Behaviour,* **46,** 257–265.

Galef, B. G., Jr. (1994). Olfactory communications about foods among rats: A review of recent findings. In B. G. Galef, Jr., M. Mainardi, & P. Valsecchi (Eds.), *Behavioral aspects of feeding: Basic and applied research in mammals* (pp. 83–102). Chur, Switzerland: Harwood Academic Publishers.

Galef, B. G., Jr., Attenborough, K. S., & Whiskin, E. E. (1990). Responses of observer rats to complex, diet related signals emitted by demonstrator rats. *Journal of Comparative Psychology,* **104,** 11–19.

Galef, B. G., Jr., & Beck, M. (1985). Aversive and attractive marking of toxic and safe foods by Norway rats. *Behavioral and Neural Biology,* **43,** 298–310.

Galef, B. G., Jr., & Beck, M. (1990). Diet selection and poison avoidance by mammals individually and in social groups. In E. M. Stricker (Ed.), *Handbook of Neurobio-*

logy (Vol. 10): Neurobiology of food and fluid intake (pp. 329–349). New York: Plenum Press.

Galef, B. G., Jr., Beck, M., & Whiskin, E. E. (1991). Protein deficiency magnifies social influence on the food choices of Norway rats (*Rattus norvegicus*). *Journal of Comparative Psychology,* **105,** 55–59.

Galef, B. G., Jr., & Clark, M. M. (1971a). Parent-offspring interactions determine time and place of first ingestion of solid food by wild rat pups. *Psychonomic Science,* **25,** 15–16.

Galef, B. G., Jr., & Clark, M. M. (1971b). Social factors in the poison avoidance and feeding behavior of wild and domesticated rat pups. *Journal of Comparative and Physiological Psychology,* **75,** 341–357.

Galef, B. G., Jr., & Henderson, P. W. (1972). Mother's milk: A determinant of the feeding preferences of weaning rats pups. *Journal of Comparative and Physiological Psychology,* **78,** 213–219.

Galef, B. G., Jr., & Heiber, L. (1976). The role of residual olfactory cues in the determination of feeding site selection and exploration patterns of domestic rats. *Journal of Comparative and Physiological Psychology,* **90,** 727–739.

Galef, B. G., Jr., Iliffe, C., & Whiskin, E. E. (1994). Social influences on rats' (*Rattus norvegicus*) preferences for flavored foods, scented nest materials and odors associated with harborage sites: Are flavored foods special? *Journal of Comparative Psychology,* **108,** 266–273.

Galef, B. G., Jr., & Kennett, D. J. (1985). Delays after eating: Effects on transmission of diet preferences and aversions. *Animal Learning and Behavior,* **13,** 39–43.

Galef, B. G., Jr., & Kennett, D. J. (1987). Different mechanisms for social transmission of diet preference in rat pups of different ages. *Developmental Psychobiology,* **20,** 209–215.

Galef, B. G., Jr., Kennett, D. J., & Stein, M. (1985). Demonstrator influence on observer diet preference: Effects of simple exposure and the presence of a demonstrator. *Animal Learning and Behavior,* **13,** 25–30.

Galef, B. G., Jr., Kennett, D. J., & Wigmore, S. W. (1984). Transfer of information concerning distant food in rats: A robust phenomenon. *Animal Learning and Behavior,* **12,** 292–296.

Galef, B. G., Jr., Mason, J. R., Preti, G., & Bean, N. J. (1988). Carbon disulfide: A semiochemical mediating socially induced diet choice in rats. *Physiology and Behavior,* **42,** 119–124.

Galef, B. G., Jr., McQuoid, L. M., & Whiskin, E. E. (1990). Further evidence that Norway rats do not socially transmit learned aversions to toxic baits. *Animal Learning and Behavior,* **18,** 199–205.

Galef, B. G., Jr., Mischinger, A., & Malenfant, S. (1987). Further evidence of an information centre in Norway rats. *Animal Behaviour,* **35,** 1234–1239.

Galef, B. G., Jr., & Sherry, D. F. (1973). Mother's milk: A medium for the transmission of cues reflecting the flavor of mother's diet. *Journal of Comparative and Physiological Psychology,* **83,** 374–378.

Galef, B. G., Jr., & Smith, M. (1994). Susceptibility of artificially reared rat pups to social influences on food choice. *Developmental Psychobiology, 27,* 85–92.

Galef, B. G., Jr., & Stein, M. (1985). Demonstrator influence on observer diet preference: Analysis of critical social interactions and olfactory signals. *Animal Learning and Behavior, 13,* 31–38.

Galef, B. G., Jr., & Whiskin, E. E. (1992). Social transmission of information about multi-flavored foods. *Animal Learning and Behavior, 20,* 56–62.

Galef, B. G., Jr., & Whiskin, E. E. (1994). Passage of time reduces effects of familiarity on social learning: Functional implications. *Animal Behaviour, 48,* 1057–1062.

Galef, B. G., Jr., & Whiskin, E. E. (1995). Learning socially to eat more of one food than of another. *Journal of Comparative Psychology, 190,* 99–101.

Galef, B. G., Jr., & Wigmore, S. W. (1983). Transfer of information concerning distant foods: A laboratory investigation of the "information-centre" hypothesis. *Animal Behaviour, 31,* 748–758.

Galef, B. G., Jr., Wigmore, S. W., & Kennett, D. J. (1983). A failure to find socially mediated taste aversion learning Norway rats (*Rattus norvegicus*). *Journal of Comparative Psychology, 97,* 358–363.

Grover, C. A., Kixmiller, J. S., Erickson, C. A., Becker, A. H., Davis, S. F., & Nallan, G. B. (1988). The social transmission of information concerning aversively conditioned liquids. *Psychological Record, 38,* 557–566.

Hall, W. G. (1975). Weaning and growth of artificially reared rats. *Science, 190,* 1313–1316.

Hepper, P. G. (1988). Adaptive fetal learning: Postnatal exposure to garlic affects postnatal preferences. *Animal Behaviour, 36,* 935–936.

Hill, W. L. (1995). On the importance of evolution to developmental psychobiology. *Developmental Psychobiology, 28,* 117–129.

Laland, K. N., & Plotkin, H. C. (1991). Excretory deposits surrounding food sites facilitate social learning about food preferences in Norway rats. *Animal Behaviour, 41,* 997–1005.

Laland, K. N., & Plotkin, H. C. (1993). Social transmission of food preferences among Norway rats by marking of food sites and by gustatory contact. *Animal Learning and Behavior, 21,* 35–41.

Martin, L. T., & Alberts, J. R. (1979). Taste aversions to mother's milk: The age-related role of nursing in acquisition and expression of learned associations. *Journal of Comparative and Physiological Psychology, 93,* 430–445.

Mineka, S., & Cook, M. (1988). Social learning and the acquisition of snake fear in monkeys. In T. R. Zentall & B. G. Galef, Jr. (Eds.), *Social learning: Psychological and biological perspectives* (pp. 51–74). Hillsdale, NJ: Erlbaum.

Posadas-Andrews, A., & Roper, T. J. (1983). Social transmission of food preferences in adult rats. *Animal Behaviour, 31,* 265–271.

Richard, M. M., Grover, C. A., & Davis, S. F. (1987). Galef's transfer of information effect occurs in a free-foraging situation. *Psychological Record, 37,* 79–87.

Richter, C. P., Holt, L., & Baralare, B. (1938). Nutritional requirements for normal growth and reproduction in rats studied by the self-selection method. *American Journal of Physiology, 122,* 734–744.

Rozin, P. (1969). Adaptive food sampling patterns in vitamin deficient rats. *Journal of Comparative and Physiological Psychology, 69,* 126–132.

Sherry, D. F., & Galef, B. G., Jr. (1984). Cultural transmission without imitation: Milk bottle opening by birds. *Animal Behaviour, 32,* 937–938.

Strupp, B. J., & Levitsky, D. E. (1984). Social transmission of food preferences in adult hooded rats (*Rattus norvegicus*). *Journal of Comparative Psychology, 98,* 257–266.

Valsecchi, P., & Galef, B. G., Jr. (1989). Social influences on the food preferences of house mice (Mus musculus). *International Journal of Comparative Psychology, 2,* 245–256.

4

Social Learning in Monkeys: Primate "Primacy" Reconsidered

DOROTHY M. FRAGASZY

Department of Psychology
University of Georgia
Athens, Georgia 30602

ELISABETTA VISALBERGHI

Istituto di Psicologia
Consiglio Nazionale delle Ricerche
Rome, Italy

INTRODUCTION

The recognition that humans share many traits with other primates can have as an unintended correlate an uncritical willingness to ascribe human traits to other primate species (Kennedy, 1992). Behavioral researchers are more likely to provide higher-order "cognitive" explanations for behaviors in primates than members of other orders, perhaps reflecting some intuitive notion that cognitive continuity extends from humans to other primates, but not to other orders. These two tendencies are as misleading for primatologists as they are for the general public (Visalberghi & Fragaszy, 1990); and they extend to our views of social learning. They are apparent, for example, in the cognitive slant to explanations for behaviors in nonhuman primates (such as dietary choices or skilled foraging actions) that we assume are either learned socially, or are socially modulated in humans. In this chapter, we make the case that the apparently natural inclination to attribute a special character to social learning in monkeys, relative to social learning in other animals, is unwarranted. This is not to say that social influences are not important to primates, as to other orders. Rather, the comparative psychological issue is whether a different set of underlying mechanisms supports social learning in primates than in other orders.

We draw on some recent research in our laboratories to illustrate how homogeneity in dietary choice among individual monkeys living in the same group may be supported in a direct way by social interactions and social predispositions, in the absence of social-learning processes that are not shared with other orders. Primate researchers have long suggested that food preferences acquired by young monkeys and apes are strongly influenced by the activities and choices of their mother in particular, and other group members in general (Goodall, 1986; Hall, 1963; Kawamura, 1959). In the general case, people observed young monkeys and apes exploring foods in the presence of their mothers, and thought that these interactions with foods might play a formative role in the infant's later choices of foods. When systematic observations were made of selected folivorous species, the hypothesis that the mother's choices were correlated with infant's choices as supported (Watts, 1985; Whitehead, 1986). For example, Watts showed that infant gorillas ingested foods chosen by their mothers much more often than they ingested other foods.

The notion that infants acquire food preferences, or at least information about what is edible, either from their mother or from other individuals through social learning has become embedded in our ideas about primates. This notion is supported indirectly by work with rodent and other mammals that has documented the pervasive role that socially provided information plays in the development of food preferences in other mammalian orders. Young animals' food choices are indeed likely to be influenced in many ways by their mothers or other group members, sometimes in nonobvious ways (such as through the taste of the mother's milk; Galef and Sherry, 1973). But there is an important difference between the ways in which social influences on feeding have been thought about in primates and in other mammalian orders. In rats, for example, it has not been suggested that individuals intend to teach others through their activities, nor has it been hypothesized that infants learn about mothers' food choices merely from watching her actions, or from mothers actively sharing food with infants. In rodents and other nonprimate orders, functional explanations are offered, without attribution of intention (Caro & Hauser, 1992). In our view, this is the level of explanation which is appropriate for nonhuman primates as well, until compelling evidence suggests otherwise. We have not yet seen such evidence for monkeys.

Sharing of food has been considered a particularly likely means for young individuals to acquire food preferences from others (Lefebvre, 1985). This behavior is present in several species of nonhuman primates as well as in other orders (Feistner & McGrew, 1989). The notion that food sharing might be a means of "educating" young monkeys about food identity or quality is problematic, however. A recent experiment by Price and Feistner (1993) illustrates why the notion is becoming untenable. Food sharing, defined as one individual handing over food to another (rather than just tolerating another taking food), is especially evident among callitrichid monkeys (marmosets and tamarins). Price and Feistner examined the extent to which tamarins shared food with their offspring under varying conditions of ease of access, novelty, and abundance. Adult tamarins shared food more often when access to food was restricted, either by physical location or by scarcity; under these conditions, infants fed themselves less often. Adults shared foods less often when novel fruit were presented. Thus, food sharing in these animals is more likely to contribute to the infant maintaining adequate nutrition in the present than to the infant acquiring useful information about foraging for the future.

The callitrichid monkeys are indeed unusual among primates in the frequency and probable nutritional importance to the infants of adult-initiated food sharing. More often, researchers propose that infants learn what to eat by watching what those around them are eating, or by sampling what others are eating after taking food from the others, in a tolerated fashion (e.g., Fedigan, 1982; see Box 1984 for general review). Granted, this is a methodologically difficult hypothesis on which to generate strong tests, because diets of most species are broad, exposure to elements in the diet occurs over a long time, and infants spend a long time ingesting small amounts of foods prior to and during weaning and in varying social circumstances. Thus there are many overlapping means by which infants could acquire food preferences and knowledge of foraging techniques, and the specific contribution of social interactions to this process, however striking the interactions may be, cannot be clearly assessed.

This problem points to another reason why people assume that social learning is more important to primates than to other animals: their relatively long period of juvenescence, combined with their intense sociality. It is still an open question whether, during their extended juvenescence (Pereira & Fairbanks, 1993), young primates learn more or different things about the environment than do other animals. Likewise, it is still an open question how much of primate learning during this period is socially mediated. A lengthy juvenescence in the natal group provides time and occasion for social learning, to be sure, but neither of these is sufficient to produce social learning.

We know now that in at least some cases having a "model" is not necessary for species-normal dietary selections to develop in primates. For example, Milton (1993a) has documented that a group of juvenile spider monkeys, raised primarily by humans and subsequently released into a natural environment lacking other spider monkeys, developed species-normal diets. As Milton points out, spider monkeys are frugivorous, and fruits are unlikely to have toxic components to the same extent as other foods (particularly leaves). Leaf-eating species might be expected to benefit more from social learning about selection of food, on the basis of the potential toxicity of their foods, than would frugivorous species. Indeed, the strongest examples of social influence on food selection in nonhuman primates from field observations involve folivorous species (Watts, 1985; Whitehead, 1986).

EXPERIMENTAL STUDIES OF FEEDING

We have recently completed laboratory studies investigating simple social tendencies influencing feeding behavior in tufted capuchin monkeys (*Cebus apella*) (Visalberghi & Fragaszy, in press; Fragaszy, Visalberghi, & Galloway, 1995; Fragaszy, Feuerstein, & Mitra, 1995). We decided to examine social influences on feeding behavior after spending several years studying social influences on tool using and problem solving in capuchin monkeys (Fragaszy & Visalberghi 1989, 1990; Fragaszy, Vitale, & Ritchie, 1994, Adams-Curtis & Fragaszy, 1995). Through this work we came to the inescapable conclusion that capuchin monkeys do not acquire instrumental behaviors (such as using a tool, solving a mechanical puzzle, or washing sandy foods) by watching a skilled demonstrator. Nevertheless, we believe that social influences were powerful in these monkeys, even if not sufficient to produce tool-using behaviors or skilled sequences of action in observers. The social processes probably of greatest importance to capuchin monkeys in these circumstances were not imitation or teaching. Rather, they were more likely to be the mundane mechanisms of social facilitation and local enhancement, which influence the rate and location of activity more than its precise form (Galef, 1988; Clayton, 1978). These aspects of social influence should be evident in feeding, and feeding is the area of behavior most often considered when social learning is discussed in nonhuman primates.

Capuchin monkeys (*Cebus* spp.) are a generalist genus exhibiting a flexible use of food resources across a broad range of habitats (Fragaszy, Visalberghi, & Robinson, 1990). Capuchin monkeys live in groups of 10 to more than 30 individuals. Groups are relatively cohesive during the day, although individuals spread out while foraging (Robinson, 1981). Social relations within the group, both in captivity and in nature, are generally pacific (Fedigan, 1993; Fragaszy et al., 1994), and individuals have occasionally been observed in the field to allow others to take food from them (Perry & Rose, 1994). Capuchin monkeys are primarily frugivorous, although animal prey also constitutes an important part of their diet. They are opportunistic and destructive foragers, exploiting many plant and animals foods (such as nuts in hard shells, and invertebrates buried in wooden substrates) that other species can not exploit. In all habitats, individuals have a wide diet (Brown & Zunino, 1990; Fragaszy & Boinski, 1995; Izawa & Mizuno, 1977; Izawa, 1979; Janson & Boinski, 1992; Robinson, 1986; Rose, 1994; Terborgh, 1983).

    Many animal species are known to exhibit neophobia toward novel (potential) foods, either avoiding them completely or eating only small quantities at first (Rozin, 1976). Caution toward novel potential foods serves to minimize an animal's exposure to toxic compounds and other deleterious substances which can be present in foods (Glander, 1982; Milton, 1993b). Nonetheless, the diversity of diet across and within populations of capuchin monkeys suggests that propensities to sample novel foods and to incorporate them into the diet must also be well developed. In opportunistic species such as capuchin monkeys, individuals ought to balance their inclination to sample novel food with caution toward potentially harmful food-stuffs. Given the tolerant nature of capuchin monkeys towards others with food, it seemed to us that social enhancement and facilitation could play a role in how these monkeys respond to novel foods and other food-related challenges.

Study 1: Social Influences on Response to Familiar and Novel Foods in Juvenile and Adult Capuchin Monkeys

Social partners can increase an individual's consumption of food through the phenomenon of social facilitation (the increased performance of a behavior while in the presence of another individual performing that behavior; see Clayton, 1978). In adult humans, social facilitation of eating has been demonstrated repeatedly (e.g., de Castro & Brewer, 1992). Social facilitation of feeding has also been demonstrated in captive pairs of rhesus macaques (Harlow & Yudin, 1933). Surprisingly, we could find no experimental studies of social facilitation of feeding in the primatological literature since 1933.

    There are several other ways in which a social partner could influence an observer's interest in feeding, or in what it is feeding on. For example, a partner's activity might draw the observer's attention to a food and to actions performed with it. Or, an observer might obtain bits of food from another animal, and thus sample the food and gain experience manipulating it. We were not sure at the outset what forms of social influence would be evident among capuchin monkeys, how social tendencies might change as young animals grew more independent in their feeding, or how social influences might be affected by the nature of the foods involved.

In both Mongolian gerbils (*Meriones unguiculatus*) and Norway rats (*Rattus norvegicus*), social facilitation of feeding is more effective when a food is novel than when it is familiar (Forkman, 1991; Galef, 1993). In rats, for example, consumption of a novel food is more affected than consumption of a familiar food, following an encounter with a conspecific that has been eating both (Galef, 1993). Galef suggests that more powerful social influences on food choices toward novel food are advantageous because they decrease the risk of enlarging the diet. One can hypothesize that the more powerful effects of social facilitation on novel foods would be evident in those species exhibiting broad diets and opportunistic foraging styles. Both rats and capuchin monkeys are generalist feeders, and therefore one can expect a pattern of differential facilitation of feeding similar to that found in rats to be present in capuchin monkeys.

In our first study, we investigated whether consumption of novel foods in capuchin monkeys might be socially facilitated to a greater extent than consumption of familiar foods (Visalberghi & Fragaszy, 1995). First, we presented abundant quantities of several familiar foods to individuals in their home cage, either while their groupmates were present (social condition), or while subjects were alone (individual condition). We tested 11 captive capuchin monkeys (ages 2–15 years) living in two groups. On average, individuals ate a similar number of pieces of food in the two conditions. Unlike humans (de Castro & Brewer, 1992) and rhesus monkeys (Harlow & Yudin, 1933), these capuchin monkeys did not eat more of familiar foods while companions were present and feeding, than when alone.

Next, we presented 20 totally unfamiliar foods to each subject in the same two conditions: social or individual. The unfamiliar foods were selected to provide a wide variety of flavors and textures, and to be similar to the kinds of foods (animal protein, fruit, legumes, flowers, fungi) that capuchin monkeys eat in natural environments. Four familiar foods were also presented to all subjects. A single food was presented during each test session. The 20 novel foods were divided into two sets; each group received one set in the social condition and the other set in the individual condition.

Acceptance of the 20 novel foods ranged from complete (all the 11 subjects ate the food in at least one sample) to unanimous rejection. Acceptance did not seem to be related to food quality, in terms of protein or carbohydrate content. Overall, an average of 7.5 individuals ate each novel food at least once (pooled conditions). In both conditions, all subjects ate the familiar foods more often than novel foods.

Two findings point to the presence of social facilitation of feeding on unfamiliar foods. First, a significant number of individuals (9 of 11) ate novel foods on more samples when tested together than when tested alone (see Fig. 1). Second, 9 of

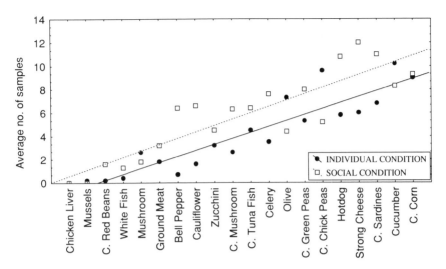

Fig. 1. The average number of samples in which an individual was eating a novel food while alone (individual condition) or with its groupmates (social condition) in Study 1. The abbreviation C in the legend stands for "canned" (e.g., canned chick peas).

10 individuals ate a larger number of novel foods in the social condition (mean = 7.8) than in the individual condition (mean = 5.9) (Wilcoxon (10), $T- = 7, p <$ .05). Thus, the monkeys ate more and more varieties of novel foods when tested together than when tested alone. However, once again, subjects ate equivalent amounts of familiar foods in both conditions.

In summary, capuchin monkeys distinguished between novel and familiar foods, treating only novel foods with caution in both social and individual conditions, i.e. eating less of them than of familiar foods. Social facilitation of feeding was only evident with novel foods. This finding supports Galef's (1993) suggestion that social facilitation is more evident in behavior toward novel foods than toward familiar foods, at least in species which feed opportunistically on a broad range of foods.

We may argue that the most important outcome of social facilitation is not consumption at the first encounter, but longer-term consequences for repeated sampling of a novel food. That is, perhaps repeated facilitation of sampling allows an individual either to incorporate a novel substance into its list of "acceptable" foods, or to reject the substance. We are now investigating the time course of facilitation during repeated encounters with an initially novel food. Of course, social facilitation of consumption of an already-familiar and acceptable food would not have any consequence for exploiting new food resources.

Study 2: Social Influences on Responsiveness to Novel Foods in Young Capuchin Monkeys

It is likely that younger individuals would obtain a greater benefit from learning from others about the palatability of unfamiliar potential foods. Therefore, one might expect social influences on feeding, such as social facilitation, to be more pronounced in infant and juvenile monkeys than in adults. We followed our first study of responses to novel foods (carried out in relatively small groups with no infants) with additional studies of young capuchin monkeys living in large groups with several matrilines present (Fragaszy, Feuerstein, & Mitra, 1995a).

We started with the general hypothesis that young individuals seek information from others about food. Social learning in this view is an active rather than passive process. More specifically, we hypothesized that young capuchin monkeys: (1) would display caution toward novel foods (i.e., eat less of novel foods, or begin to eat them later, than familiar foods), (2) would express greater interest in novel than familiar foods that others are eating, and (3) would be more likely to eat a novel food following inspection of that food while it was eaten by another than before inspection. These predictions follow from the functional view that infants would both seek and make use of available information provided by the actions of others (that is, socially provided information) to improve their competence in feeding themselves.

We presented 17 novel foods, once each, to two groups. We also presented two kinds of familiar foods in other sessions. Eleven infants ranging in age from 4 to 12 months served as subjects. We followed each infant in the group during a 10-min focal sample in each test day. On each test day, we conducted three consecutive focal observations in each group, replenishing the supply of food (spread liberally on the floor of the cage) before each focal observation period. Behaviors toward others which we scored included sniffing the mouth, touching food, showing visual interest in another's food, and attempting to take food. A food owner's response was scored as tolerant or resistant. Figure 2 illustrates the typical form in which interest in another's food was expressed: Individuals approached another very closely, leaning into the other's field of view to see what it was handling or had in its mouth.

The basic findings surprised us. First, infants did not display caution toward novel foods; if anything, they were more attracted to them than to familiar foods. Infants did not delay picking up food or eating it as a function of whether it was familiar or novel, and 8 out of 11 ate novel foods more frequently than familiar foods. Second, 9 of 11 infants showed more interest in another's food when that food was novel, indicating that they did discriminate familiar and novel foods.

Fig. 2. A young monkey expressing interest in another individual's food, and the food possessor displaying characteristic tolerance of the youngster's approach. Photograph by Jennifer Feuerstein and Devjani Mitra.

However, infants expressed interest toward other's food equally often before and after they had eaten some of the novel food themselves. Thus, infants did not reliably seek information from others before eating a novel food.

Study 3: Young Monkeys' Behavior toward Nuts

The second study with infants concerned a rather different kind of opportunity for social influence (Fragaszy, Feuerstein, & Mitra, 1996). The problem for the infant (as defined by the experimenter) was not whether to eat a novel food, but rather how to obtain a familiar food when the food was encased in a hard shell. In this study, we presented unshelled pecan nuts, *Carya illinoensis,* to two groups. Capuchin monkeys typically break nuts open by a combination of banging and biting, and adults can readily open pecan nuts.

Using the same design as in the previous study, we presented abundant quantities of pecans and commercial chow to two groups of capuchin monkeys. In control trials, only commercial chow was presented. We studied 11 infants from 2 to 8 months of age and two juveniles 21 and 22 months of age at the start of testing. We noted whenever infants picked up nuts or pieces of nut from the floor, collected pieces which had dropped near another eating nuts, or obtained nuts directly from another.

We found that young capuchin monkeys displayed a high degree of interest in the nuts that others were eating; in fact, more interest than they displayed in either novel foods or in familiar chow (Fig. 3). Infants attempted to take nuts from others roughly 14 times per hour. Surprisingly, even with this highly desirable food, infants were nearly universally tolerated when they approached others with nuts (about 90% of cases). When they did encounter resistance, it overwhelmingly took the form of movement away from the infant, rather than assertive retaliation. Infants were readily able to watch others, and to inspect and obtain nuts that others were holding. We even saw infants eating from their mother's hand. Similar

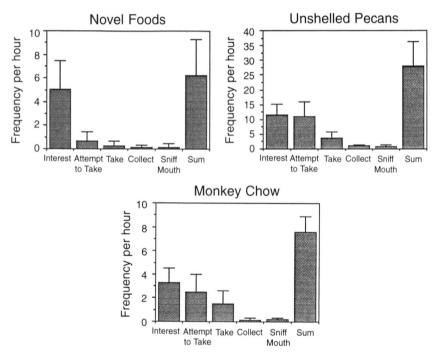

Fig. 3. The rate of social interactions per subject during feeding in young capuchin monkeys when presented with abundant foods in their home groups. The figures given here for sessions with monkey chow come from the control sessions during Study 3, with unshelled pecans. Note the difference in scale along the Y axis in the middle panel (Pecans). All forms of social interactions involving food were more frequent with unshelled pecans than with novel foods or monkey chow.

behaviors were observed with chow, although less frequently. Adults were equally tolerant toward infants' activities with chow as with nuts, but the infants were simply less likely to approach another holding chow than another holding a nut.

Infants did not learn how to open nuts either from watching others or from interacting with them. Infants younger than 6 months old did not open any nuts; half of the infants between 7 and 12 months old opened nuts, although not in the first few sessions. The two juveniles were able to open the nuts from the start of testing. Thus, opening pecans is a skill that infants are likely to acquire near the end of their first year.

The skill of opening nuts did not appear to be a technically difficult one. The behavior toward nuts by animals that could open them was not appreciably different from the behavior of individuals that could not open them. All animals treated nuts similarly: They bit them and banged them in no particular orientation against any available substrate. The secret to success seemed to be the force with which the nuts are struck against the substrate, and for adults, the ability to get the jaws around enough of the shell to crack it by biting. Youngsters' jaws were too small to bite the nuts open, however, and they had to rely on banging. Banging is a behavior that is deeply embedded in the repertoire of capuchin monkeys. Hand-reared individuals display this behavior prior to any contact with conspecifics, and banging is surely one of the most characteristic ways that capuchin monkeys interact with objects (Izawa & Mizuno, 1977; Fragaszy & Adams-Curtis, 1991, Visalberghi, 1988).

Adults' behavior toward infants was not affected by infants' competence or incompetence at opening the nuts. Infants could and did take nuts from others even if they were capable of opening them themselves. Adults did not appear to monitor infants' behavior at all.

These studies give us insight into the tolerant nature of social interactions during feeding in captive capuchin monkeys (see also de Waal, Luttrell, & Canfield, 1993). The findings suggest that the social environment could support learning about difficult or unfamiliar foods. On the other hand, social learning by infants is likely to be a rather casual and nondirected affair. Social interactions occur frequently during feeding, but they are not timed so as to provide information at an optimal time (prior to consumption of novel foods; prior to attempts to open nuts). Infants appear to be particularly attracted to others (of all ages) when the others are acting on an object. Whether they do so to obtain information relevant to a current uncertainty must still be determined, but our findings suggest this is unlikely. Adults are oblivious to infants' competence with the food at hand, and do not modify their behavior in any way that would enhance an infant's possibility of acquiring useful information.

In summary, a strongly functional interpretation of social interactions during feeding (that infants are doing these things in order to learn something) seems unwarranted. Altogether, rather simple social tendencies are evident that could, under the right circumstances, inform an infant about food. Much of the time, however, they seem to provide no immediately useful information at all. They are normal activities occurring in a stream of tolerant social interactions among group members, and they may have no more significance than that.

COMPARISON WITH ANOTHER SPECIES OF MONKEY: BABOONS

Our data concern well-fed monkeys living in a risk-limited captive environment, where toxic items are never presented and where there are no deleterious consequences for inefficient foraging. It might be that reliance on socially available information, or the importance of that information, would be more evident among monkeys in natural environments, where there is a higher priority on accuracy in food selection and efficiency in feeding, and not all individuals obtain optimal nutrition. A similar phenomenon has been found by Galef, Beck, and Whiskin (1991) in laboratory rats: The food choices of protein-deprived juvenile rats were more affected by social interaction than were those of their better-fed peers.

King (1994) considers the possibility that social interactions provide infant savanna baboons (*Papio cynocephalus*) living in Amboseli, Kenya, with useful information about foods and foraging. She noted several kinds of behaviors in baboons (termed cofeeding, tolerated scrounging, and muzzle-muzzle behavior) that could plausibly contribute to infants' learning about foods and foraging from others. Cofeeding involves the infant and another feeding synchronously on the same food item within one meter of each other. Cofeeding occurs most often when the mother is feeding on corms, a preferred food which is nutritious and available all year, but one that is difficult for infants to obtain on their own. Corms must be extracted from beneath the surface of the soil, and infants lack the strength and dexterity to achieve this. Tolerated scrounging (taking small bits of another's food, either from the food owner or from nearby, while the owner takes no action to prevent it) is also common. King's qualitative notes suggest that scrounging, like cofeeding, occurs selectively with corms, whether scrounging involves the mother or the nearest neighbor. Others have observed a similar pattern. Stein (1984) reported that infants scrounge corms more often than any other food from adult males. Muzzle-muzzle behavior involves one baboon placing its muzzle within 2.5 cm of another baboon's muzzle, presumably to smell and look at the muzzle of the other. Infant baboons initiated this behavior proportionately more often than did other age groups, and initiated it both proportionately more often to adults, and proportionately more often while the other was feeding.

All these findings suggest, in King's view, that baboon infants actively seek information from adults about foods—corms in particular. These findings also are consistent with the view that the social dynamics of baboon groups (including strong tolerance of young infants) promotes acquisition by infants of some foods that are difficult for them to obtain directly (an immediate benefit), and promotes acquisition of species-typical foraging skills (a future benefit). Distinguishing among these alternatives is not yet possible; in any case, they are not mutually exclusive.

If infants are acquiring information from others, the process is definitely one-sided. King noted that adult baboons showed no concern for infants' foraging activities, even when an infant ingested an item not eaten by adults in the near presence of her mother, and vomited 3 min later. This observation is a striking example of what nonhuman primates do *not* do: They do not exhibit behavioral negation (interference or punishment to prevent activity) (Visalberghi & Fragaszy, in press) in circumstances in which we would expect they should, if they were attentive to the risks facing another (see also Cheney & Seyfarth, 1990). Moreover, adults did not demonstrate techniques other than in the course of their own foraging activity, and they did not give food to infants, leading King to conclude that adults did not "donate" information to infants.

King's observations are strikingly similar to ours, despite the differences in species and setting. In both captive capuchin monkeys and wild baboons, young animals acquired foraging skills with no direct assistance from adults, but in a socially tolerant setting in which they could occasionally obtain certain hard-to-process foods from others. The infants were responsible for initiating interactions during feeding, and initiated interactions at different rates according to food value or difficulty of obtaining the food. Infant capuchin monkeys, like infant baboons, initiated interactions differentially toward those that had food, and initiated such interactions more often when the other was eating something more preferred than when it was eating something less preferred. Social information is available in these conditions, but it is probably not as strongly related to infants' feeding as earlier researchers thought it might be, and it is certainly not affected by adult attention to infant activity. It is not clear whether or to what extent infants are "seeking" information through their activities, even though they are clearly the initiators of the interactions and the transfers of food that occur. Infant baboons may simply be trying to obtain some of what adults are eating, especially when it is a favored food that they cannot obtain themselves, as corms are. We are more inclined to view the social interactions occurring between infants and others as immediately motivated by the infants' interest in the food, and desire to eat it themselves. This parsimonious interpretation does not rule out the possibility that acquisition of information may be a byproduct of the process, but it deemphasizes acquisition of information as a proximate motivator of the infant's behavior.

POTENTIAL GROWING POINTS FOR STUDIES OF SOCIAL LEARNING IN MONKEYS

If social learning among monkeys is not marked by special characteristics which differentiate it from social learning among other animals, it may still vary across species in illuminating ways, and comparative research with monkeys, as with other animals, ought to address these differences. Comparative research on social learning already has two clear growing points which apply as much to primates as to other orders, and these should be pursued. These are ecological correlates, and evolutionary implications. As an example of the first theme, Giraldeau, Caraco, and Valone (1994) analyze how aspects of social and ecological conditions, such as group size and the distribution of food resources, influence the opportunities, costs, and benefits of social learning. Laland, Richerson, and Boyd's (1993) analysis of the contribution of social learning to behavioral evolution is an example of the second theme. However, a social-behavioral level of analysis has not been prominent in comparative research; we think that analysis of species differences at this level will complement the ecological and evolutionary perspectives now gaining prominence. In particular, links among social learning, social dynamics, and emotional expressiveness (including the richness of emotional signaling, and the sensitivity of others to these signals) deserve systematic attention from comparative researchers.

The first theme, social dynamics, is familiar in primatological research, and an extensive data base supports a variety of predictions linking species differences in social dynamics with social learning (Coussi-Korbel & Fragaszy, 1995). In general, the extent to which individuals are able to approach others busy with interesting activities or objects or to match their activity with others, both of which allow social learning to occur, depends on the tolerance that exists among individuals in the group. An obvious initial prediction is that information is likely to spread more quickly and more widely within a tolerant group than a despotic group.

The second theme, emotional expressiveness, is a less familiar topic in systematic comparative research. It is time to acknowledge that emotions are intimately connected with cognition, including learning, in nonhumans, as they are in humans

(Niedenthal & Kitayama, 1994; Izard, 1993), and time to move forward with rigorous empirical study of the contributions of emotion to behavior in animals. Since Darwin (1872/1965), it has been recognized that some species exhibit a greater range of emotional expressions than others. Expression of emotion in nonhumans can be assumed to be functional signals—that is, to affect others, as they do in humans (Campos, Mumme, Kermoian, & Campos, 1994). Humans exhibit exquisitely sensitive, automatic, and rapid responsiveness to the expressions of emotion in others (Hatfield, Cacioppo, & Rapson, 1994), and this capacity is intimately connected to the human pattern of social learning from early in life (Trevarthen, 1993). Humans respond unconsciously to the expression of affect in others with milliseconds. The phenomenon has been termed "emotional contagion," and defined as "the tendency to automatically mimic and synchronize facial expressions, vocalizations, postures, and movements with those of another person and, consequently, to converge emotionally." (Hatfield, Cacioppo, & Rapson, 1992). Humans differ in their sensitivity to the emotional expressions of others, and their ability to make appropriate responses to others' expressions (Salovey & Meyer 1989/1990). There is accumulating evidence that those who are more sensitive to others' emotional expressions are better at negotiating social affairs, and better social learners in general (Hatfield et al., 1994, and reviewed in Salovey & Meyer, 1989/1990). In other words, emotional convergence seems to result in more efficient interpretation of others' behavior, perhaps serving as a kind of fifth gear which allows the cognitive engine to work more efficiently in the domain of social learning.

Nonhuman species may be susceptible to emotional contagion in the same fashion as humans are, and with the same result. If they are, this process would be central to the propensity of one animal to produce behavior convergent with that observed in another, or to approach a place or object where another has recently been active—in other words, to display local enhancement and social facilitation. Differences in emotionality are as likely to contribute to behavioral differences among species of nonhuman primates and among species of other orders, as they are to contribute to individual differences in humans.

It will be an important task to determine the extent of species and individual differences in emotional contagion, and the relation of emotional contagion to learning in various circumstances. It is premature to spell out a large cast of related predictions in this domain, but consider at least the following initial expectations for variations across species and within group: Species exhibiting greater emotional contagion and more nuanced displays of emotion should also exhibit more frequent or more effective social learning than do species with less emotional contagion and impoverished emotional displays. Within species, emotional responsiveness of individuals will likely be significantly affected by social dynamics, because individuals will be differentially attentive to particular others, and thus likely to respond differentially to expressions of emotions in others. This should produce a mosaic of social learning opportunities within a group, leading to heterogeneity of timing, at least, of social learning, and perhaps also to heterogeneity in what is learned. We look forward to rigorous experiments designed to illuminate these, and related, predictions.

CONCLUSION

We began this chapter by claiming that it is easy, for laymen and scientists alike, to look at nonhuman primates and to envisage human-like cognition as being behind their behavior. We have argued that this attitude has stood in the way of an objective assessment of social learning capacities in nonhuman primates, particularly monkeys. We showed that in the case of capuchin monkeys dealing with novel or difficult foods, simple social tendencies are sufficient to explain what is observed.

   Although we have studied only one species, the literature describing social learning in other species of monkeys is consistent with our findings (e.g., baboons; King, 1994). Social learning processes such as social facilitation that are important across many taxa are also the important ones in monkeys. Behavioral features such as social dynamics and the richness of emotional signaling, as well as ecological characteristics, such as characteristics of the diet or resource distribution in the particular case of feeding, are more likely to predict important characteristics of social learning in a given species than taxonomic status. We need to understand social learning in these terms better than we do now before we can decide whether monkeys exhibit features of social learning that set them apart from other animals.

ACKNOWLEDGMENTS

We thank the editors for inviting us to participate in the conference on which this volume is based. Research was supported by a Bilateral Italy-USA grant of the Consiglio Nazionale delle Richerche, Italy, to E. Visalberghi, and by a Research Scientist Development Award from the National Institutes of Health, USA, to D. Fragaszy. We thank L. Toll, J. Feuerstein, A. Galloway, D. Mitra, and M. Valente for their help in collecting and analyzing data.

REFERENCES

Adams-Curtis, L. E., & Fragaszy, D. (1995). Influence of a skilled model on the behavior of conspecific observers in tufted capuchin monkeys (*Cebus apella*). *American Journal of Primatology, 37,* 65–71.

Box, H. O. (1984). *Primate behaviour and social ecology.* London: Chapman and Hall.

Brown, A. D., & Zunino, G. E. (1990). Dietary variability in *Cebus apella* in extreme habitats: evidence for adaptability. *Folia Primatologica, 54,* 187–195.

Campos, J. J., Mumme, D., Kermoian, R., & Campos, R. (1994). A functionalist perspective on the nature of emotion. In N. Fox, (Ed.), The development of emotion regulation (pp. 284–303). *Monographs of the Society for Research in Child Development 59* (2-3, Serial No. 240).

Caro, T. M., & Hauser, M. D. (1992). Teaching in nonhuman animals. *Quarterly Review of Biology, 67,* 151–174.

Cheney, D., & Seyfarth, R. (1990). Attending to behaviour versus attending to knowledge: Examining monkeys' attribution of mental states. *Animal Behaviour, 40,* 742–753.

Clayton, D. A. (1978). Socially facilitated behavior. *Quarterly Review of Biology, 53,* 373–392.

Coussi-Korbel, S., & Fragaszy, D. M. (1995). On the relation between social dynamics and social learning. *Animal Behaviour, 50,* 1441–1453.

de Castro, J., & Brewer, E. M. (1992). The amount eaten in meals by humans is a power function of the number of people present. *Physiology and Behaviour, 51,* 121–125.

Darwin, C. (1872–1965). *The expression of the emotions in man and animals.* Chicago: University of Chicago Press.

Fedigan, L. M. (1982). *Primate paradigms: Sex roles and social bonds.* Montreal: Eden Press.

Fedigan, L. M. (1993). Sex differences and intersexual relations in adult white-faced capuchin monkeys. *International Journal of Primatology, 14,* 853–877.

Feistner, A.T.C., & McGrew, W. C. (1989). Food-sharing in primates: A critical review. In P. K. Seth & S. Seth (Eds.), *Perspectives in primate biology,* (*Vol. 3;* pp. 21–36). New Delhi: Today and Tomorrow's Printers.

Forkman, B. (1991) Social facilitation is shown by gerbils when presented with novel but not with familiar food. *Animal Behaviour, 42,* 860–861.

Fragaszy, D. M., & Adams-Curtis, L. E. (1991). Generative aspects of manipulation in tufted capuchin monkeys (*Cebus apella*). *Journal of Comparative Psychology, 105,* 387–397.

Fragaszy, D. M., & Boinski, S. (1995). Patterns of individual diet choice and foraging efficiency in wedge-capped capuchin monkeys (*Cebus olivaceus*). *Journal of Comparative Psychology, 109,* 339–348.

Fragaszy, D., Feuerstein, J., & Mitra, D. (1996). Social and asocial behaviors of young capuchin monkeys in the presence of difficult foods. Unpublished manuscript. University of Georgia, Georgia.

Fragaszy, D. M., & Shaffer, D. (1994). Developmental foundations of social learning in humans and other primates. In J. J. Roeder, B. Thierry, J. R. Anderson, & N. Herrenschmidt, (Eds.), *Current primatology,* Vol. III. *Social development, learning, and behaviour* (pp. 329–337). Strasbourg: University Louis Pasteur Press.

Fragaszy, D. M., Visalberghi, E., & Galloway, A. (1995). Behavior of infant capuchin monkeys with novel foods. Unpublished manuscript. University of Georgia, Georgia.

Fragaszy, D. M., Visalberghi, E., & Robinson, J. G. (1990). Variability and adaptability in the genus *Cebus. Folia Primatologica,* **54,** 114–118.

Fragaszy, D. M., & Visalberghi, E. (1989). Social influences on the acquisition of tool-using behaviors in tufted capuchin monkeys (*Cebus apella*). *Journal of Comparative Psychology,* **103,** 159–170.

Fragaszy, D. M., & Visalberghi, E. (1990). Social processes affecting the appearance of innovative behaviors in capuchin monkeys. *Folia Primatologica,* **54,** 155–165.

Fragaszy, D. M., Vitale, A., & Ritchie, B. (1994). Variation among juvenile capuchins in social influences on exploration. *American Journal of Primatology,* **32,** 249–260.

Galef, B. G., Jr. (1988). Imitation in animals: History, definition, and interpretation of data from the psychological laboratory. In T. Zentall and B. G. Galef, Jr. (Eds.), *Social learning: Psychological and biological perspectives,* (pp. 3–28). Hillsdale, NJ: Erlbaum.

Galef, B. G., Jr. (1993). Function of social learning about food: A causal analysis of effects of diet novelty on preference transmission. *Animal Behaviour,* **46,** 257-265.

Galef, B. G., Jr., Beck, M., & Whiskin, E. E. (1991). Protein deficiency magnifies social influence on the food choices of Norway rats (*Rattus norvegicus*). *Journal of Comparative Psychology,* **105,** 55–59.

Galef, B. G., Jr., and Sherry, D. F. (1973). Mother's milk: A medium for the transmission of cues reflecting the flavor of mother's diet. *Journal of Comparative and Physiological Psychology,* **83,** 374–378.

Giraldeau, L.-A., Caraco, T., & Valcone, T. (1994). Social foraging: Individual learning and cultural transmission of innovations. *Behavioral Ecology,* **5,** 35–43.

Glander, K. E. (1982). The impact of plant secondary compound on primate feeding behavior. *Yearbook of Physical Anthropology,* **25,** 1–18.

Goodall, J. (1986). *The chimpanzees of gombe.* Cambridge, MA: Harvard University Press.

Hall, K. R. L. (1963). Observational learning in monkeys and apes. *British Journal of Psychology,* **54,** 201–226.

Harlow, H. F., & Yudin, H. C. (1933). Social behavior of primates. *Journal of Comparative Psychology,* **16,** 171–185.

Hatfield, E., Cacioppo, J. T., & Rapson, R. L. (1992). Emotional contagion. In M. S. Clark (Ed.), *Review of personality and social psychology, Vol. 14. Emotion and social behavior* (pp. 151–177). Newbury Park, CA: Sage.

Hatfield, E., Cacioppo, J. T., & Rapson, R. L. (1994). *Emotional contagion.* Cambridge: Cambridge University Press.

Izard, C. E. (1993). Four systems of emotion activation: Cognitive and noncognitive processes. *Psychological Review,* **100,** 68–90.

Izawa, K. (1979). Foods and feeding behavior of wild black-capped capuchin monkeys (*Cebus apella*). *Primates,* **20,** 57–76.

Izawa, K., & Mizuno, A. (1977). Palm fruit cracking behavior of wild black-capped capuchin (*Cebus apella*). *Primates,* **18,** 773–792.

Janson, C., & Boinski, S. (1992). Morphological versus behavioral adaptation for foraging in generalist primates: The case of the cebinae. *American Journal of Physical Anthropology,* **88,** 483–498.

Kawamura, S. (1959). The process of sub-culture propagation among Japanese macaques. *Primates,* **2,** 43–60.

Kennedy, J. S. (1992). *The new anthropomorphism.* Cambridge: Cambridge University Press.

King, B. J. (1994). *The information continuum.* Santa Fe, NM: SAR Press.

Laland, K. N., Richerson, P. J., & Boyd, R. (1993). Animal social learning: toward a new theoretical approach. In P. P. Klopfer, P. P. Bateson, & N. Thomson (Eds), *Perspectives in ethology* (pp. 249–276). New York: Plenum Press.

Lefebvre, L. (1985). Parent-offspring food-sharing: A statistical test of the early weaning hypothesis. *Journal of Human Evolution,* **14,** 255–261.

Milton, K. (1993a). Diet and social organization of a free-ranging spider monkey population: The development of species-typical behavior in the absence of adults. In M. Pereira & L. Fairbanks (Eds.), *Juvenile Primates* (pp. 173–181). New York: Oxford University Press.

Milton, K. (1993b). Diet and primate evolution. *Scientific American,* **269,** 70–77.

Niedenthal, P. M., & Kitayama, S. (Eds.) (1994). *The heart's eye.* New York: Academic Press.

Pereira, M., & Fairbanks, L. (1993). *Juvenile primates: Life history, development, and behavior.* New York: Oxford University Press.

Perry, S., & Rose, L. (1994). Begging and transfer of coati meat by white-faced capuchin monkeys, *Cebus capucinus. Primates,* **35,** 409–415.

Price, E. C., & Feistner, A.T.C. (1993). Food-sharing in lion tamarins: Tests of three hypotheses. *American Journal of Primatology,* **31,** 211–221.

Robinson, J. G. (1981). Spatial structure in foraging groups of wedge-capped capuchin monkeys (*Cebus nigrivittatus*). *Animal Behaviour,* **29,** 1036–1056.

Robinson, J. G. (1986). Seasonal variation in use of time and space by the wedge-capped capuchin monkey, *Cebus olivaceus:* Implications for foraging theory. *Smithsonian Contributions in Zoology,* **431,** 1–60.

Rose, L. (1994). Sex differences in diet and foraging behavior in white-faced capuchins, *Cebus capucinus. International Journal of Primatology,* **15,** 95–114.

Rozin, P. (1976). The selection of foods by rats, humans, and other animals. In: D. Lehrman, R. A. Hinde, & E. Shaw (Eds.), *Advances in the study of behavior* (Vol. 6; pp. 21–76). New York: Academic Press.

Salovey, P., & Mayer, J. D. (1989/1990). Emotional intelligence. *Imagination, Cognition and Personality,* **9,** 185–211.

Stein, D. M. (1984). *The sociobiology of infant and adult male baboons.* Norwood, NJ: Ablex.

Terborgh, J. (1983). *Five new world primates.* Princeton, NJ: Princeton University Press.

Trevarthen, C. (1993). The function of emotions in early infant communication and development. In J. Nadel & L. Camaioni (Eds.), *New perspective in early communicative development* (pp. 48–81). New York: Routledge.

Visalberghi, E. (1988). Responsiveness to objects in two social groups of tufted capuchin monkeys (*Cebus apella*). *American Journal of Primatology,* **15,** 349–360.

Visalberghi, E., & Fragaszy, D. (1990). Do monkeys ape? In: S. T. Parker & K. T. Gibson (Eds.), *Language and intelligence in monkeys and apes* (pp. 247–273). New York: Cambridge University Press.

Visalberghi, E., & Fragaszy, D. (in press). Pedagogy and imitation in monkeys: yes, no, or maybe? In: D. R. Olson (Ed.) *Handbook of education and human development: New models of learning, teaching and schooling.* Oxford: Blackwell.

Visalberghi, E., & Fragaszy, D. (1995). The behaviour of capuchin monkeys (*Cebus apella*) with food: The role of social context. *Animal Behaviour,* **49,** 1089–1095.

de Waal, F. B. M., Luttrell, L. M., & Canfield, M. E. (1993). Preliminary data on voluntary food sharing in brown capuchin monkeys. *American Journal of Primatology,* **29,** 73–78.

Watts, D. P. (1985). Observations on the ontogeny of feeding behavior in mountain gorillas (*Gorilla gorilla beringei*). *American Journal of Primatology,* **8,** 1–10.

Whitehead, J. M. (1986). Development of feeding selectivity in mantled howling monkeys, *Alouatta palliata.* In: J. G. Else & P. C. Lee (Eds.), *Primate ontogeny, cognition, and social behaviour* (pp. 105–117). Cambridge: Cambridge University Press.

5

Copying and Mate Choice

LEE A. DUGATKIN

Department of Biology
University of Louisville
Louisville, Kentucky 40292

INTRODUCTION

Whereas animal psychologists have been studying social learning for close to a century, with the exception of work on "imprinting," ethologists and evolutionary ecologists have only recently begun studying this subject in earnest. This lag time may in part be due to the lack of a sound theoretical framework for the evolution of culture. Over the last 15 years, a framework has begin to emerge, as population geneticists, anthropologists and evolutionary ecologists have begun developing models for the evolution of cultural behavior, including social learning (Cavalli-Sforza & Feldman, 1981; Boyd & Richerson, 1985; Findlay, Lumsden, & Hansell, 1989a,b; Findlay, 1991; Laland, 1994a,b; Kirkpatrick & Dugatkin, 1994). Cultural transmission is ubiquitous in humans, but has also been found in nonhumans in the context of foraging in rats (see Galef, 1976 for a review) and pigeons (see LeFebvre & Palameta, 1988 for a review), song learning in birds (Slater, Eales, & Clayton, 1988), mate choice in fish (Dugatkin, 1992; Dugatkin & Godin, 1992, 1993), and a variety of other situations (see Bonner, 1980; Cavalli-Sforza & Feldman, 1981; Boyd & Richerson, 1985; Zentall & Galef, 1988; and other chapters in this volume).

In their "dual inheritance" model, Boyd and Richerson (1985) argue that all of the forces that lead to changes in gene frequencies—natural selection, drift, mutation and migration—have analogs within the realm of cultural evolution. Boyd and Richerson's models demonstrate how cultural change can be studied with techniques similar to those developed by population geneticists. Furthermore, they demonstrate how cultural evolution and genetic evolution can operate in the same or opposite directions, and how either can be the predominate force—depending on the particular scenario (Richerson & Boyd, 1989).

One well-studied suite of behaviors that is almost certainly influenced by cultural, as well as innate factors, is a female's choice of mates. Copying the mate choice of others is conceptually intriguing, because genetic models of sexual selection indicate that female mate choice may coevolve with the male trait being chosen (see Kirkpatrick, 1987a; Bradbury & Andersson, 1987; Pomiankowski, 1988; Kirkpatrick & Ryan, 1991; Moller, 1994; and Andersson, 1994 for reviews). If copying plays a role in mate choice, then the dynamics of sexual selection (e.g., the coevolution of female preference and male trait) may be influenced by cultural evolution in ways that may be distinct from genetic evolution (Boyd & Richerson, 1985). Furthermore, studying culture in the context of female mate choice may also allow us to examine experimentally the evolution of a trait (female preference) when both innate and cultural factors operate simultaneously (see below).

Most work on female mate choice has assumed that a female's preference for a particular male trait is under some sort of genetic control (sensu Fisher, 1958). It is, however, almost certainly true that social, nongenetic factors also play a role in mate choice (Boyd & Richerson, 1985). One area of sexual selection that remains relatively unexplored is to what extent a female's preference is affected by the preference of other females, i.e., do females copy the mate choice of others, and if so, under what conditions? Furthermore, how do genetic factors and social/cultural factors (such as copying) interact to affect the dynamics of sexual selection, particularly the coevolution of female mate choice and male traits? One place to begin answering these questions is a definition of mate-choice copying.

A Definition

To date, the only formal definition of (female) mate-choice copying is that from Pruett-Jones (1992). According to Pruett-Jones' definition, mate copying occurs when:

> The conditional probability of choice of a given male by a female is either greater or less than the absolute probability of choice depending on whether that male mated previously or was avoided, respectively.
>
> *Pruett-Jones (1992)*

One strong feature of this operational definition is that conditional and absolute probabilities can be measured, and hence a mate-choice copying hypothesis can be supported or refuted by available data. However, Pruett-Jones (1992) fails to capture a critical aspect of mate-choice copying in that individuals need not observe others choose mates for a mate-choice copying to have occurred under this definition (observation here is meant in the most general terms and includes not just vision, but any sensory input that allows one to know that a mating is taking place). For example, imagine a species in which males defend territories that differ in their defensability against predators. Females in this species choose to live in the safest territories and are further attracted to areas that have other individuals (in this case females) in them, because of the benefits associated with the selfish herd (Hamilton, 1971) and confusion effects (Landeau & Terborgh, 1986). Further imagine that moving between territories often is a dangerous activity. So, females choose territories based on how safe they are, and then they mate with the male on such a territory.

It is the above scenario mate-choice copying? If one were to go and measure conditional and absolute probabilities of mating, they would certain differ, and hence this qualifies as a case of mate-choice copying according to Pruett-Jones (1992). Yet the fact remains that observing the choice of others plays no causal role in the difference between absolute and conditional probabilities of males mating, and this scenario might be depicted as an independent choice of safe territories that is reinforced by what behavioral ecologists call conspecific cueing (Keister, 1979). To avoid this problem, I propose modifying Pruett-Jones' (1992) definition as follows. Mate-choice copying occurs when:

> The conditional probability of choice of a given male by a female is greater than the absolute probability of choice depending upon whether that male mated previously. Further, the information about a male's mating history (or some part of it) must be obtained by the female via observation.

This definition then retains the measurable properties that make Pruett-Jones' (1992) definition testable, but stresses the importance of observation per se in any definition of mate-choice copying. It also restricts copying to the case when a male's conditional probability of being chosen is greater after he mates, which makes some intuitive sense.

It is worth noting, however, that adding the element of observation to this definition raises the possibility that some types of what might be called "weak" mate choice copying will be overlooked. For example, imagine a species that at some time in the past met the conditions of the more stringent definition of mate-choice copying. In this species, females observed a male territory holder mate with another female and were then more likely to mate with that male themselves. Suppose that females then stayed on a particular territory after mating. Now, a new "rule of thumb" emerges which instructs females to look at how many, if any, other females are on a male's territory. Since this information correlates very well with how many females have mated with a male, females can get a quick, but accurate assessment of how many others have mated with the territory holder without actually taking the time (or being lucky enough to view) such matings. This rule of thumb should increase in frequency, and eventually females need not observe any matings, but are nonetheless in some sense mate-choice copying. It is difficult, however, if not impossible, to distinguish this case from the one in which females were choosing territories based on safety (as outlined above) and so it is perhaps best (or at least reasonable) to simply not include this type of mate-choice copying in our definition.

Theoretical Approaches to Female Mate-Choice Copying

There are at least three, nonmutually exclusive questions that one can model with respect to female mate-choice copying and sexual selection:

1. What affect does female mate-choice copying have on variance in male reproductive success ("the opportunity for selection" approach)?
2. Under what circumstances will copying strategy invade a population composed of noncopiers ("the game theory" approach)?
3. How will female mate copying affect the coevolution of innate female preferences for males and the male trait that is the subject of that preference ("the population genetics" approach)?

"The Opportunity for Selection" Approach

One obvious affect of female mate-choice copying is to change the distribution of matings among males in a population. Using computer simulations, Bradbury, Vehrencamp, and Gibson (1985) found that independent choice of mates on the part of females was unable to account for the skew in male reproductive success that is common in lekking species. Wade and Pruett-Jones (1990) use the parameter I—the opportunity for selection—to address how copying might influence variance in male reproductive success. I is a measure of the variance in relative fitness and sets an upper limit on the amount of phenotypic change possible by natural selection (Wade, 1979; Wade & Pruett-Jones, 1990). Following Bradbury et al. (1985), Wade and Pruett-Jones note that the same mating systems in which variance in male reproductive success is most skewed (lek breeding systems) are also those most likely to have females employing a mate-choice copying rule because of the abundant chance to observe the preferences of others. While some of the variance in male reproductive success in most species can be accounted for, "excess" variance has been attributed to "noise" in the system.

Wade and Pruett-Jones (1990) examined a population in which females bred sequentially and a male's probability of mating with the i_{th} female was a function of how many females had selected him as a mate in the past. They found that:

> Female copying increases the frequency of extreme values in the mating distribution. Whenever there is female copying there are more males that do not mate and more males that obtain large numbers of matings.
>
> *Wade and Pruett-Jones (1990)*

In many ways this result is not surprising (but see Marks, Deutsch, & Clutton-Brock, 1994 for evidence for how other factors might affect the above results). The strength of Wade and Pruett-Jones' (1990) model is, however, that it provides a means for looking at the distribution of matings and inferring the extent of mate-choice copying.

> . . . when the number of males is large and the sex ratio is approximately unity, there is a simple and sensitive test for the possible existence of female copying. First, testing the observed distribution of matings among males with the expected under random mating (from the binomial distribution) will test for nonrandom mating. Second, calculating the ratio of the observed proportion of males that do not mate with those that mate once estimates s (*the copying parameter*). This estimates the degree of female copying that could account for the observed deviation from random expectation, assuming that female copying is the only factor responsible for nonrandom mating.
>
> *Wade and Pruett-Jones (1990)*

It is worth noting that this model does not require that females observe males mating per se, but rather that prior male success affects current probability of mating. As such, observation of mating is not necessary for the model to work, and hence this approach does not meet the criteria of the definition put forth earlier. That is not to say that the model can not accommodate females observing males, but rather that it does not require it.

The Game Theory Approach

Game theory models of animal behavior look at the evolution of strategies, when the fitness of one individual is affected not only by its own actions, but by the actions of others (Maynard Smith, 1982). The simplest possible game involving mate-choice copying involves a copying strategy (copy the mate choice of other females) and a chooser strategy (independent assessment of male quality) competing with one another in an infinitely large population. Pruett-Jones (1992) models such a game in which all individuals have a baseline fitness, W. Mate assessment has a cost, k, but those who assess males receive some benefit, f. Analysis of this game shows that when $f > k$, copying and choosing coexist in a population at frequencies k/f and $1 - (k/f)$, respectively.

Losey, Stanton, Telecky, & Tyler (1986) built a computer simulation that examined copying and choosing strategies in a structured environment which simulated a lek breeding area. The distribution of male genetic quality was set in any given run of the simulation (but could be changed across runs). A second distribution, that of female choosing ability, was also preset. This distribution determined the probability that a female using the choosing strategy mated with a male of a given genetic quality. Females using the copying strategy were allowed a fixed number of "peeks" at the choice made by other females (copiers and choosers alike). If a copier had not observed any males mate she either selected just as chooser would (the "smart copier" model) or she selected males randomly (the "dumb" copier model); however, if she had seen males mate, she simply chose the male that she had seen mate the greatest number of times.

For most of the parameter space explored, Losey et al. (1986) found that a mixture of copiers and choosers, with copiers in the minority, was the "evolutionarily stable stage." This frequency dependence was due to copiers copying each other rather than choosers, as the frequency of copying surmounted a critical threshold. Their model generates four other testable predictions about female mate-choice copying in lekking species: (1) copying fares best when choosers are able to correctly identify high quality males, (2) copiers should have an advantage when competition between young plays a minor role, (3) copiers should have an advantage when fecundity is low, and (4) copying should fair well when females visit leks for either long periods of time, on numerous occasions, or both.

Using a simulation technique similar to Losey et al. (1986), Dugatkin and

Hoglund (1995) examined mate-choice copying, when the tendency to copy is a continuous variable, and when female reproduction is time constrained (i.e., if females spent too much time assessing males, they suffered reduced fecundity). In this model, females moved from territory to territory in a random fashion, and only a single female was assumed to be on a given male's territory at once. All females assessed "male quality" and used a rule that instructed them to mate with a male if its perceived genetic quality was greater than some value (V), but female assessment was subject to some error. Dugatkin and Hoglund assumed that a copier's, but not a noncopier's, perceived value of male quality is affected by the choice other females make. Females have a limited breeding season and consequently the more time units that pass before they mate, the lower their fitness at mating. The tendency to copy can thus have contradictory effects. While copiers may avoid the costs associated with mating late, because all females make errors in assessing a male's quality, copiers may copy others who themselves are making "bad" choices—i.e., choosing males whose real quality is less than V.

Results indicate that the only variable that had clear-cut interpretable effects on the evolution of copying was V, the minimum perceived male quality for female mating (Dugatkin & Hoglund, 1995). As V increased, the probability that a copying strategy invaded the population increased. Increasing V probably favored copying, because as V increased fewer males had the perceived quality score required by females. When any female, however, did mate, copiers increased their quality assessment of that male, making it more likely that they would mate early in the sampling process and thus avoid the costs of late mating. Unlike the case of Losey et al.'s (1986) model, the copying strategy in Dugatkin and Hoglund (1995), for the most part, did better as the frequency of copiers increased (except for high values of V and many copiers). The positive frequency-dependent effect found was tied to the cost of mating late in the breeding season. Basically, some copiers (as well as noncopiers) mated early in the season. The more copiers there were, the more early matings copiers saw, and the more likely they were to lift their assessment of male qualities and thus to mate earlier themselves—hence the positive frequency dependence.

The "Population Genetics" Approach

One shortcoming of the above techniques for modeling female mate choice is that the affect of such choice on male traits, and more importantly on the coevolution of female mate choice and male trait, are not considered. Kirkpatrick and Dugatkin (1994) and Laland (1994a,b) have recently begun developing such models, in which the male trait is transmitted genetically, but the female preference is transmitted via "oblique" transmission (Kirkpatrick & Dugatkin) and vertical transmission in the latter (Laland); see Laland and Richerson, this volume, for more on these models). Oblique transmission occurs when young females' preferences are influenced by adults, but the adults need not be their genetic parents. In essence, oblique transmission allows juveniles to have many "cultural parents" (Boyd & Richerson, 1985). Oblique transmission is likely much closer to the mode found in species where no postbirth parental care exists, juveniles are often found in mixed-age groups, and juveniles seem to make no distinctions between adults (e.g., genetic parent vs nonparent).

Kirkpatrick and Dugatkin (1994) developed two models in which female preferences were affected by an "innate" preference (which cannot evolve in the model) and a culturally transmitted preference (which can evolve). The male trait is inherited via a haploid autosomal or a Y-linked locus and has no cultural component. In the "mass copying" model, immature females observed many adults make choices among potential mates. When these females matured, they had a stronger preference for the male type they most frequently saw mating. In the "single mate-copying" model, an immature female observed a single randomly chosen mating [that may or may not have involved her genetic parent(s)] and augmented her preference for the type of male she observed mating. Results indicate that culturally determined female preference and the male trait can coevolve, with a positive frequency-dependent advantage to the more common male trait allele (i.e., "runaway" cultural sexual selection). This can cause a male trait with low viability to increase in frequency. Under some conditions, both models can lead to two alternative stable equilibria for the male trait, but neither model supports variation at the male trait locus. These models also make the somewhat counterintuitive prediction that cultural transmission, via copying, makes it more difficult for a novel male trait to spread in a population.

Given the little data available on copying and female mate choice, it is somewhat surprising that both models Kirkpatrick and Dugatkin (1994) developed found that copying makes it more difficult for a novel male trait to spread in the population. For example, female mate copying appears to play a role in the mating system of the sage grouse (*Centrocerus urophasianus*). Gibson, Bradbury, and Vehrencamp (1991) found that not only did the trait that females prefer in males change across years at the same lek location (with many of the same individuals), but that in leks that were only a few kilometers apart, with females were choosing males based on very different sets of criteria. While it is always possible that conditions were dramatically different across leks and across years, and thus different male traits were "optimal" at different locations and times, the data do not support such a conclusion. Gibson et al. (1991) hint that copying, per se, may have led to the diversity of traits that females preferred. Similarly, experiments on the strength of copying on mate-choice decisions in guppies (*Poecilia reticulata*) suggests that it plays some role in maintaining genetic variation among the numerous traits that females use to determine potential mates (Kodric-Brown, 1993).

One potential reason for the discrepancy between Kirkpatrick and Dugatkin's (1994) model's prediction regarding variation in male traits and the little data available, is that the model assumes a very simplistic population structure—a single, infinitely large population. The black grouse and guppies described above live in metapopulations: a cluster of populations in which individuals can disperse and immigrate. For example, guppies living in the Northern Mountain ranges of Trinidad are often found in slow-moving third and fourth order tributaries (Seghers, 1973: Dugatkin, personal observation). During the dry season, there is little immigration or emigration between these isolated subgroups, or between subgroups and the fish in the main sections of the stream. During the rainy season, however, the situation is likely reversed, with much emigration and immigration between groups (Shaw, Carvalho, Magurran, & Seghers, 1991). What effects such population structure has on the evolution of mate copying, and the co-evolution of female choice and male traits, however, remain unexplored. For example, suppose the "birth" of new groups (e.g., new lek sites open for black grouse, new ponds are colonized by guppies) is common. Drift (both genetic and cultural drift) in small, newly founded populations, in conjunction with female mate copying, might allow novel male traits to take hold in such populations, thus allowing variation in male traits to exist within the metapopulation. It is difficult to know the answer to such questions without formal mathematical models.

Empirical Studies

Mate-Choice Copying in Guppies

The strongest evidence to date for female mate-choice copying comes from the guppy (*P. reticulata*) (Dugatkin, 1992; Dugatkin & Godin, 1992, 1993). The guppy is an ideal species for examining female mate copying for at least five reasons: (1) female mate choice has been extensively studied in this species (see Kodric-Brown, 1990 for a review). For example, in addition to studies that examine how innate preferences may be co-evolving with male traits (i.e., color patterns; Houde, 1988; Houde & Endler, 1990; Breden & Hornaday, 1994; Wilkinson & Reillo, 1994), much is known about the genetics of male color patterns (Winge, 1927; Haskins, Haskins, McLaughin, & Hewitt, 1961; Houde, 1992). (2) The guppy's social structure typically consists of mixed-sex shoals (Magurran & Seghers, 1991; Seghers, 1973), within which females likely have opportunities to view (and potentially copy) the mate choice of nearby conspecifics. (3) Guppies are native to the streams of the northern mountain ranges of Trinidad, West Indies. Within these streams, a series of waterfalls divide guppy populations into areas in which they are under high predation pressure (downstream sites) and other areas (often only kilometers away) in which they are under slight predation pressure (upstream sites). Thus, the opportunity exists to directly examine how selection pressure (here, predation) affects mate copying. (4) Guppies exhibit normal courtship when placed in small aquaria, and thus are ideal for manipulative laboratory experiments. (5) Guppies breed in laboratory tanks quite readily and their generation time is short, thus making work on the interaction of genetic and cultural factors influencing mate choice feasible.

Dugatkin (1992) examined mate-choice copying using a 10-1 aquarium situated between two end chambers constructed of clear Plexiglas. A single male was placed into each of these end chambers. The "focal" female—the individual that potentially copies the behavior of other females—was placed in a clear Plexiglas canister in the center of the aquarium. A removable glass partition also created a section of the aquarium into which another female was placed. This individual will subsequently by referred to as the "model" female, i.e., the female whose behavior may be copied by the focal female.

At the start of a trial, males were placed into the end chambers and the focal and model females were placed in their respective positions in the aquaria. Fish were given 10 min during which the focal female could observe the model female near one of the two males. The model female and the glass partition were then removed, and the focal female was released from her canister and given 10 min to swim freely and choose whichever male she preferred. In these trials, the focal female chose the male which had been beside the model female 17 out of 20 times (G test $p < 0.005$).

While the results of this experiment are consistent with the hypothesis that females copy the mate choice of others, there are several alternative explanations. First, since guppies are schooling fish (Magurran & Seghers; Seghers, 1973), the focal female might simply be choosing the area which recently had the largest group of fish (in this case, two). A second treatment was conducted to test this "schooling hypothesis." It was identical to the above protocol, except that females were placed in the end chambers. In this case, the focal female chose the female in the end chamber closest to the model in only 10 out of 20 trials. The schooling hypothesis is therefore insufficient to explain the results.

Dugatkin (1992) also examined three other alternatives to female mate choice copying. Other treatments in this set of experiments found that : (1) If male courtship was removed from the experiment (via one-way mirrors), but all else was held constant, focal females chose randomly between males. That is, if lighting was adjusted so that males saw their own images in one-way mirrors, and hence did not court females, focal females showed no preferences for the male near the model. Thus, females were not simply going to the area that recently contained both a male and a female. (2) When choosing between males, focal females were not basing their decision on the activity patterns of males. 'Naive' females, who did not observe one male near the model (and the other away from it) showed no preference across males. In addition, Dugatkin (unpublished) has shown that if both males in the side canisters were courting females (and hence courtship activity did not differ across male), but the focal could only see a female near one male, she consistently preferred that male. (3) Focal females remembered the identity of the male chosen by the model, i.e., if the focal saw male 1 near the model, but the position of the end chambers containing male 1 and 2 were switched before that focal herself chose, she still preferred male 1. As such, copying by females is the only explanation compatible with the results of Dugatkin (1992).

In the absence of mate-copying opportunities, female guppies will choose between males on the basis of a number of phenotypic traits, such as size, tail length, and color patterns (Kodric-Brown, 1985; Bischoff, Gould, & Rubenstein, 1985; Breden & Stoner, 1987; Houde, 1988; Stoner & Breden, 1988; Houde & Endler, 1990); but if females prefer certain males over others, can their preference be reversed by social cues, such as information on the mate preference of other females? Dugatkin and Godin (1992) examined this question in a "reversal" experiment with two treatments. In both treatments, focal females were allowed individually to choose between two males, on two separate occasions. The treatments differed mainly in that Treatment II provided the focal female with the opportunity to copy (imitate) the experimentally staged mate preference of another female (the model), whereas such an opportunity was not available in Treatment I.

In Treatment I, each trial consisted of paired consecutive preference tests, in which a female chose between the same two males on two separate occasions (30 min apart). No model was present in Treatment I. Treatment II examined whether a female's initial choice of a mate can be altered by an opportunity to copy the mate preference of another female. The protocol of this treatment was similar to that of Treatment I, but with one major difference. In Treatment II, after a focal female initially chose between two males in the first paired preference test, she was returned to the canister for another 30-min acclimation period, during which time opaque partitions blocked her view of the males in the end chambers. After this, a second female was placed behind a clear Plexiglas partition in the quarter of the experimental tank adjacent to the male not chosen by the focal female in the first preference test. The opaque partitions were then removed, thereby allowing the focal female to see both males and the model female for 10 min. Following this "viewing period," the model female was removed from the arena and the focal female was then allowed to choose between the two males a second time. Twenty such trials were conducted with different females.

In Treatment I, females consistently preferred the same male across preference tests, when compared to a null model of random choice. Thus, in the absence of mate-copying opportunities, females did not change their preference for a given male. When compared with the low frequency of reversals observed in Treatment I, female guppies in Treatment II reversed their initial preference for particular males significantly more often, by copying the mate choice of a nearby (model) female. Thus, this 'reversal' experiment shows clearly that social cues can alter female choice, as individual females switched their initial preference for a particular male after they had observed another female choose differently.

In the above work, the focal and model females were matched for size. It was assumed that females of equal size had similar prior experience in choosing between males. Dugatkin and Godin (1993) examined the case in which focal and model females differed in size. They wondered whether smaller females were more likely to copy the mate choice of larger females than vice versa. Assuming older females have more experience in choosing mates than their younger counterparts, then one would predict that younger, relatively inexperienced, females would copy the choice of older females more frequently than older females copy the choice of younger ones. Results indicate that this is in fact the case, as younger females did copy the mate choice of older females, but not vice versa.

Mate-Choice Copying or Attraction to Nests with Eggs in Fishes?

A number of studies of female choice in fishes indicate that females prefer to mate with males which already have broods from prior matings (Ridley & Rechten, 1981; Constanz, 1985; Marconato & Bisazza, 1986; Sikkel, 1988, 1989; Gronell, 1989; Knapp & Sargent, 1989; Unger & Sargent, 1988; Jamieson & Colgan, 1989; Goldschmidt, Bakker, & Feuth-De Bruijn, 1993). Results from these studies, however, are ambiguous with respect to copying behavior. First, females do not actually observe the choice of others in these cases, and so these examples are not mate-choice copying under the definition presented earlier. Second, it is possible, however, that a female may be attracted to nests with eggs because this is a clear indication that a male has been chosen by other females and hence such a rule could act as a rule of thumb. This interpretation is also somewhat suspect. Females may prefer to spawn with males for other reasons (Sargent, 1988)—for example, eggs in a nest may indicate that the male is of "high" quality, e.g., able to defend a nest against predators. Another possibility is that females choose such nests to spawn because of increased courtship rates of males guarding nests with many eggs, rather than because of the presence of the eggs per se (Jamieson & Colgan, 1989); but see Goldschmidt et al. (1993) for evidence that this is not the case for sticklebacks, *Gasterosteus aculeatus*). Yet another alternative evolutionary explanation to female mate-choice copying is the "dilution" hypothesis which argues that females choose such nests because their eggs are less likely to be eaten by an egg predator or the guardian male (Rohwer, 1978). This dilution hypothesis is supported experimentally as males also prefer nests with eggs (Unger & Sargent, 1988).

Deer

In many lek-breeding species of mammals, females often join males that have the largest harem (Clutton-Brock, Hiraiwa-Hasegawa, & Robertson, 1989). Clutton-Brock and McComb (1993) examined whether fallow deer (*Dama dama*) are attracted to such harems because of the male, or because they are attracted to other females per se; perhaps because being in a large group reduces the risks associated with predation. To address this, Clutton-Brock and McComb (1993) performed two experiments. In the first experiment, pairs of oestrous females were given a choice among 4 paddocks. Two paddocks contained solitary males, one contained a male with a moderate number of females (8), and one a male with a large number of females (either 32 or 38). In the second, pairs of females were again given a choice among four paddocks. This time the paddocks contained a male with 9 females, 9 females with no male, a male with 19 females, and 19 females with no male present.

In the first experiment, females showed a strong preference for males with harems, rather than solo males. The second experiment, however, indicates that this preference was not due to copying the mate choice of others, as females showed no preference for paddocks containing a male plus females over paddocks containing females alone. Given that anoestrous females do not show a preference for associating with females (over mixed-sex groups), and that actually seeing a male mate does not increase the probability that a female observing this will join his harem, McComb and Clutton-Brock (1994) argue that estrous female fallow deer copy each other's movements in an attempt to increase group size, to decrease per capita predation, and most importantly, to avoid harassment by males in mixed-sex groups.

Mate-Choice Copying in Grouse

One of the first suggestions of female mate-choice copying in the literature comes from Lill's (1974) work on sexual selection in the lek breeding white-bearded manakin, *Manacus manacus*. In many lekking bird species, a single male often receives a large proportion of the available mating opportunities (see Wiley, 1991 for a review). It seems reasonable then to ask whether some of the success of males is due to female mate-choice copying (Bradbury & Gibson, 1983; Bradbury, Vehrencamp, & Gibson, 1985; Wade & Pruett-Jones, 1990), particularly since females in these lekking species often visit a lek numerous times and in groups, and thereby have the opportunity to copy each others choice of mates (Hoglund, Alatalo, & Lundberg, 1990). Much of the best evidence on this question in birds comes from Hoglund, et al.'s (1990) study of black grouse (*Tetrao tetrix*) and Gibson, Bradbury, & Vehrencamp's (1991) work on sage grouse (*C. urophasianus*).

Recognizing the inherent difficulties of studying mate-choice copying in the field, Hoglund, Alatalo, and Lundberg (1990) tried nonetheless to see if at least the prerequisites for copying exist in black grouse. They examined sequences of male matings, the assumption being that copying would increase the probability that consecutive, or near consecutive matings would be secured by males. As expected, Hoglund et al. (1990) found a skewed distribution of male mating success, with most males obtaining zero, but some males obtaining as many as 24 matings. They also found, however, that males were more likely to have whatever copulations they did obtain come nearly consecutively, suggesting that female mate-choice copying may indeed play a significant role in this system. The natural history of black grouse and the data collected by Hoglund et al. also suggest that young females are the most likely to employ mate-choice copying strategies.

Gibson, Bradbury, and Vehrencamp (1991) examined mate choice of sage grouse females (*C. urophasianus*) in two different leks over a 4-year period. Somewhat surprisingly, they found the traits that females preferred in males differed within and between leks (where within lek variation occurs over the four breeding seasons studied). Gibson, et al. (1991) examined mate-choice copying within leks on two different time scales: within and between day. For within-day copying, they hypothesized that the unanimity of female mate choice would increase as more hens mated on a given day, because more opportunities to observe and imitate would exist on such days. The data support this hypothesis.

Female Aggregations in Isopods

In an attempt to test the strength of Wade and Pruett-Jones' (1990) model, Shuster and Wade (1991) examined female mate choice in the marine isopod

Paracerceis sculpta. Breeding occurs in sponges in this species, and males exist in three distinct genetic morphotypes, labeled α, β, and γ (Shuster, 1987, 1989). α males defend breeding territories, β males resemble breeding females, and small γ males enter into α male territories and engage in sperm competition with any males present (Shuster, 1987, 1989).

Shuster (1987, 1990, 1991) has shown that "variance in harem size among males in this species appears significantly influenced by the tendency for females to prefer breeding aggregations consisting of more than one female" and that "in the field, large breeding aggregations contain a disproportionate number of recently inseminated females, suggesting that large harems are particularly attractive to sexually receptive females." (Shuster & Wade, 1991). Shuster and Wade determined how much variance in male reproductive success in *P. sculpta* could be accounted for by females preferring sponges that contained other females. They found that for much of the year, the copying parameter (s) was significantly different from random and that this deviation from randomness corresponded with an increase in I, the opportunity for sexual selection.

Why are females attracted to sponges that contain other females in this species? Shuster and Wade (1991) suspect that:

> The simplest explanation is provided by considering the influence of natural selection on female life history characteristics in this species (Shuster, 1991). To reproduce successfully, females must migrate from breeding sites in coralline algae to breeding sites in sponges. During this trip, females experience considerable predation pressure from reef fish (Thompson & Lehner, 1976; Shuster, 1991). Risks incurred during mate choice or breeding site selection are thought to diminish opportunities for females to discriminate between potential mates (Kirkpatrick, 1987b). Thus, it may be costly to females in this species, owing to predation, to move extensively to discriminate male and/or sponge characteristics. Females may instead use the scent of other females in spongocoels as an indication of the quality of breeding sites. This tactic would permit females to move towards established (and thus perhaps relatively safe) breeding sites with minimal exposure to predation.
>
> *Shuster and Wade (1991)*

While Shuster and Wade (1991) clearly show that what they refer to as copying has the effect of increasing variance in male reproductive success, it is not clear whether females are using the mate choice of others as a factor in their own decision-making process, as they do not seem to be observing the choice of others (again, using observation in the broad sense). Until further work is done in this system, it might best be thought of as an example of "female aggregation" or "conspecific cueing" (Keister, 1979) rather than female mate-choice copying.

DISCUSSION

The study of copying and mate choice is clearly still in its infancy. This subject has been examined in a only a handful of animals (in very few taxa) and under artificial conditions. Although work to date has uncovered some interesting results, many fundamental questions remain completely unexplored. Below, I address some areas that I believe to be particularly interesting for future work on female mate choice.

In some species it should be possible to examine (quantitatively) the relative importance of genetic and social factors on the evolution of female mate choice. If evidence exists that female preference for a male trait has a heritable component, we can then "titrate" this heritable preference against social factors. For example, in the guppy, females from some populations prefer males with lots of orange coloration over males that are not as orange, and this preference has a heritable component (Houde 1988; Houde & Endler, 1990). That being the case, experiments could be constructed to determine whether a female prefers a drabb male, but one who has a model female(s) by his side, over a brighter orange male with no model (Dugatkin 1996).

As evidence by the distribution of authors in this volume, social learning is of interest to both biologists and psychologists. Experiments on female mate-choice copying may serve as a bridge between these fields. For example, although behavioral factors that juvenile females experience during their development into adults play a large role in general models of cultural evolution (Boyd & Richerson, 1985), such factors are rarely investigated experimentally, particularly in the case of female mate copying. The ontogeny of behavior is of interest to both animal psychologists and behavioral ecologists, and mate-choice copying experiments that have elements of interest for both fields are quite feasible. Furthermore, researchers from both fields are interested in how an animal's internal "state" (e.g., hunger level) affects it decision-making process (Mangel & Clark, 1988). Since one possible advantage of female mate-choice copying is to avoid the costs of assessing a male's suitability as a potential mate, state variables may play a role in mate-copying decisions. For example, a hungry female searching for food may be more likely to mate copy than a satiated female, who is under less time budget constraints. This hypothesis can be easily tested and provide information on both proximate and ultimate aspects of copying and female mate choice.

There are many other interesting questions that need to be addressed regarding mate choice and copying. For example, what are the costs and benefits of copying? Do some individuals in a given cohort copy and others assess mate quality independently? Does the frequency of mate-choice copying differ across taxa? If so, is this difference correlated with the cognitive abilities of individuals? My hope is that these and similar questions will be addressed across a wide gamut of species and that such work will spur others to give more thought to the way in which female mate-choice copying can be linked with other fundamental questions in the field of behavioral ecology.

REFERENCES

Andersson, M. (1994). *Sexual selection*. Princeton, NJ: Princeton University Press.

Bischoff, R. J., Gould, J. L., & Rubenstein, D. I. (1985). Tail size and female choice in the guppy (*Poecilia reticulata*). *Behavioral Ecology and Sociobiology,* **17,** 252–255.

Bonner, J. T. (1980). *The evolution of culture in animals*. Princeton, NJ: Princeton University Press.

Boyd, R., & Richerson, P. J. (1985). *Culture and the evolutionary process*. Chicago: The University of Chicago Press.

Bradbury, J. W., & Andersson, M. B. (Eds). (1987). *Sexual selection: Testing the alternatives*. New York: Wiley-Interscience.

Bradbury, J. W., & Gibson, R. M. (1983). Leks and mate choice. In P. Bateson, (Ed.), *Mate choice* (pp. 109–138). Cambridge: Cambridge University Press.

Bradbury, J. W., Vehrencamp, S. L., & Gibson, R. M. (1985). Leks and the unanimity of female choice. In P. J. Greenwood, P. H. Harvey, & M. Slatkin (Eds.), *Essays in honour of John Maynard Smith* (pp. 301–314). Cambridge: Cambridge University Press.

Breden, F., & Stoner, G. (1987). Male predation risk determines female preference in the Trinidadian guppy. *Nature,* **329,** 831–833.

Breden, F., & Hornaday, K. (1994). Test of indirect models of selection in the Trinidad guppy. *Heredity,* **73,** 291–297.

Cavalli-Sforza, L. L., & Feldman, M. W. (1981). *Cultural transmission and evolution: A quantitative approach*. Princeton, NJ: Princeton University Press.

Clutton-Brock, T. H., Hiraiwa-Hasegawa, M., & Robertson, A. (1989). Mate choice on fallow deer leks. *Nature,* **340,** 463–465.

Clutton-Brock, T. H., & McComb, K. (1993). Experimental tests of copying and mate choice in fallow deer (*Dama dama*). *Behavioral Ecology,* **4,** 191–193.

Constanz, G. (1985). Alloparental care in the tessellated darter, *Etheostoma olmstedi* (Pieces:Percidae). *Environmental Biology of Fish,* **14,** 175–183.

Dugatkin, L. A. (1992). Sexual selection and imitation: females copy the mate choice of others. *American Nature, ***139,** 1384–1389.

Dugatkin, L. A. (1996). The interface between culturally-based preferences and genetic preferences: Female mate choice in *Poecilia reticulata. Proceedings of the National Academy of Science, USA,* in press.

Dugatkin, L. A., & Godin, J.-G., J. (1992). Reversal of female mate choice by copying. *Proceedings of the Royal Society of London: Series B,* **249,** 179–184.

Dugatkin, L. A., & Godin, J.-G. J. (1993). Female mate copying in the guppy, *Poecilia reticulata:* age dependent effects. *Behavioral Ecology,* **4,** 289–292.

Dugatkin, L. A., & Hoglund, J. (1995). Delayed breeding and the evolution of mate copying in lekking species. *Journal of Theoretical Biology,* **174,** 261–267.

Findlay, C. (1991). The fundamental theorem of natural selection under gene-culture transmission. *Proceedings of the National Academy of Science USA* **88,** 4874–4876.

Findlay, C., Lumsden, C., & Hansell, R. (1989a). Behavioral evolution and biocultural games: Vertical cultural transmission. *Proceeding of the National Academy of Science USA,* **86,** 568–572.

Findlay, C. S., Hansell, R.I.C., & Lumsden, C. J. (1989b). Behavioral evolution and biocultural games: Oblique and horizontal cultural transmission. *Journal of Theoretical Biology,* **137,** 245–269.

Fisher, R. A. (1958). *The genetical theory of natural selection* (2nd ed.). New York: Dover.

Galef, B. G. (1976). Social transmission of acquired behavior: A discussion of tradition and social learning in vertebrates. *Advances in the Study of Behavior,* **5,** 77–100.

Gibson, R. M., Bradbury, J. W., & Vehrencamp, S. L. (1991). Mate choice in lekking sage grouse: The roles of vocal display, female site fidelity and copying. *Behavioral Ecology,* **2,** 165–180.

Goldschmidt, T., Bakker, T. C., & Feuth-De Bruijn, E. (1993). Selective copying in mate choice of female sticklebacks. *Animal Behavior,* **45,** 541–547.

Gronell, A. M. (1989). Visiting behaviour by females of the sexually dichromatic damelsfish, *Chrysiptera cyanea* (Teleostei: Pomacentradae): a probable method of assessing male quality. *Ethology,* **81,** 89–122.

Hamilton, W. D. (1971). Geometry of the selfish herd. *Journal of Theoretical Biology* **31,** 295–311.

Haskins, C. P., Haskins, E. F., McLaughin, J. J., & Hewitt, R. E. (1961). Polymorphism and population structure in *Lebistes reticulatus,* a population study. In W. F. Blair (Ed.), *Vertebrate Speciation* (pp. 320–395). Austin: University of Texas.

Hoglund, J., Alatalo, R. V., & Lundberg, A. (1990). Copying the mate choice of others? Observations on female black grouse. *Behaviour,* **114,** 221–236.

Houde, A. E. (1988). Genetic difference in female choice in two guppy populations. *Animal Behavior,* **36,** 510–516.

Houde, A. E. (1992). Sex-linked heritability of a sexually selected character in a natural population of *Poecilia reticulata. Heredity,* **69,** 229–235.

Houde, A. E., & Endler, J. A. (1990). Correlated evolution of female mating preference and male color pattern in the guppy. *Poecilia reticulata. Science,* **248,** 1405–1408.

Jamieson, I. G., & Colgan, P. W. (1989). Eggs in the nests of males and their effect on mate choice in the three spined stickleback. *Animal Behavior,* **38,** 859–865.

Keister, R. (1979). Conspecifics as cues: a mechanism for habitat selection in the Panamanian grass anole (*Anolis auratus*). *Behavioral Ecology and Sociobiology,* **5,** 323–330.

Kirkpatrick, M. (1987a). Sexual selection by female choice in polygynous animals. *Annual Review of Ecological System,* **18,** 43–70.

Kirkpatrick, M. (1987b). The evolutionary forces acting on female mating preferences in polygenous animals. In J. Bradbury & M. Andersson (Eds.), *Sexual Selection Testing the Alternatives* (pp. 67–82). New York: Wiley-Interscience.

Kirkpatrick, M., & Dugatkin, L. A. (1994). Sexual selection and the evolutionary effects of mate copying. *Behavioral Ecology and Sociobiology,* **34,** 443–439.

Kirkpatrick, M., & Ryan, M. (1991). The evolution of mating preferences and the paradox of the lek. *Nature,* **350,** 33–38.

Knapp, R. A., & Sargent, R. C. (1989). Egg-mimicry as a mating strategy in the fantail darter, *Etheostoma flabellare:* females prefer males with eggs. *Behavioral Ecology and Sociobiology,* **25,** 321–326.

Kodric-Brown, A. (1985). Female preference and sexual selection for male coloration in the guppy. *Behavioral Ecology and Sociobiology,* **17,** 199–205.

Kodric-Brown, A. (1990). Mechanisms of sexual selection: insights from fishes. *Annales Zoologici Fennici.* **27,** 87–100.

Kodric-Brown, A. (1993). Female choice of multiple male criteria in guppies: interacting effects of dominance, coloration and courtship. *Behavioral Ecology and Sociobiology,* **32,** 415–420.

Laland, K. N. (1994a). On the evolutionary consequences of sexual imprinting. *Evolution,* **48,** 477–489.

Laland, K. N. (1994b). Sexual selection with a culturally transmitted mating preference. *Theoretical Population and Biology,* **45,** 1–15.

Landeau, L., & Terborgh, J. (1986). Oddity and the "confusion effect" in predation. *Animal Behavior,* **34,** 1372–1380.

LeFebvre, L., & Palameta, B. (1988). Mechanisms, ecology and population diffusion of socially learned, food-finding in feral pigeons. In T. Zentall & B. G. Galef (Eds.), *Social learning: Psychological and biological perspectives* (pp. 141–164). Hillsdale, NJ: Erlbaum.

Lill, A. (1974). Sexual behaviour of the lek-forming white-bearded manakin (*Manacus manacus Hartert*). *Zeitschrift für Tierpsychologie,* **36,** 1–36.

Losey, G. S., Stanton, F. G., Jr., Telecky, T. M., & Tyler, W. A. (1986). Copying others, an evolutionarily stable strategy for mate choice: A model. *American Naturalist* **128,** 653–664.

Magurran, A. E., & Seghers, B. H. (1991). Variation in schooling and aggression amongst guppy populations in Trinidad. *Behaviour,* **118,** 214–234.

Mangel, M., & Clark, C. (1988). *Dynamic modeling in behavioral ecology.* Princeton, NJ: Princeton University Press.

Marconato, A., & Bisazza, A. (1986). Males whose nests contain eggs are preferred by female *Cottus gobio. Animal Behavior,* **34,** 1580–1582.

Marks, A., Deutsch, J. C., & Clutton-Brock, T. H. (1994). Stochastic influences, female copying and the intensity of sexual selection on leks. *J. Theo. Biol.,* **170,** 159–162.

Maynard Smith, J. (1982). *Evolution and the Theory of Games.* Cambridge: Cambridge University Press.

McComb, K., & Clutton-Brock, T. H. (1994). Is mate choice copying or aggregation responsible for skewed distributions of females on leks? *Proceedings of the Royal Society of London: Series B,* **255,** 13–19.

Moller, A. P. (1994). *Sexual selection and the barn swallow.* Oxford: Oxford University Press.

Pomiankowski, A. (1988). The evolution of female mate preferences for male genetic quality. In P. Harvey & L. Partridge (Eds.), *Oxford Surveys in Evolutionary Biology* (Vol. 5, pp. 136–184). Oxford: Oxford University Press.

Pruett-Jones, S. G. (1992). Independent versus non-independent mate choice: Do females copy each other? *American Nature,* **140,** 1000–1009.

Richerson, P., & Boyd, R. (1989). The role of evolved predispositions in cultural evolution. Or, Human sociobiology meets Pascal's wager. *Ethology and Sociobiology* **10,** 195–219.

Ridley, M., & Rechten, C. (1981). Female sticklebacks prefer to spawn with mate whose nests contain eggs. *Behavior* **16,** 152–161.

Rohwer, S. (1978). Parental cannibalism of offspring and egg raiding as a courtship strategy. *American nature,* **112,** 429–440.

Sargent, R. C. (1988). Paternal care and egg survival both increase with clutch size in the fathead minnow, *Pimephales promelas. Behavioral Ecology and Sociobiology,* **23,** 33–37.

Seghers, B. H. (1973). *An analysis of geographic variation in the antipredator adaptations of the guppy, Poecilia reticulata.* Unpublished Doctoral dissertation, University of British Columbia.

Shaw, P. W., Carvalho, G. R., Magurran, A. E., & Seghers, B. H. (1991). Population differentiation in Trinidadian guppies (*Poecilia reticulata*): patterns and problems. *Journal of Fish Biology,* **39** (Suppl.), 203–209.

Shuster, S. M. (1987). Alternative reproductive behaviours: three distinct male morphs in *Paracerceis sculpta,* an intertidal isopod from the northern gulf of Mexico. *J. Crustacean Biology,* **7,** 318–327.

Shuster, S. M. (1989). Female sexual receptivity associated with molting and differences in copulatory behaviour among the three male morphs in *Paracerceis sculpta. Biological Bulletin,* **117,** 331–337.

Shuster, S. M. (1990). Courtship and female mate selection in a semelparous isopod crustacean, *Paracerceis sculpta. Animal Behavior,* **40,** 390–399.

Shuster, S. M. (1991). The ecology of breeding females and the evolution of polygyny in *Paracerceis sculpta*, a marine isopod crustacean. In R. Bauer & J. Martin (Eds.), *Crustaceaa Sexual Biology* (pp. 91–110). New York: Columbia University Press.

Shuster, S. M., & Wade, M. J. (1991). Female copying and sexual selection in a marine isopod crustacean, *Paracerceis sculpta. Animal Behavior,* **42,** 1071–1078.

Sikkel, P. (1988). Factors influencing spawning-site choice by female garabaldi, *Hypsypops rudicundus* (Pieces: Pomacentradae). *Copeia, 1988,* 710–788.

Sikkel, P. (1989). Egg presence and developmental stage influence spawning-site choice by female garabaldi. *Animal Behavior,* **38,** 447–456.

Slater, P., Eales, L., & Clayton, N. (1988). Song learning in zebra finches: Progress and prospects. *Advances in the Study of Behavior,* **18,** 1–33.

Stoner, G., & Breden, F. (1988). Phenotypic differentiation in female preference related to geographic variation in male predation risk in the Trinidad guppy. *Behavioral Ecology and Sociobiology,* **22,** 285–291.

Thompson, D. A., & Lehner, C. E. (1976). Resilience of a rocky intertidal fish community in a physically unstable environment. *Journal of Experimental Marine Biology and Ecology,* **22,** 1–29.

Unger, L. M., & Sargent, R. C. (1988). Alloparental care in the fathead minnow, *Pimephales promelas:* Females prefer males with eggs. *Behavioral Ecology and Sociobiology,* **23,** 27–32.

Wade, M. J. (1979). Sexual selection and variance in reproductive success. *American Nature,* **114,** 742–747.

Wade, M. J., & Pruett-Jones, S. G. (1990). Female copying increases the variance in male-mating success. *Proceeding of the National Academy of Science USA,* **87,** 5749–5753.

Wiley, R. H. (1991). Lekking in birds and mammals: behavioral and evolutionary issues. *Advances in the Study of Behavior,* **20,** 201–291.

Wilkinson, G. S., & Reillo, P. (1994). Female choice response to artificial selection on an exaggerated male trait in a stalk-eyed fly. *Proceedings of the Royal Society of London: Series B,* **255,** 1–6.

Winge, O. (1927). The location of eighteen genes in Lebises reticulatus. *Journal of Genetics,* **18,** 1–43.

Zentall, T. R., & Galef, B. G. (1988). *Social learning: Psychological and biological perspectives.* Hillsdale, NJ: Erlbaum.

6

Is Social Learning an Adaptive Specialization?

LOUIS LEFEBVRE

Department of Biology
McGill University
Montréal, Québec, Canada H3A 1B1

LUC-ALAIN GIRALDEAU

Department of Biology
Concordia University
Montréal, Québec, Canada H3G 1M8

INTRODUCTION

L earning, like other behavioral or structural traits, may vary in its usefulness according to the particular environmental problems an animal faces.
In psychology, Rozin and Kalat (1971) were the first to propose explicitly
that some learning abilities could be seen as adaptive specializations molded by
natural selection to cope with particular ecological demands. Three major assumptions underlie this view: (1) learning is not a single, general, set of rules for the
modification of behavior, but an assemblage of discrete abilities that may be oriented in different directions in different contexts; (2) because different species face
different ecological contexts, learning abilities can be expected to vary across species; and (3) the origin of ecologically correlated learning differences is divergent
natural selection. Rozin and Kalat's (1971) views have since been expanded and
applied to particular types of learning by, among others, Roper (1983), Sherry and
Schacter (1987), Rozin and Schull (1988), and Shettleworth (1993).

As Shettleworth (1993) points out, adaptive specialization is part of a wider,
ecological, program for the study of learning. Several logical (Plotkin & Odling-
Smee, 1979; Johnston, 1982) and mathematical (Stephens, 1991) models for the
evolution of learning also fall within this ecological program, as does recent comparative work on spatial memory in birds and mammals (Balda & Kamil, 1989;
Hilton & Krebs, 1990; Sherry, Jacobs, & Gaulin, 1992) and flower exploitation skills
in hymenoptera (Laverty, 1994; Laverty & Plowright, 1988; Dukas & Real, 1991).

Like any other learning ability, social learning can also be seen as an adaptive specialization to particular environmental demands. In fact, more than a decade before Rozin and Kalat (1971), Klopfer (1959, 1961) explicitly proposed a set of ecological predictions for interspecific differences in social learning (or "empathic" learning, in his terminology). Klopfer (1959) suggested that "the degree of social flocking will be found to show a positive correlation with the ability to learn avoidance and other responses through empathic processes." Subsequently, Klopfer (1961) restated this prediction with respect to solitary and social animals ("solitary species should be much less likely to have their characteristic behavior established by observation") and added a second prediction concerning opportunism: "birds with highly conservative and restricted habits would show a different type of observational effect than more tolerant species." Several authors have since emphasized the potential role of group living (Krebs, MacRoberts, & Cullen, 1972; Emlen & Oring, 1977; Mason & Reidinger, 1981; Strupp & Levitsky, 1984; Altmann, 1989) and of a generalist-opportunist lifestyle (Gandolfi, 1975; Strupp & Levitsky, 1984; Laland & Plotkin, 1990) in the evolution of social learning.

At the empirical level, Klopfer (1961) and Sasvàri (1979, 1985a) have provided comparative data that addresses Klopfer's prediction concerning the difference in social learning in conservative and opportunistic species. Experiments by Cambefort (1981) and colleagues (Jouventin, Pasteur, & Cambefort, 1976) are relevant to the effects of group living. In all three cases, interspecific differences in social learning appear to support the ecological view. Mandrills are quicker at social learning of an avoidance response than are baboons, which are, in turn, more rapid than are vervet monkeys; these differences are in the same direction as species differences in gregariousness (Cambefort, 1981; Jouventin et al., 1976). Opportunistic great tits learn an avoidance discrimination more easily in social conditions than conservative greenfinches (Klopfer, 1961; note, however, that this study does not use the standard trained tutor/naive observer design, but tests pairs of naive birds learning together). Great tits and blackbirds, more opportunistic than their respective congenerics the marsh tit and songthrush, also socially learn a new food searching behavior more rapidly (Sasvàri, 1979, 1985a). At first glance, the comparative literature therefore seems to suggest that social learning is an adaptive specialization to opportunist and gregarious lifestyles.

PROBLEMS WITH COMPARATIVE TESTS OF LEARNING

Confounding Variables

When these comparative learning tests are examined more closely, however, several problems become apparent. First, any number of confounding variables could have caused the interspecific differences observed. Some of these potential confounds have been singled out by MacPhail (1982): In any learning task, there are procedural biases that are likely to favor one species over another and, thus, to cause apparent interspecific differences that are, in fact, attributable to the task itself or to motivational effects caused by differential response to pretest deprivation.

Opportunism and anthropophilia can also have major confounding effects on interspecific tests of adaptive specialization (Lefebvre, in press). If a species is more tolerant of humans, less neophobic (Greenberg, 1984, 1989), less stressed by captivity and laboratory handling, and actually adapts well in the field to environmental changes brought about by human activity (e.g., urbanization), this species may perform better in any captive learning test for these reasons only, not because it has any adaptive specialization for the abilities being tested. In fact, many ecological views of learning suggest a general link between opportunism and all forms of learning (Johnston, 1982). It is therefore important to control this general effect before any conclusions are made on specialization for a particular learning ability.

Adaptive Departures from Interspecific Trends

Secondly, contrary to what is done in other areas of evolutionary biology, adaptive predictions concerning learning are usually made on interspecific differences in the trait itself and not on adaptive departures from phyletic variation in the trait and its covariates. If, for instance, we wish to test whether food caching is associated with a larger hippocampus (Sherry, Vaccarino, Buckenham, & Herz, 1989; Krebs, Sherry, Healy, Perry, & Vaccarino, 1990), copulation by females with several males is associated with larger testes (Harcourt, Harvey, Larson, & Short, 1981), or polygyny is associated with a greater degree of sexual dimorphism (Clutton-Brock & Harvey, 1977; Alexander, Hoogland, Howard, Noonan, & Sherman, 1979), we do not ask whether, at the absolute level, Clark's nutcrackers have a larger hippocampus than canaries, chimpanzees have larger testes than gorillas, or female–male size differentials are greater in polygynous baboons than they are in monogamous gibbons. We first take out the confounding effects of body size on the relevant variables and ask whether the residual deviation of hippocampus size, testes weight, or male–female differential, regressed against body weight, is larger in caching, multimale breeding or polygynous species, respectively. Looking at departures from normal phyletic variation in a trait rather than at the trait itself solves both problems mentioned above: It effectively controls for confounding variables (a testis or hippocampus or male–female differential can be larger in a given species simply because that species is large) and it focuses the comparative prediction at the appropriate level, i.e., adaptive departures from conservative covariates that may be strongly constrained phyletically. In the case of structural traits, this constraint is often allometric: Body parts and life history traits change on a predictable scale as animals become larger (Peters, 1983). It is not the allometric change per se that is interesting for most adaptive hypotheses, but ecologically correlated departures from allometric scaling. Any absolute species difference is just as likely to reflect phyletic scaling as it is to express the specialization predicted by an ecological hypothesis. Both components are worth studying. Scaling, for instance, has long been the focus of comparative psychologists and evolutionary biologists interested in the relationship between brain size and learning (Jerison, 1973; Riddell & Corl, 1977; Wyles, Kunkel, & Wilson, 1983; Stephan, Baron, & Frahm, 1991). It should be emphasized, however, that adaptive specialization deals essentially with the second, residual, component of interspecific variation.

STATISTICAL PROCEDURES FOR THE REMOVAL OF CONFOUNDING VARIABLES

This section reviews a series of statistical procedures that control the effects of confounding variables in comparative tests of learning and make adaptive predictions at the appropriate level of interspecific variation. The examples are based on hypothetical data sets generated for the purpose of this chapter and on actual data taken from Sasvàri's (1985 a,b) work on *Parus* and *Turdus* species, Klopfer's (1961) work on great tits and greenfinches and Lefebvre and colleagues' (Lefebvre, Palameta, & Hatch, 1996) work on pigeons and Zenaida doves. Both hypothetical and empirical examples deal specifically with social learning, but the technical and theoretical issues raised can be generalized to other learning abilities.

Two families of techniques will be discussed: linear regressions and analysis of variance. In both cases, we argue that the appropriate comparative test on a particular learning ability lies not in the standard prediction of differences between species, but in the relationship between the performance of each species on control tasks and tasks relevant to adaptive specialization. For linear regressions, this relationship can be expressed, for instance, by the residual deviation from the plot of the specialization task against the control task. For ANOVAs, this relationship can be found in the interaction term of a species-by-task design.

In both approaches, the standard procedure in a comparative social-learning test should include five steps: (1) choose two or more closely related species that differ on an ecological covariate that is predicted to be associated with the learning specialization (Domjan & Galef, 1983); (2) make sure *a priori* that the species do not also differ on an obvious confounding variable (e.g., the species should not differ both in opportunism and gregariousness); (3) test the species on one or more tasks that measure the learning ability predicted to vary in an adaptive way (Kamil, 1988); and (4) on one or more control tasks for which there is no adaptive prediction of interspecific differences; and (5) test the interspecific prediction on the relative, not the absolute, species differences on the tasks.

LINEAR REGRESSIONS

Linear Regressions on Nonindependent Groups

The most straightforward regression technique for the removal of confounding variables applies to sets of subjects in two or more species measured sequentially (repeated measures) on two types of tasks: one that assesses the presumed specialized ability (in the present case, social learning), and another (control tasks) that deals with potential confounding variables. The regression approach to a design of this type is illustrated in Figs. 1A and 1B. Each data point in both panels of the figure represents a separate individual, who is tested sequentially on the control and social tasks (sequential testing may be counterbalanced over tasks, but this may have serious consequences on the results, producing for instance a spurious negative correlation between performance on different tasks; Beauchamp, personal communication). In Fig. 1A, each subject's score (e.g., latency to learning, trials-to-criterion) on the control task is plotted on the x-axis and its score on the social task is plotted on the y-axis. Statistically, the adaptive hypothesis then applies to the intertask regressions that characterize each species; these regressions are predicted to differ significantly in the direction indicating, for instance, better social-learning performance (i.e., lower latencies) by the more gregarious species. In the example illustrated, the regression for the territorial species falls, as predicted, above the one for the gregarious species. When more than one control variable is available, an extra

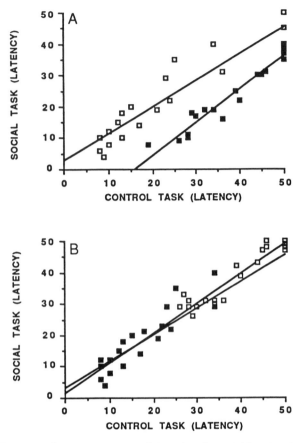

Fig. 1. Latency to learning on a control (*x*-axis) and a social learning (*y*-axis) task for individuals from two hypothetical species predicted (gregarious species: dark squares) or not predicted (territorial species: open squares) to show the learning specialization. (A) The species differ in the direction predicted and show significantly different regressions. (B) The species differ equally on the control and specialization tasks and their regressions do not differ.

check against confounding variables is the comparison of species regressions on two control variables. In this case, the species regressions should not differ in the direction predicted by the adaptive hypothesis.

If two species differ in the same direction in both control and specialization tasks, this will produce different mean scores and different species distributions on each task, but it will have no effect on the regressions. As seen in Fig. 1B, subjects from the gregarious species have lower latencies on both the social and control tasks and are therefore concentrated closer to the origin. The clouds of points representing the two species are at distinct ends of the graph, but their regressions do not differ. It should be noted that the examples given in Figs. 1A and 1B feature positive correlations between tasks, but that the procedure outlined here also applies to cases where the tasks are negatively correlated; an example of a positive intertask correlation is provided by Beauchamp, Giraldeau, and Dugatkin, (1994), but negative correlations between social and nonsocial tasks can be predicted when a frequency-dependent relationship exists between public and private forms of information gathering. For instance, individuals that rely more often on scrounging may, as a consequence, be relatively poor at the individual learning of a foraging skill, which may force them to learn socially from the producers they selectively follow (Giraldeau, Caraco, & Valone, 1994; Giraldeau & Lefebvre, 1987).

Linear Regressions on Independent Groups: Several Species and Two Tasks

When control and adaptive specialization tasks are performed on different individuals, interspecific comparisons cannot be done on the primary data obtained for each subject. As in other areas of comparative biology, regressions must be calculated on species averages for each variable, with each data point in the regression representing one species. If data are available for several species, mean performance of each species on the social task can be plotted on one axis (in Fig. 2, the y-axis) and mean performance on the control task on the other axis (Fig. 2, x-axis; note that standard errors are plotted around the species means to indicate within-species individual variation). The linear regression for the intertask relationship across species can then be calculated for the entire data set and the test of the adaptive hypothesis conducted on the residuals from this regression. Species predicted to show the adaptive specialization should have a significantly higher mean residual value than species that have no ecological reason to show the specialized ability. If, for instance, we want to test the idea that social learning is a specialized ability which is more useful in an opportunistic lifestyle, then the residuals of the more opportunistic species should tend to be negative and the residuals of the conservative species should tend to be positive; in other words, opportunistic species should fall below the regression line and conservative ones should fall above it (Fig. 2). Alternatively, if the number of species tested is very large, a separate regression can be run for species likely to show the specialization and species not likely to do so. In this case, the adaptive prediction is for significantly different regressions, with the regression of the specialized species (in this case, opportunistic) falling below that of the unspecialized ones (in this case, conservative).

In the social-learning literature, the only data set featuring several species in which each one is tested on both a social and a nonsocial task is found in Sasvàri's (1979, 1985a,b) work on *Parus* and *Turdus* species. For the social task, Sasvàri (1979, 1985a) tested 20 adult individuals each from three species of titmice (the great tit *P. major,* the blue tit *P. caeruleus,* and the marsh tit *P. palustris*) and two species of thrushes (the blackbird *T. merula* and the songthrush *T. philomelos*). Each bird was required to watch a conspecific find food by lifting a piece of linen (Sasvàri also tested juvenile observers and heterospecific demonstrators, but these data are not used in our reanalysis). For the nonsocial task, Sasvàri (1985b) tested 21 individuals from the same five species (presumed to be different individuals with respect to the social-task data). In this case, the birds were first conditioned to peck a key at one locus in a feeding apparatus, then to shift their pecking to a second location.

It should be noted that Sasvàri (1985a) makes no explicit claim that his demonstration of an ecologically based species difference in social learning should be interpreted as an adaptive specialization; he simply documents ecologically correlated species differences in social (Sasvàri, 1979, 1985a) and nonsocial (Sasvàri, 1985b) tasks and relates both sets of interspecific differences to opportunism. His work is nevertheless useful in testing key features of the specialization hypothesis and is the largest available data set published in subject-by-subject detail (Tables 1 and 2 in Sasvàri, 1985a; Table 1 in Sasvàri, 1985b) for social and nonsocial tasks in several species. Furthermore, Sasvàri's ecological variable, opportunism, is one of the two predicted correlates of social learning proposed by Klopfer (1959, 1961). If Klopfer's prediction is supported and social learning is an adaptive specialization correlated with opportunism, then the more opportunistic species of *Parus* and *Turdus* should, in Sasvàri's experiments, show a lower latency to social learning than the less opportunistic species, taking into account species differences in the nonsocial task (Sasvàri, 1985b). In our linear regression approach, the more opportunistic and urbanized species, *P. major* and *T. merula,* should both fall on the same side of the regression and their more conservative congenerics, *P. palustris* and *T. philomelos,* should both fall on the other. *P. caeruleus,* which is intermediate in terms of opportunism, should fall on or close to the regression.

Two potential problems with Sasvàri's data (as well as with other comparative-learning tests) are that several individuals from some of the species fail to learn within the limits of the experiment and that the dependent variables of the two tasks are different. The social task is measured in trials to criterion (with a maximum of 100 trials) and the nonsocial task in latency to learning (with a maximum

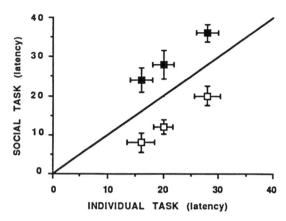

Fig. 2. Latency to learning on an individual (*x*-axis) and a social (*y*-axis) task for six hypothetical species predicted (more opportunistic species: open squares) or not predicted (more conservative species: dark squares) to show the learning specialization. Standard errors on both tasks are given by the *x* and *y* error bars. The line represents the best-fitting regression for the sample of species.

of 15 min). Because interspecific variation is likely to occur both in terms of the number of subjects who succeed in learning and the latency required for this learning, it is important to keep both sources of variation and assign a maximum latency score to the subjects who fail. This introduces a ceiling effect on the data, however, while the difference in dependent variables for social and nonsocial tasks leads to heterogeneous variances. Both problems can be minimized with transformations of the raw data: ceiling effects can be normalized by reciprocal transformations ($1/x$), while data taken on different scales can be transformed to z scores. For the z scores to be useful in interspecific work, each standardization must include all the species involved in a comparison for a given test. For example, an overall mean and standard deviation are calculated for all birds of species A and B on one task, and then each subject from either species is given its z score based on its deviation from the interspecific mean. If species A does better than species B on a particular task, individuals from A will tend to have positive z score, while individuals from B will tend to have negative z scores; if the two species do not differ, the positive and negative z scores will tend to be distributed randomly between A and B individuals. For Sasvàri's data, a separate standardization is conducted for *Parus* and *Turdus* species, but all five species are plotted together on the same graph. Note that the transformations now reverse the direction of our interspecific predictions, since the low latencies of opportunistic species become positive z scores based on the inverse of latency.

As Fig. 3 illustrates, opportunistic species are distributed on either side of the regression and there is thus no systematic difference in the residuals of the two more conservative species (studentized residuals: *T. philomelos* = 0.737; *P. palustris* = −1.330) and the two more opportunistic ones (*T. merula* = −.762; *P. major* = 1.989). The only result that fits the prediction is the intermediate position of *P. caeruleus* (−0.280). Compared to conservative species, opportunistic Passerines therefore do not seem to perform differently in social and nonsocial learning tasks.

Linear Regressions on Independent Groups: Two Species and Several Tasks

In cases where only two species are tested (these cases are frequent in the comparative literature on learning), a different regression approach has to be used. Instead of learning tasks representing each axis and species representing each data point, the *x*- and *y*-axes can now each represent one of the two species. The mean performance of each species on the different tasks then yields the data points for the regression (Lefebvre, in press). For this approach to work, several learning tasks

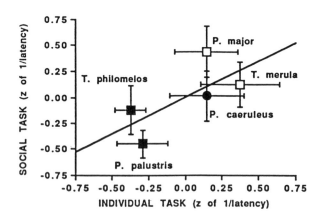

Fig. 3. Transformed (z score of reciprocal) latency to learning on an individual (*x*-axis) and a social (*y*-axis) task for five species of Passerines (data reanalyzed from Sasvàri, 1985a,b). Standard errors on both tasks are given by the x and y error bars. More opportunistic species are indicated by open squares, more conservative species by dark squares, and intermediate ones by a circle. The line represents the best-fitting regression through the sample of species.

must be used and the individuals tested from each species must be different for each task. In cases such as these, tests of the adaptive hypothesis either predict a significant difference between the regression linking the various control groups and the regression linking the specialization groups. Alternatively, if there are several control groups, but only a limited number of specialization groups (therefore precluding a regression of the latter), the regression can be calculated on the control groups only and the specialization groups tested against the 95% confidence limits of the control regression.

An example of the latter approach is provided by comparative work on Columbids involving two opportunistic species, three independent control groups per species, and two social groups (Lefebvre et al., 1996). The study tests the second of the two adaptive predictions made by Klopfer (1959), that group-living species should be better at social learning than more solitary species. The two species tested are the gregarious feral pigeon, *Columba livia,* and a tropical dove, *Zenaida aurita,* that is as urbanized in Barbados as the pigeon is in temperate regions. Unlike the pigeon, however, Zenaida doves forage alone on year-round territories over almost all the island of Barbados (Pinchon, 1963). A total of 50 pigeons and 63 Zenaida doves were used in five different conditions featuring different individuals in each group. In two conditions, the information required to find a food reward was provided by a conspecific tutor, while in three conditions, only nonsocial cues coming from the apparatus allowed the bird to solve the task. As as the case with Sasvàri's (1985a,b) data, trials to learning were transformed to z scores of the reciprocals to normalize variances and ceiling effects caused by some birds in some tasks reaching the maximum number of trials without learning.

As illustrated by Fig. 4, the two social tasks did not depart from the regression line fit through the three nonsocial tasks. There is, thus, no evidence that gregarious pigeons and territorial Zenaida doves respond differently to social and nonsocial tasks, beyond the general interspecific difference that applies to all tasks. Pigeons learn more rapidly than territorial Zenaida doves whether the learning situation is social or not.

ANALYSIS OF VARIANCE

As mentioned earlier, the ANOVA approach to a comparative test of adaptive specialization predicts a significant species-by-task interaction, whether or not the species and task main effects are also significant. The ANOVA approach differs from other comparative ones, which either feature no control tasks or, in cases

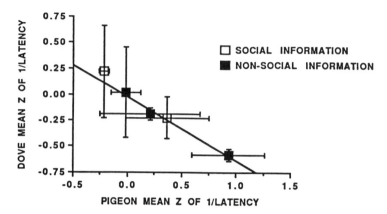

Fig. 4. Transformed (z score of reciprocal) latency to learning on social (open squares) and nonsocial (dark squares) tasks for gregarious pigeons (x-axis) and territorial Zenaida doves (y-axis). Standard errors on both tasks are given by the x and y error bars. The line represents the best-fitting regression through the nonsocial tasks.

where controls are run, expect no baseline species differences in the direction predicted by the adaptive hypothesis; in such cases, interspecific differences in the baseline are expected either to be nonsignificant (Hilton & Krebs, 1990) or opposite to those of the adaptive specialization (Shettleworth, 1993). In some ways, the approach we suggest here is more conservative than its alternatives: it requires a significant interaction, not a simple nonsignificant baseline in cases where sample sizes are often small and the probability of type-2 error is consequently fairly high. In other aspects, the approach is more liberal: the species may very well differ in the same direction on both the control and specialization tasks, provided that the predicted interspecific difference on the adaptive test is sufficiently larger than the interspecific difference on the baseline test for the interaction to be significant. By focusing on the species-by-task interaction, there is no a priori requirement that the species not differ on the baseline or else differ in a direction that is opposite to the adaptive prediction.

Three possible outcomes of a comparative test analyzed with our approach are presented in Fig. 5. Each section of the figure is produced with a hypothetical data set based on two species tested on both a social-learning task and a nonsocial-control task. As in the empirical examples used in the previous section, tasks need not be measured on similar scales and may also feature several subjects in one species that failed to learn one task within the limits imposed on the test, if the data are transformed into z scores of the reciprocal of trials-to-learning. Fig. 5A illus-

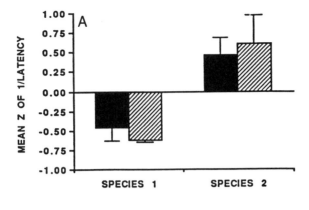

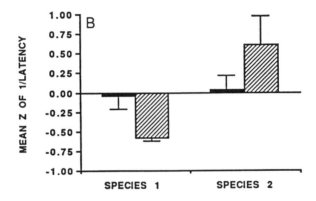

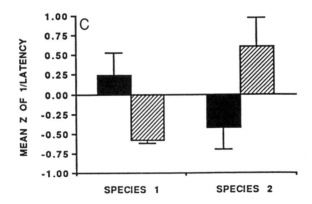

Fig. 5. Mean transformed (z score of reciprocal) latency to learning for two hypothetical species tested on a nonsocial control (dark histograms) and a social (hatched histograms) task. Error bars represent standard errors of the mean. (A) Species differ in the same direction on both tasks. (B) Species differ slightly on the control task, but strongly on the social one. (C) Species differ in opposite directions on the two tasks.

trates a case where the two species differ in the same direction on both specialization and control tasks; in this case, an ANOVA would reveal a significant species effect, but no significant species-by-task interaction. In Fig. 5B, the species differ slightly on the control, nonsocial, task, but differ strongly on the social (specialization) task; an ANOVA here would reveal both a significant species effect and a significant species-by-task interaction. In Fig. 5C, the species effect of the ANOVA is nonsignificant, but the interaction term is highly significant: The two species differ on the two tasks, but in opposite directions. The adaptive prediction would thus be rejected for Fig. 5A and supported in Figs. 5B and 5C.

The ANOVA approach can be applied to Klopfer's (1961) data on great tits and greenfinches (*Carduelis chloris*) learning to discriminate between palatable and unpalatable food in social (learning in pairs) and nonsocial conditions (learning alone). Unlike Sasvàri's (1985a,b), Klopfer's (1961) data do not need to be transformed into z scores, since the social and nonsocial conditions are measured on the same scale. Klopfer (1961) reports that, compared to learning latencies in nonsocial conditions, greenfinches appear to be less efficient at learning in pairs than great tits, with several greenfinches failing to learn within the 12 trial limit. When Klopfer's data are reexamined with an ANOVA, however, a weakly significant species-by-task interaction appears only when the subjects that failed to learn by Trial 12 are removed from the analysis (Fig. 6C: $F(1,28) = 4.27, p = 0.048$). On both transformed ($z$ scores of $1/x$) and untransformed data that includes non-learners, the interaction is nonsignificant (transformed: Fig. 6A, $F(1,34) = 1.40, p = 0.24$; untransformed: Fig. 6B, $F(1,34) = 3.19, p = 0.08$); in both these cases, however, the species main effect reaches borderline levels of significance (both $p = 0.05$).

Figure 7 illustrates the ANOVA approach as it applies to Sasvàri's (1985a,b) data. As in the regression presented earlier, trials-to-learning in the social task and latency to learning in the nonsocial task have been transformed to z scores of their reciprocals; in this analysis, *Turdus* and *Parus* spp are graphed separately, since, contrary to the regression approach, the ANOVA does not require that the number of species included in each analysis be as large as possible. The conclusions of the ANOVA are identical to those of the linear regression: The three *Parus* (Fig. 7A) and two *Turdus* species (Fig. 7B) differ in the same direction on the two tasks, but the species-by-task interactions fail to reach significance whether nonlearners are included or not in the analysis (*Parus:* with nonlearners: $F(2,117) = 1.34, p = 0.5$; without: $F(2,89) = 2.16, p = 0.12$; *Turdus:* with nonlearners: $F(1,78) = 1.46, p = 0.23$; without: $F(1,67) = 0.04, p = 0.84$). The species main effect is significant in all cases but one (p ranges from 0.002 to 0.04, except for *Turdus* without nonlearners,

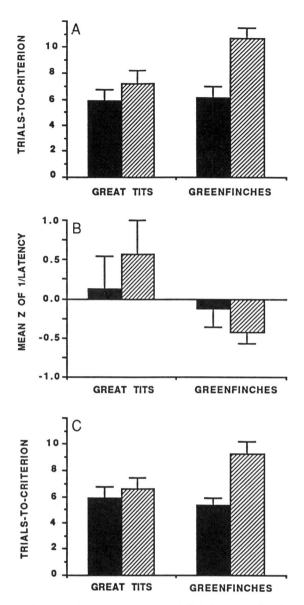

Fig. 6. Mean latency to learning in a nonsocial (dark histograms) and a social (hatched histograms) task for great tits and greenfinches (data reanalyzed from Klopfer, 1961). Error bars represent standard errors of the mean. (A) Untransformed trials-to-criterion, including subjects who approach the feeders but fail to learn. (B) Transformed (z scores of the reciprocal) trials-to-criterion, including nonlearners. (C) Untransformed trials-to-criterion, excluding nonlearners.

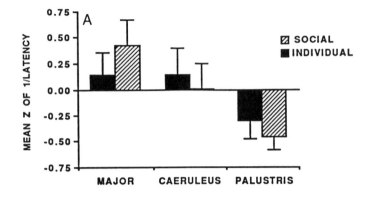

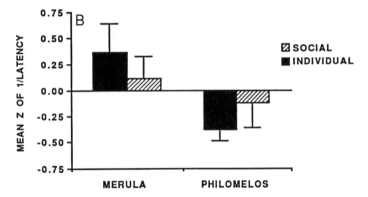

Fig. 7. Mean transformed (z score of reciprocal) latency to learning in a nonsocial (dark histograms) and a social (hatched histograms) task for (A) *Parus* and (B) *Turdus* species (data reanalyzed from Sasvàri, 1985a,b). Error bars represent standard errors of the mean.

where $p = 0.58$). Similar conclusions apply to the feral pigeon vs Zenaida dove comparison of Lefebvre et al. (1996). The species main effect is again significant, but not the species-by-task interaction.

For the eight Passerine and Columbid species tested up to now, consistent interspecific differences therefore appear to characterize both social and nonsocial tasks. Thus, there is, for the moment, little comparative evidence that social learning is an adaptive specialization to group living and/or opportunism. Instead, these ecological variables seem to be associated with interspecific differences in both individual and social forms of learning, a conclusion that can be linked to the general ecological view proposed by Johnston (1982). The conclusion is also relevant to Heyes' (1994) suggestion that social and asocial forms of learning share the same mechanisms. At the intraspecific level, a similar association between individual and social learning has also been found by Beauchamp et al. (1994), who report a positive correlation between the learning latencies of pigeons on a social and a nonsocial task.

Our failure to support specialization in the case of social learning should not be taken as a general rejection of the adaptive view. In fact, the same data used to question specialization for social tasks can be used to support adaptive theory for another important learning ability, spatial memory (Sherry & Schacter, 1987). Two of the *Parus* species tested by Sasvàri (1985a,b) differ in food-storing behavior, one of the major ecological correlates of spatial learning: *P. palustris* caches food, while *P. major* does not. As predicted by the specialization hypothesis, *P. palustris* does better than *P. major* in spatial learning tasks (Healy & Krebs, 1992), a difference which is opposite to the one reported by Sasvàri (1979, 1985a,b) for tasks that do not require spatial memory. Since *P. major* is both more opportunistic and gregarious than *P. palustris* (Barnes, 1975), the spatial memory difference cannot be due to the confounding variables we have examined in this chapter, but may instead be a true adaptive specialization to food storing.

CONCLUSIONS

Given the small number of species yet examined, more comparative work is clearly needed on ecological determinants of social learning. To be conclusive, this work must abide by the same methodological standards that have been developed in other areas of comparative biology (Harvey & Pagel, 1991). For what little data have been tested up to now, there are no obvious ecological factors uniquely correlated with quantitative interspecific differences in social learning. Once confounding variables have been removed, both Columbid and Passerine species that differ in opportunism or gregariousness show learning differences that seem to apply both to social and nonsocial tasks.

The fact that opportunism is related to general differences in learning is not extremely surprising and may even be slightly tautological, given that both rapid learning and opportunism can be defined as an ability to modify behavior quickly in the face of environmental novelty. Relating gregariousness to social learning is just as obvious (learning by observing others requires having others to observe), but the link found in Columbids between group living and individual learning seems more counterintuitive.

Scramble competition occurring in groups may be a key factor shaping both social and nonsocial forms of learning in foraging situations. Palameta (1989; Lefebvre & Palameta, 1988) and Dolman (1991) have proposed that scramble competition may lead to a "mental arms race," in which rapid learning of a feeding technique is favored to reduce the temporary advantage that knowledgeable foragers in a group have over naive ones. Speed is the major determinant of success in scramble competition, where food is removed from competitors by being ingested as quickly as possible. Both social and nonsocial forms of learning may be useful in reducing the temporary speed advantage held by a knowledgeable competitor. On the other hand, speed of individual and social learning appear unimportant for success in interference competition, where competitors are removed from the vicinity of food through aggression.

Throughout this chapter, we have discussed comparative methods that can be used to test the hypothesis that social learning is an adaptation to specific ecological conditions. An alternative to the comparative approach would involve testing predictions concerning the use of social learning derived from optimality models (Harvey & Pagel, 1991). For instance, even though pigeons are able to learn food-finding skills socially, they do not always do so; when subjects obtained (scrounged) shares of the tutor's food discoveries, social learning was not used to acquire food-finding skills (Giraldeau & Lefebvre, 1987; Giraldeau & Templeton, 1990). This result may be interpreted as an adaptation by which pigeons refrain from learning a skill that is economically unprofitable. Testing the specialization hypothesis therefore could also involve the formulation of economic models predicting the ecological circumstances under which individuals should use their social-learning ability.

In closing, we suggest that if social learning occurs, it likely accomplishes some biologically relevant function. To date, too few adaptive hypotheses have been proposed and fewer still have been tested to allow us to reject this view. We wish to encourage students of social learning to formulate new and testable functional hypotheses, while promoting a more stringent means of testing them through comparative analyses.

ACKNOWLEDGMENT

This research was supported by grants from NSERC to L.L. and L.-A.G.

REFERENCES

Alexander, R. D., Hoogland, J. L., Howard, R. D., Noonan, K. M., & Sherman, P. W. (1979). Sexual dimorphism and breeding systems in Pinnipeds, Ungulates, Primates and Humans. In N. A. Chagnon & W. Irons (Eds.), *Evolutionary Biology and Human Social Behavior* (pp. 402–435). North Scituate, MA: Duxbury Press.

Altmann, S. A. (1989). The monkey and the fig. *American Scientist, 77,* 256–263.

Balda, R. P., & Kamil, A. C. (1989). A comparative study of cache recovery in three corvid species. *Animal Behaviour,* **38,** 486–495.

Barnes, J.A.G. (1975). *The titmice of the British Isles.* Newton Abbott: David & Charles.

Beauchamp, G., Giraldeau, L.-A., & Dugatkin, L. A. (1994). The relationship between individual and social learning in pigeons. Human Frontier Science Program Workshop on Social Learning, Madingley, UK.

Cambefort, J. P. (1981). Comparative study of culturally transmitted patterns of feeding habits in the Chacma baboon (*Papio ursinus*) and the vervet monkey (*Cercopithecus aethiops*). *Folia Primatologica,* **36,** 243–263.

Clutton-Brock, T. H., & Harvey, P. H. (1977). Primate ecology and social organization. *Journal of Zoology (London),* **183,** 1–39.

Dolman, C. (1991). Preference for a heterospecific demonstrator in a territorial dove. Unpublished master's thesis, McGill University, Montréal.

Domjan, M., & Galef, B. G., Jr. (1983). Biological constraints on instrumental and classical conditioning: retrospect and prospect. *Animal Learning and Behavior,* **11,** 151–161.

Dukas, R., & Real, L. A. (1991). Learning foraging tasks by bees: a comparison between social and non-social species. *Animal Behaviour,* **42,** 269–276.

Emlen, S. T., & Oring, L. W. (1977). Ecology, sexual selection and the evolution of mating systems. *Science,* **197,** 215–223.

Gandolfi, G. (1975). Social learning in non-primate animals. *Bollettino di Zoologia,* **42,** 311–329.

Giraldeau, L.-A., Caraco, T., & Valone, T. J. (1994). Social foraging: Individual learning and cultural transmission. *Behavioral Ecology,* **5,** 35–43.

Giraldeau, L.-A., & Lefebvre, L. (1987). Scrounging prevents cultural transmission of food-finding behaviour in pigeons. *Animal Behaviour,* **35,** 387–394.

Greenberg, R. (1984). Differences in feeding neophobia in the tropical migrant warblers, *Dendroica castanea* and *D. pennsylvanica. Journal of Comparative Psychology,* **98,** 131–136.

Greenberg, R. (1989). Neophobia, aversion to open space and ecological plasticity in Song and Swamp sparrows. *Canadian Journal of Zoology,* **67,** 1194–1199.

Harcourt, A. H., Harvey, P. H., Larson, S. G., & Short, R. V. (1981). Testis weight, body weight and breeding system in primates. *Nature,* **293,** 55–57.

Harvey, P. H., & Pagel, M. D. (1991). *The comparative method in evolutionary biology.* Oxford: Oxford University Press.

Healy, S. D., & Krebs, J. R. (1992). Comparing spatial memory in two species of tits: Recalling a single positive location. *Animal Learning and Behavior,* **20,** 121–126.

Heyes, C. M. (1994). Social learning in animals: Categories and mechanisms. *Biological Reviews,* **69,** 207–231.

Hilton, S. C., & Krebs, J. R. (1990). Spatial memory of four species of Parus: performance on an open field analogue of a radial maze. *Quarterly Journal of Experimental Psychology,* **42B,** 345–368.

Jerison, H. J. (1973). *Evolution of brain and intelligence.* New York: Academic Press.

Johnston, T. D. (1982). The selective costs and benefits of learning: an evolutionary analysis. *Advances in the Study of Behavior,* **12,** 65–106.

Jouventin, P., Pasteur, G., & Cambefort, J. P. (1976). Observational learning of baboons and avoidance of mimics: Exploratory tests. *Evolution,* **31,** 214–218.

Kamil, A. C. (1988). A synthetic approach to the study of animal intelligence. In D. W. Leger, (Ed.), *Comparative perspectives in modern psychology. Nebraska symposium on motivation,* (Vol. 35, pp. 257–308). Lincoln, NE: University of Nebraska Press.

Klopfer, P. H. (1959). Social interactions in discrimination learning with special reference to feeding behavior in birds. *Behaviour,* **14,** 282–299.

Klopfer, P. H. (1961). Observational learning in birds: The establishment of behavioral modes. *Behavior,* **17,** 71–80.

Krebs, J. R., MacRoberts, M. H., & Cullen, J. M. (1972). Flocking and feeding in the great tit *Parus major*—an experimental study. *Ibis,* **114,** 507–530.

Krebs, J. R., Sherry, D. F., Healy, S. D., Perry, V. H., & Vaccarino, A. L. (1990). Hippocampal specialization in food-storing birds. *Proceedings of the National Academy of Science (USA),* **86,** 1388–1392.

Laland, K. N., & Plotkin, H. C. (1990). Social learning and social transmission of foraging information in Norway rats (*Rattus norvegicus*). *Animal Learning and Behavior,* **18,** 246–251.

Laverty, T. M. (1994). Bumble bee learning and flower morphology. *Animal Behaviour,* **47,** 531–545.

Laverty, T. M., & Plowright, R. C. (1988). Flower handling by bumble bees: a comparison of specialists and generalists. *Animal Behaviour,* **36,** 733–740.

Lefebvre, L. (in press). Ecological correlates of social learning: Problems and solutions for the comparative method. *Behavioural Processes,* Special issue on cognition and evolution.

Lefebvre, L., & Palameta, B. (1988). Mechanisms, ecology, and population diffusion of socially learned, food-finding behavior in feral pigeons. In T. R. Zentall & B. G. Galef, Jr. (Eds.), *Social learning: Psychological and biological perspectives* (pp. 141–164). Hillsdale, NJ: Erlbaum.

Lefebvre, L., Palameta, B., & Hatch, K. K. (1996). Is group-living associated with social learning? A comparative test of a gregarious and a territorial Columbid. *Behaviour,* **69,** 1–21.

MacPhail, E. M. (1982). *Brain and intelligence in vertebrates.* Oxford: Clarendon Press.

Mason, J. R., & Reidinger, R. F. (1981). Effects of social facilitation and observational learning on feeding behavior of the Red-winged Blackbird (*Agelaius phoeniceus*). *Auk,* **98,** 778–784.

Palameta, B. (1989). The importance of socially transmitted information in the acquisition of novel foraging skills by pigeons and canaries. Doctoral dissertation, King's College, Cambridge.

Peters, R. H. (1983). *The ecological implications of body size*. Cambridge: Cambridge University Press.

Pinchon, R. (1963). *Faune des Antilles françaises: le Oiseaux*. Fort-de-France: Muséum National d'Histoire Naturelle.

Plotkin, H. C., & Odling-Smee, F. J. (199). Learning, change and evolution: an enquiry into the teleonomy of learning. *Advances in the Study of Behavior,* **10,** 1–41.

Roper, T. J. (1983). Learning as a biological phenomenon. In T. R. Halliday & P. J. B. Slater (Eds.), *Animal Behaviour* (Vol. 3: Genes, Development and Learning, pp. 178–212). Oxford: Blackwell.

Rozin, P., & Kalat, J. W. (1971). Specific hungers and poison avoidance as adaptive specializations in learning. *Psychological Review,* **78,** 459–486.

Rozin, P., & Schull, J. (1988). The adaptive-evolutionary point of view in experimental psychology. In R. C. Atkinson, R. J. Hernnstein, G. Lindzey, & R. D. Luce (Eds.), *Stevens's handbook of experimental psychology* (2nd ed., Vol. 1, pp. 503–546). New York: Wiley.

Riddell, W. I., & Corl, K. G. (1977). Comparative investigation of the relationship between cerebral indices and learning abilities. *Brain, Behavior and Evolution,* **14,** 305–308.

Sasvàri, L. (1979). Observational learning in Great, Blue & Marsh Tits. *Animal Behaviour,* **27,** 767–771.

Sasvàri, L. (1985a). Different observational learning capacity in juvenile and adult individuals of congeneric bird species. *Zeitschrift für Tierpsychologie,* **69,** 293–304.

Sasvàri, L. (1985b). Keypeck conditioning with reinforcements in two different locations in thrush, tit and sparrow species. *Behavioural Processes,* **11,** 245–252.

Sherry, D. F., Jacobs, L. F., & Gaulin, S. J. C. (1992). Spatial memory and adaptive specialization of the hippocampus. *Trends in Neuroscience,* **15,** 298–303.

Sherry, D. F., & Schacter, D. L. (1987). The evolution of multiple memory systems. *Psychological Review,* **94,** 439–454.

Sherry, D. F., Vaccarino, A. L., Buckenham, K., & Herz, R. S. (1989). The hippocampal complex of food-storing birds. *Brain Behavior, and Evolution,* **34,** 308–317.

Shettleworth, S. J. (1993). Where is the comparison in comparative cognition? Alternative research programs. *Psychological Science,* **4,** 179–184.

Stephan, H., Baron, G., & Frahm, H. (1991). *Insectivora*. Berlin: Springer-Verlag.

Stephens, D. W. (1991). Change, regularity, and value in the evolution of animal learning. *Behavioral Ecology,* **2,** 77–89.

Strupp, B. J., & Levitsky, D. E. (1984). Social transmission of food preferences in adult hooded rats (*Rattus norvegicus*). *Journal of Comparative Psychology,* **98,** 257–266.

Wyles, J. S., Kunkel, J. G., & Wilson, A. C. (1983). Birds, behavior and anatomical evolution. *Proceedings of the National Academy of Science USA,* **80,** 4394–4397.

Developing a Theory of Animal Social Learning

KEVIN N. LALAND

Sub-Department of Animal Behaviour
University of Cambridge
Madingley, Cambridge CB3 8AA
United Kingdom

PETER J. RICHERSON

Division of Environmental Studies
Center for Population Biology
University of California
Davis, California 95616

ROBERT BOYD

Department of Anthropology
University of California
Los Angeles, California 90024

Over the last century, several hundred empirical studies have generated significant amounts of information about social learning in animals. While these studies have spawned a number of intuitive schemes designed to collate and categorize information on animal social learning, development of a formal theoretical framework is still in its infancy. One might expect many benefits from such a framework. Formal theory can structure and discipline thinking, tighten hypotheses, clarify mechanisms, identify key parameters, and raise questions that inspire empirical research.

The field of animal social learning has not yet reached the stage of development where theoretical and empirical projects guide and inform each other. Much of the theory that does exist has been adapted from models of human culture rather than specifically designed for the analysis of social learning in other animals. Although existing theory successfully addresses some very general evolutionary questions concerned with the adaptive properties of social learning, and models the transmission of socially learned variants through populations, it does not inform our understanding of within-individual learning processes. It was never designed to do so.

In contrast, the majority of experimental studies of animal social learning are concerned with learning processes occurring within individuals. The traditional laboratory study of animal social learning pairs a naive observer animal with a trained demonstrator conspecific in an attempt to establish the nature of the information transmitted between them. Such experiments say little about the diffusion of socially learned behavior through populations, or the maintenance of cultural differences among groups; they were never designed to do so. There is now ample evidence that demonstration of social learning is not sufficient grounds for the conclusion that social transmission is of sufficient penetration to result in diffusion of behavior through a population (Galef & Allen, in press; Laland & Plotkin, 1990, 1993). For this reason, traditional laboratory studies rarely generate information that can guide construction of mathematical models of the diffusion of socially transmitted information.

Recently, experimental studies of how social learning propagates behavior patterns in groups of nonhuman animals have begun to emerge, the findings of which may greatly inform the modeling exercise. Diffusion studies of mammals and birds in captive populations have given fresh insight into transmission dynamics, and established that social interaction can interfere with, as well as mediate social transmission (Giraldeau & Lefebvre, 1987; Lefebvre, 1986; Palameta & Lefebvre, 1985). Transmission chain studies have demonstrated that diet preferences, foraging, and predator information can be transmitted along chains of individuals, and can shed light on factors that affect the stability of transmission (Curio, Ernst, & Vieth, 1978; Galef & Allen, 1995; Laland & Plotkin, 1990, 1992, 1993).

Most transmission chain studies have employed designs using a chain of successive, single demonstrators. We have suggested that this approach could be extended to "multigeneration" experiments in small groups that had varying degrees of overlap in group membership across generations (Laland, Richerson, & Boyd, 1993). Such an approach was recently adopted by Galef and Allen (1995) to explore factors important in the propagation of traditions of food preference in small groups of Norway rats. Founder groups were taught an arbitrary food preference, and then individual members of the founder group were slowly replaced with naive subjects. Several generations of replacements after the last founder had been removed, the arbitrary food preference was still evident. Galef and Allen also found that restricting access to foods to a few hours per day significantly enhanced the stability of the arbitrary tradition. This is a good illustration of the fact that the transmission process may have emergent properties that the study of single demonstrator to observer social-learning episodes will not uncover.

Field studies suggest that cultural transmission may be common. Traditional laboratory experimentation has established something of the nature of the information transmitted between individuals. However, diffusion and transmission chain studies provide the most useful information for development of a theory of transmission dynamics. For instance, the diffusion studies by Lefebvre and Giraldeau have shown that scrounging (exploitation of food discovered by others) can be an important factor affecting diffusion of food-related behavior. Such studies have also demonstrated that migration between groups can affect the stability of socially transmitted traits in populations. Transmission chain studies have taught us that foraging information can be both gained and lost throughout transmission (Laland & Plotkin, 1990, 1992), and that different social learning mechanisms can combine to reinforce the stability of a transmission (Laland & Plotkin, 1993). Multigeneration, multiple-individual experiments are beginning to capture some of the population-level properties of animal social learning. The onus is now on theoreticians to incorporate such findings into transmission models.

We have previously argued that empirical and theoretical work suggests two very different conceptions of social learning (Laland et al., 1993). At one extreme, many mathematical models assume that social learning may result in stable traditions, transmitted across generations, and resembling human culture. A few field studies, such as those focusing on bird song, suggest that animal populations may exhibit stable behavioral traditions (Baker & Jenkins, 1987; Kroodsma & Baylis, 1982). At the other extreme, the results of many field and laboratory studies suggest that much nonhuman social learning may function to enhance foraging efficiency, allowing individuals rapidly to home in on appropriate behavior, with the transmitted information usually of only transient value (Lefebvre & Palameta, 1988). These studies often suggest that foraging information is transmitted through animal populations horizontally, between unrelated individuals, (rather than vertically, from parent to offspring). The well studied rat food preference system (Galef, 1988) is an example. We have referred to these two contrasting views as "traditional" and "highly horizontal" social transmission, and have suggested that they may be adaptations to quite different evolutionary problems (Laland et al., 1993).

Most social learning in most animals appears to fall in the highly horizontal category (Galef, 1988; Lefebvre & Palameta, 1988). Transmission may be ephemeral in such a system, spreading rapidly but being rapidly replaced, often on time scales far shorter than a generation. While traditional social transmission would make individuals track short-term environmental variability less effectively than individual learning, since the social learners' behavior is correlated with that of previous generations, highly horizontal transmission often seems to allow individuals to track environmental variability more efficiently than individual learning, as social learners can rapidly home in on appropriate behavior by sharing up-to-date foraging and predatory information.

For theoreticians, the distinction between traditional and highly horizontal social transmission is important because it determines the kind of mathematical analysis that is appropriate. To date, mathematical models of social learning have been developed to address, broadly, three questions: How do socially transmitted traits spread through populations? What are the adaptive advantages of social learning? And, what are the evolutionary consequences of a capacity for social learning? The following three sections review the progress made by the models in generating answers to these questions. It is important to note, however, that most theoretical analyses of the transmission dynamics and adaptive properties of social learning, including those applied to animals (e.g., Boyd & Richerson, 1988), have been concerned with traditional social transmission (Giraldeau, Caraco, & Valone's 1994 study is one exception). In later sections we attempt to address this imbalance by developing mathematical models of highly horizontal transmission systems, tailored to the specific properties of animal social learning. We use mathematical models to illustrate how the diffusion dynamics of a socially transmitted trait may be influenced by processes such as scrounging and bystander interference with demonstration. We also present formal theoretical analyses of the adaptive advantages of social learning in a highly horizontal context, such as at an information center.

The utility of any mathematical model rests critically on its assumptions, and hence it is vital that models of social learning are able to utilize relevant empirical findings. For instance, most mathematical models of social learning assume that interaction with an individual exhibiting a particular trait will increase the probability that an observer will exhibit the same trait. This means, with the exception of models of guided variation which explicitly incorporate individual learning, the models assume that traits may be maintained in populations through social interaction. In contrast, Heyes (1994) has argued that the stability of a socially transmitted trait depends primarily on reinforcement. After Dawkins (1976), Heyes maintains that the spread and maintenance of a behavior in a population depends on the fidelity with which the behavior is transmitted between individuals, the longevity of the behavior in an individual, and the fecundity of the behavior, the probability that the individual will transmit the trait to others. Heyes argues that since all three are necessary for a stable tradition, and since longevity depends on the extent to which the trait is reinforced, then the stability of a trait depends only on the stability of the environment. We also have suggested that the long-term stability of socially transmitted behavior patterns may depend more on the constancy of the local environment, or on genetic constraints, than any intrinsic properties of the transmission process (Laland et al., 1993). For this reason it is important that the effects of individual experience are explicitly incorporated into future models of animal social learning.

There may, however, be some important exceptions to Heyes's hypothesis. First, high fidelity and fecundity could offset the need for longevity, and the behavior could spread like an epidemic, lasting only briefly in an individual. Food preferences may spread in this way among populations of social foragers. Second, equally reinforced alternatives may be arbitrarily preserved in a population as a consequence of social learning. Empirical support for this hypothesis comes from transmission chain studies (Galef & Allen, 1995; Laland & Plotkin, 1990, 1992, 1993). Third, Heyes's position may rest on an overly static view of the organism. The stimuli to which an animal is exposed depend to a large extent on what the animal does, and where it goes—both of which can be strongly influenced by social interaction. It may be that it is social learning that ensures that an animal is continuously exposed to the same patterns of reinforcement. Social interaction may prevent behavior patterns from drifting out of an individual's repertoire, by rein-troducing behavioral variants (Galef & Whiskin, 1994). Fourth, arbitrary, nega-tively reinforced, or maladaptive traits may be preserved in an individual's reper-toire if they increase reproductive success through sexual selection (Laland, 1994). If any of these exceptions are valid, it creates the possibility that arbitrary, suboptimal, and even maladaptive patterns of behavior may be maintained in a population through social transmission. Ultimately, the extent to which social transmission can be predicted by patterns of reinforcement is an empirical issue, which can only be addressed with the appropriate population-level studies. In the short term, how-ever, it is important that we think carefully about the assumptions of mathematical models, and tailor them to what is known about social learning in animals.

The rest of this chapter reviews the theory of animal social learning as it presently stands, pointing out its strengths and its shortfalls, and discussing the extent to which it sheds light on empirical questions. Ultimately, our goal is to foster the further integration of empirical and theoretical approaches.

MODELS OF TRANSMISSION DYNAMICS

Much of the modeling of social transmission in animal populations has emerged from a related theoretical tradition concerned with the development of mathemati-cal models of human culture. Cavalli-Sforza and Feldman (1981) developed models of the spread of cultural variants under three types of transmission: vertical (from parent to offspring), oblique (from parental generation to offspring), and horizontal (within-generation transmission). Variants can increase in frequency as a result of selection operating at the individual level (this is cultural selection: one trait is more likely to be adopted than the alternatives, for instance, wearing blue jeans), or biological level (this is natural selection: individuals that express one trait are more likely to survive or leave offspring than individuals who express alternatives, for example, nonsmokers). Since this theory explores the diffusion of cultural variants through populations, it can be informed and tested by diffusion and transmission chain studies.

Boyd and Richerson (1985) have developed a similar body of theory. They consider how a trait can spread through a population by a process called "guided variation," in which individuals acquire behavioral traits culturally (typically, although not necessarily, from their parents), and then modify them on the basis of their personal experience. They also consider the effects of various forms of "biased transmission," which is a similar mechanism to Feldman and Cavalli-Sforza's "cultural selection." Biased transmission occurs when, given a choice between alternative modeled variants, individuals are more likely to adopt some variants than others (Boyd & Richerson, 1985). To bias their acquisition of cultural traits, individuals must be able to compare the behavior of at least two different individuals and evaluate them. Various types of bias exist, depending on what information individuals use to bias their acquisition. In direct bias, individuals might try out two alternative behavior patterns and use their own experience to choose the preferred one for adoption. Food preference experiments frequently show pronounced bias effects in that rats seem to innately prefer some flavors relative to others (Chou & Richerson, 1992). In the case of frequency-dependent bias, the commonness or rarity of a trait among a set of models might affect the probability of transmission nonlinearly. The scrounging effect noted by Lefebvre (1986) is an example of negative frequency dependence, and both Galef, Attenborough, and Whiskin (1990) and Chou and Richerson (1992) document an example of positive frequency-dependent bias with multiple demonstrators in the rat food preference paradigm. Social learners might also use cues about some traits in demonstrators, for example, health, to choose which demonstrator to attend to in order to acquire other traits, such as food preferences. Boyd and Richerson call this indirect bias. For example, among redwing blackbirds the social learning of a food preference is affected by whether the model bird becomes sick or remains well (Mason, 1988). Indirect bias may play a role in the diet acquisition of this species.

The cultural forces in transmission models (cultural selection, guided variation, bias) are motivated by the idea that social learners are active "decision makers" that combine information available from both social sources and individual experience according to some mental algorithm or another. In the simplest case, guided variation, Boyd and Richerson (1985) suppose that learners acquire their behavior using a weighted average of individual experience and social cues according to the rule

$$Y = aX + (1 - a)(H + \varepsilon)$$

where Y is the mature behavior of the individual after both social and individual learning have taken place, X is the behavior learned socially, H is the state of the environment expressed in terms of what behavior individual learning favors, ε represents the effects of errors in the individual learning process, a is the proportionate weight given to social cues, and $1 - a$ the weight given to personal experience. If a is close to one, the individual depends mostly on social cues and nearly ignores its own experience. As a approaches zero, the individual depends almost entirely on its own experience and ignores social cues. Although very simple, this model of mixed individual and social learning is fairly general. The same qualitative model can be derived assuming a very simple learning process (the linear-learning model from behavioral psychology) or a very advanced decision-making capabilities (a Bayesian decision maker). Qualitatively similar models also result from quantitative character models like the one above where Y can vary continuously (e.g., the amount of a given food eaten per day) and from models where the choice is between discrete, gene-like units (i.e., whether to forage in water or not). Direct bias is similar to guided variation, except that social learners use individual experience to sort among preexisting, socially acquired behavioral variants rather than generating their own variation.

A recent experiment by Lefebvre and Giraldeau (1994) provides a good illustration of why an investigator might want to move beyond the simplest and most general sort of transmission model. Models of social transmission generally predict that a novel variant will exhibit a sigmoidal (S-shaped) diffusion through the population under bias, or a saturating (inverted J-shaped) pattern of increase under guided variation (Boyd & Richerson, 1985; Cavalli-Sforza & Feldman, 1981). In both cases, rates of evolution slow as the favored trait approaches maximum penetration. Although there is considerable evidence to suggest that innovations spread through human populations with a pattern that resembles a sigmoid curve (Rogers, 1983), until recently (Lefebvre, 1995a,b), there has been very little information about diffusion in animal populations. Lefebvre and Giraldeau presented pigeons with the task of opening a stoppered test tube to obtain seeds in the presence of variable numbers of trained tutors and naive bystanders. They found that the rate of learning increased with tutor number, and decreased with bystander number. This is a significant finding, since it suggests that the spread of some innovations through nonhuman animal populations will not be sigmoidal as predicted by the models. In fact, Lefebvre and Giraldeau suggest diffusion will be exponential.

Why do the models predict sigmoidal diffusion? The most appropriate models for nonhuman animal populations are the oblique and horizontal diffusion models developed by Cavalli-Sforza and Feldman (1981) and Boyd and Richerson (1985), which predict sigmoidal diffusion. For Cavalli-Sforza and Feldman, the change in the frequency of a trait due to either oblique or horizontal transmission (the expressions are the same in either case) depends on (i) the proportion of individuals that do not have the trait, and (ii) the probability that these individuals will adopt it. If the frequency of the trait prior to a transmission episode is u, then (i) is given by $1 - u$, and (ii) by fu, where f is a constant (Cavalli-Sforza & Feldman, 1981). It makes sense that the probability that a naive individual will adopt a trait is proportional to the frequency of that trait in the population, because the frequency of a trait is the probability that any individual that the naive individual contacts will exhibit the trait. Cavalli-Sforza and Feldman developed a model in which unbiased vertical transmission (which does not change the frequency of the trait) operates in conjunction with biased oblique (or horizontal) transmission. After each generation the new frequency is $u' = u + fu(1-u)$, and the change in frequency $\Delta u = fu(1-u)$ (Cavalli-Sforza & Feldman, 1981). Boyd and Richerson reach the same expression (1985). Consideration of the possibility that individuals may lose the target trait on contact with a naive observer merely changes the value of the constant f. Since the change in frequency (Δu) is proportional to the product of the frequency of individuals with the trait (u) and the frequency of naive observers ($1-u$), the theory predicts a sigmoid curve. Initially the trait spreads slowly, since there are few demonstrators, and as it approaches maximum frequency, the trait again spreads slowly, because there are few individuals left to adopt the trait. However, at intermediate frequencies, the trait spreads quickly, since neither the number of demonstrators nor observers is limiting.

Lefebvre and Giraldeau's experiment suggests that at least two refinements of the theory are required to explain the dynamics of social transmission in pigeons. First, the assumption that the probability of adopting a trait is solely proportional to the frequency of the trait in the population is inappropriate. Lefebvre and Giraldeau suggest that the probability of adopting a trait depends on two factors: (i) the probability of contacting a demonstrator, or u, and (ii) the probability that bystanders do not interfere with the observation and demonstration of the trait. If the probability that bystanders do not interfere is proportional to the number of noninterfering individuals, and if most or all noninterfering individuals are demonstrators with the trait, then the probability of a naive individual adopting the trait by oblique or horizontal transmission is given by gu^2, where g is a constant, and the change in frequency with each transmission episode is $\Delta u = gu^2(1-u)$. Figure 1 plots the diffusion of traits that spread (i) sigmoidally, and (ii) by this kind of oblique transmission. The new curve is a distorted S-shape, with the saturating effect apparent only toward the end of diffusion. Note that, in this case, interference does not block diffusion of the trait through the population, although it does slow down transmission considerably when the favored trait is rare.

Lefebvre and Giraldeau's work also suggests that some individuals will not learn the test-tube-opening task, or will reject opening as a strategy, if eating food providing by others is a more successful strategy. We can generate a simple model of this scenario if we assume that some individuals that have acquired a trait will reject it, with a probability proportional to its frequency, or hu, where h is a constant. In other words, the more individuals there are producing food, the

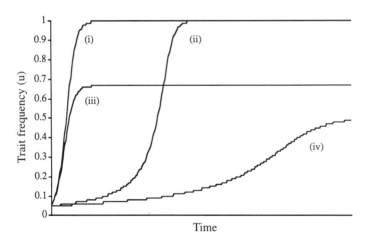

Fig. 1. Diffusion of a trait in an animal population. (i) The standard sigmoidal diffusion, $\Delta u = f(1-u)u$; (ii) with the effect of interference in trait demonstration, $\Delta u = g(1-u)u^2$; (iii) with scrounging, $\Delta u = [f(1-u)-hu]u$ and the equilibrium frequency is $f/(f+h)$; (iv) with interference and scrounging, $\Delta u = [g(1-u)-h]u^2$ and the equilibrium frequency is $1-h/g$. With interference the diffusion resembles an exponential curve more than a standard sigmoid. The effect of scrounging is to prevent the trait from spreading completely through the population.

greater the probability that some stop producing, and start scrounging. Now the change in frequency with each transmission episode is given by $\Delta u = fu(1-u) - hu^2 = [f(1-u) - hu]u$, and the equilibrium frequency of the trait, $\hat{u}$, is given by $\hat{u} = f/(f+h)$. In this case, a socially transmitted trait will only diffuse part way through the population, as shown in Fig. 1(iii). With both interference in the demonstration process, as in (ii), and scrounging, as in (iii), then the change in frequency with each transmission episode is given by $\Delta u = gu^2(1-u) - hu^2 = [g(1-u) - h]u^2$, and the equilibrium frequency of a trait, $\hat{u}$, is given by $\hat{u} = 1 - h/g$. Again, the trait only diffuses part way through the population, as shown in Fig. 1(iv).

It would be premature, on the basis of Lefebvre and Giraldeau's finding, to assume that the diffusion of all traits in nonhuman animal populations will be nearly exponential, and hence that the current theory is inappropriate. Scrounging and other forms of interference will not play a significant role for all socially transmitted traits, many of which involve no direct competition for limited resources. Those field studies that provide sufficient information to plot transmission curves suggest that the diffusion of some traits resembles exponential diffusion, others sigmoidal, and still others linear (Lefebvre, 1995a,b). Since data from field studies are rarely powerful enough to allow compelling curve fitting, it may be difficult to distinguish curve (ii) (diffusion with interference) from an exponential curve, and curve (iv) (diffusion with interference and scrounging) from a linear curve. These simple models may help to explain why apparently linear and exponential curves have been reported. Further experiments are required which, like Lefebvre and Giraldeau's insightful study, explore the transmission dynamics of other traits, in additional species.

Despite over a century of research, surprisingly little is known about the population-level dynamics of social learning. Both modeling and experimental studies are little developed compared to the complexity and diversity that is probably present in systems of social learning, human, and nonhuman alike. Several questions remain unanswered. For instance: What factors affect the stability, fidelity, and diffusion rate of a trait? How do different social-learning mechanisms affect diffusion dynamics? How do unlearned predispositions affect the stability of a transmitted behavior in a population? Further diffusion and transmission chain studies will answer some of these questions.

THE EVOLUTION OF CULTURE

Over the last decade, theoreticians have developed "dual-inheritance" models to investigate the complex interaction between culturally and genetically transmitted information (e.g., Boyd & Richerson, 1985; Feldman & Cavalli-Sforza, 1989; Feldman & Zhivotovski, 1992). Much theoretical analysis has focused on the question of the adaptive advantages (and disadvantages) of the social transmission of information (Aoki & Feldman, 1987; Boyd & Richerson, 1985, 1988; Cavalli-Sforza & Feldman, 1983). Boyd and Richerson have investigated the ecological conditions likely to lead to the evolution of cultural transmission, of genetically biased cultural transmission, and of different patterns of cultural transmission (Boyd & Richerson, 1976; Richerson & Boyd, 1978, 1989). One important feature of dual-inheritance models is that processes that guide and bias the acquisition of culture need not have very strong effects at the individual level in order to dominate the multigenerational dynamics of a cultural system. Even if each individual's own learning has the effect of making only rather marginal adjustments in what that individual learned socially, the effects of such individual learning summed over many individuals and a few generations can act as a powerful supplement to natural selection in generating adaptations.

The adaptive advantages of social learning are assessed by conducting a calculus of the costs and benefits of different systems of acquiring adaptive information. The genetic system for information transmission is cheap and accurate, but inflexible in the face of environmental variation. Individual capacities for phenotypic flexibility, ranging from inducable enzyme systems to individual learning, are supplementary systems for acquiring adaptive information (Plotkin & Odling-Smee, 1981), and have the advantage that they allow organisms to take account of short-term environmental contingencies. Individual learning is a costly way to acquire information, however, since simple systems are likely to be error prone, whereas more accurate systems may require costly nervous systems. A system of social learning may reduce the costs of individual learning by allowing individuals to rely partly on other experienced individuals rather than engaging in costly trials, exploration, and/or calculations of their own. In changing or heterogeneous environments, however, erroneous information may be obtained by social learning. Natural selection should act on the complex of genetic transmission, transmission by social learning and individual decision making to optimize over the costs and benefits of each.

Boyd and Richerson (1985, 1988) have explored the circumstances under which natural selection favors a system of social learning. Their analysis considered first the optimisation of a dependence on social vs. individual learning (the "a" and "$1-a$" weights and their analogues in a discrete-character model). In the quantitative-character model, fitness is optimized when the costs of making errors due to social learning equal those due to individual learning. In a variable environment, too little use of experience and too great a tendency to depend on social learning would cause the cultural system to track change poorly, and would favor individuals that increase their checking. However, assuming that it is easier to acquire a behavior by social than by individual learning, individuals that check too much and depend too little on socially learned traditions will pay excessive learning costs. Environmental change affects the results substantially. If rates of environmental change are slow enough, an uncritical dependence on social traditions is safe, since natural selection alone can keep the tradition current in this case. When environments change rapidly relative to the frequency of transmission, a tradition cannot track environmental change even given quite strong individual learning, and complete dependence on individual learning is favored. The model suggests there exists a broad range of intermediate rates and magnitudes of environmental change in time or space that favor some mixture of social and individual learning. The neurological and energetic costs and accuracy of individual learning also play a role in the analysis. As individual learning becomes less accurate or more costly a greater dependence on tradition is favored.

Next consider adding the opportunity to transmit information genetically rather than by social learning. Boyd and Richerson measured the fitness difference between populations with the optimal mix of individual learning and genetic transmission and populations with an optimal mix of social and individual learning. The results are quite intuitive. When environments change very slowly, all information should be transmitted genetically, since the modest demands for updating are easily met by the genetic system responding to selection. When environmental change is very rapid, tracking by pure individual learning is favored. At intermediate rates of change, social learning is an advantage.

In the case of human culture, these results have some appeal. Culture has made humans a very successful species, and our culture capacities have grown under the Pleistocene regime of great environmental fluctuation. For social learning in other animals, Boyd and Richerson's analysis raises more questions than it answers. The results seem to suggest that vertical social transmission should be very widespread. This corresponds to Darwin's and other 19th-century naturalists' enthusiasm for inheritance of acquired variation, including imitation (Galef, 1988). Yet humans appear to be unique in the degree of their reliance on tradition.

There are several possible answers to this puzzle. Perhaps the theoretical analysis described here does not represent all the costs of a system of tradition. If the psychological mechanisms that underlie stable vertical transmission require costly neural apparatus, perhaps extensive cross-modal connection in the brain, the hominid lineage might represent an adaptive revolution that somehow worked its way across a threshold. Hypotheses that invoke some form of adaptive revolution are the most popular explanation for the apparent difference between human culture and limited animal capacities for social learning (Landau, 1984). Or perhaps tradition is more common in animal populations than we currently believe, and we simply have had difficulty in demonstrating it. Most empirical studies of animal social learning have focused on foraging behavior. If we look among the traits known to be vertically transmitted in human populations, we find fears, language, social status, and occupational roles. It may be no coincidence that animal analogues of these traits (fear of snakes and other predators in primates, bird song, vocalization frequency in bats, and inherited ranks among primates) represent the best evidence for vertical transmission currently available. Perhaps, by looking at traits that are transmitted vertically among humans, we can suggest where to look for the "missing" vertical transmission in animal populations.

The models referred to thus far effectively assume that individuals interact with a single individual at random, and subsequently use the information they acquire to adopt a behavioral variant. Boyd and Richerson (1989) have extended this analysis to the situation in which individuals sample a number of models. They found that when an equal number of models use each behavior, individuals should rely completely on individual learning, but as the number of models exhibiting one behavior increases, natural selection should increase the relative weighting attributed to social learning. This is consistent with empirical findings. Galef et al. (1990) found that the greater the proportion of Norway rat demonstrators in a group that had eaten a diet, the greater the proportion of that diet the observers ate. Other investigators have found similar effects in other species (Lefebvre & Giraldeau 1994; Sugita 1980). This not only suggests that animals attend to cues from many potential models, but also that the strength of preference for particular variants is affected by the number of demonstrators exhibiting a behavior.

The very general models of the type pioneered by Cavalli-Sforza and Feldman, and Boyd and Richerson, have focused mainly on traditional social transmission. Although these models give some insight into animal social learning, they take no account of the differences between social learning in humans and in other animals. The best studied examples of social learning in nonhuman animals appear to be characterized by highly horizontal social transmission. In such a system short-term diet preferences might evolve over a few nights, guided by individual experience that transforms the transmitted information on time scales considerably shorter than a generation. Horizontal transmission has been found among social foragers (pigeons and rats) living in seemingly the most unpredictable and changeable environments (Lefebvre & Palameta, 1988), a finding that seems at variance with Boyd and Richerson's (1988) prediction that individual learning should be more important than social learning in variable environments. To date, the theoretical analysis has assumed that there may be costs associated with social learning in an unpredictable environment, as outdated, or bad information might be acquired. However, empirical studies suggest that mechanisms exist which allow animals to identify toxic diets, either directly, by preventing preferences for toxic foods or traits that reduce fitness from spreading, or indirectly, through the social transmission of preferences for valuable alternatives (Bond, 1982; Galef, 1985, 1986; Lavin, Feise, & Coombes, 1980; Mason, 1988).

An alternative explanation for the prevalence of horizontal social transmission in animal populations is that selection may have favored a facultative reversion from social to individual learning when it is most important to track major environmental shifts, which may, in turn, have allowed an increase in the routine dependence on social cues. Theoretical analyses have found that social learning is less effective than individual learning in tracking environmental variability, partly because transmission acts as a conservative force that lags behind environmental variability (Boyd & Richerson, 1985). In reality, individuals may often be forced by environmental catastrophes to temporarily ignore social cues (for example, those indicating which foods to eat) when the conflict with the state of the environment (i.e., if the food supply is exhausted).

One theoretical study which is tuned to the specific properties of animal social learning is Giraldeau et al.'s (1994) exploration of the costs and benefits of group foraging in the face of direct competition for resources (i.e., scrounging). Giraldeau et al. concentrate on within-generation social learning, assuming that both individual and social learning of a trait enhancing resource production depend on the trait's frequency, the former because of scrounging behavior. The acquired trait results in an increased ability to find resource clumps, relative to a baseline rate. They found that social learning increased the expected number of individuals foraging at the elevated rate, relative to the number expected if individuals were solitary foragers, but with no social learning there was a fitness cost to group foraging. Giraldeau et al.'s (1994) hypothesize that the adaptive function of social learning may be to allow individuals to circumvent some of the inhibitory effects that scrounging has on individual learning of a foraging skill. It is clear that there are circumstances where reliance on social learning may be favored in a variable environment.

Giraldeau et al.'s (1994) analysis focuses on the adaptive role of social learning in populations of highly competitive social foragers, such as pigeons and starlings. An explicit model of the evolution of diet breadth based on biases in favor of adding novel foods eaten by demonstrators, rapid horizontal transmission of traits within a generation, strong testing by individual experience, and rapid forgetting, would also be useful. More generally, there is a need for an analysis of the adaptive function of social learning at an information center, where a social population can exchange information about diet choice and resource location, as observed in populations of rats, bats, or gulls. We present such a model in the next section.

The Evolution of Horizontal Social Transmission

The model described below is designed to explore the adaptive value of horizontal social transmission in a population of foragers which exchange information at an information center, but where there is comparatively little direct competition for resources. We begin by classifying the individuals in an isolated population as either learners or social learners, according to whether they have genotype L or S. (This model assumes haploid genetics, but we have analyzed a diploid model and we find it gives qualitatively similar results). Learning and social learning are not viewed as alternatives, however, since here all individuals engage in individual learning, but in addition social learners are influenced by their conspecifics.

Individuals exhibit one of two phenotypic states (1 or 0), representing the diet they consume. Diet 0 represents a baseline diet of low nutritional value, which is continuously available. In reality, diet 0 could be composed of a number of different food items, and can also represent a lack of foraging success. Diet 1 is a novel diet of high nutritional value, but which is variable in its geographical and temporal distribution. Consumption of diet 1 confers a fitness advantage.

There are four classes of individual, or phenogenotype, L_1, L_0, S_1, and S_0, which we define as having frequencies $x_1 - x_4$ at time t, and express as the vector $\mathbf{x}^t$,

$$\mathbf{x} = \begin{bmatrix} x_1 \\ x_2 \\ x_3 \\ x_4 \end{bmatrix}.$$

We assume each individual experiences the following cycle on a daily, or more frequent, basis: (i) *Social Interaction:* social learners are influenced by their conspecifics diet choice, and can develop a preference for diet 1 as well as learn one of its previous locations. Individual learners are unaffected by social interaction. (ii) *Foraging:* both learners and social learners forage for food. If they do not locate diet 1, they revert to consumption of diet 0. (iii) *Selection:* diet 1 eaters have a fitness advantage.

(i) Social Interaction

Each individual interacts with a conspecific, but only social learners learn from the experience. The effects of social learning are represented by the transition probability c. An S_0 individual who interacts with an L_1 or S_1 conspecific will become S_1 with probability c. The transition from state S_0 to S_1 could represent the acquisition of a dietary preference (e.g., rats picking up cues on the breath of a conspecific), or the learning of the location of a food site (e.g., unsuccessful bats following a conspecific to food). It is assumed that there is no scrounging. Social interaction introduces novel variants (i.e., diet 1) into an individual's repertoire, but does not increase the probability that familiar foods (i.e., the baseline diet 0) will be consumed (Galef, 1993; Galef & Whiskin, 1994). Table 1 presents the probability than an individual will be L_1, L_0, S_1, and S_0, given the category of individual with which it interacts. After social interaction, time t', the phenogenotype frequencies, $\mathbf{x}^{t'}$, are given by multiplication of $\mathbf{x}^t$, by the transition matrix $\mathbf{C}$ which describes the effect of social learning,

$$\mathbf{x}^{t'} = \mathbf{C}\,\mathbf{x}^t, \qquad \text{where } \mathbf{C} = \begin{bmatrix} 1 & 0 & 0 & 0 \\ 0 & 1 & 0 & 0 \\ 0 & 0 & 1 & (x_1 + x_3)c \\ 0 & 0 & 0 & 1 - (x_1 + x_3)c \end{bmatrix}. \tag{1}$$

TABLE 1

The Probability That an Individual Will Be L_1, L_0, S_1 and S_0 after Social Interaction, Given the Individual's Current State, and the Type of Individual It Interacts with

Social interaction			Final state			
Self	Other	Frequency	L_1	L_0	S_1	S_0
L_1	L_1, L_0, S_1, S_0	x_1	1			
L_0	L_1, L_0, S_1, S_0	x_2		1		
S_1	L_1, L_0, S_1, S_0	x_3			1	
S_0	L_1, S_1	$x_4(x_1 + x_3)$			c	$1 - c$
S_0	L_0, S_0	$x_4(x_2 + x_4)$				1

TABLE 2

The Probability That an Individual Will Be L_1, L_0, S_1 and S_0 after Foraging, Given the Individual's Initial State

Initial state	Frequency	Final state			
		L_1	L_0	S_1	S_0
L_1	x'_1	$1-e$	e		
L_0	x'_2	ε	$1-\varepsilon$		
S_1	x'_3			$1-e$	e
S_0	x'_4			ε	$1-\varepsilon$

(ii) Foraging

All individuals forage for food. If they locate and consume diet 1 they commit to memory its location, and return the next foraging episode (Table 2). L_1 and S_1 individuals have a preference for diet 1, and know where it was located in the previous foraging episode. They will find and consume diet 1 with probability $1-e$, where e represents the probability that diet 1 is no longer at that site. Thus e is a measure of environmental variability. Prior to foraging, L_0 and S_0 do not have a preference for diet 1 or know where it is located. They will forage in a subsection of the local environment, and find and consume diet 1 with probability ε. Thus ε depends on the proportion of the local environment that an individual can search, the amount of the environment containing diet 1, and the probability that an unfamiliar diet will be consumed. After foraging, time t'', the phenogenotype frequencies, $\mathbf{x}^{t''}$, are given by multiplication of $\mathbf{x}^{t'}$, by the transition matrix $\mathbf{F}$ which describes the effects of foraging,

$$\mathbf{x}^{t''} = \mathbf{F}\,\mathbf{C}\,\mathbf{x}^{t}, \quad \text{where } \mathbf{F} = \begin{bmatrix} 1-e & \varepsilon & 0 & 0 \\ e & 1-\varepsilon & 0 & 0 \\ 0 & 0 & 1-e & \varepsilon \\ 0 & 0 & e & 1-\varepsilon \end{bmatrix}. \quad (2)$$

(iii) Viability Selection

Consumers of nutritious diet 1 have fitness advantage s over consumers of the baseline diet 0. We assume that there is no additional cost to social learning, either in terms of neural circuitry, or through the transmission of maladaptive information. Humans, perhaps, aside, there is little evidence that social learning relies on underlying processes, or brain structures, which are fundamentally different from those employed in individual learning and nonlearned communication. There is evidence, discussed in the above section, for the existence of mechanisms which, directly or indirectly, prevent maladaptive information from spreading. The final set of recursions, at time $t+1$, is given by multiplication of $\mathbf{x}^{t''}$ by the transition matrix $\mathbf{V}$ which describes the effect of viability selection

$$\mathbf{x}^{t+1} = \mathbf{V}\,\mathbf{F}\,\mathbf{C}\,\mathbf{x}^{t}, \qquad \text{where } \mathbf{V} = \frac{1}{1 + (x_1'' + x_3'')s}\begin{bmatrix} 1+s & 0 & 0 & 0 \\ 0 & 1 & 0 & 0 \\ 0 & 0 & 1+s & 0 \\ 0 & 0 & 0 & 1 \end{bmatrix} \quad (3)$$

The frequency of diet 0 and diet 1 consumers, f_0 and f_1 are approximately given by

$$f_0 \approx 1 - \varepsilon - \phi(1-e-\varepsilon) - s[\varepsilon + \phi(1-e-\varepsilon)], \text{ and}$$
$$f_1 \approx \varepsilon + \phi(1-e-\varepsilon) + s[1 - \varepsilon - \phi(1-e-\varepsilon)], \text{ where } \phi = (1+x_4c)(x_1+x_3). \quad (4)$$

The three terms, which together give the frequency of diet 1 consumers, are readily interpreted as: the proportion that acquire the trait anew by learning, or ε; previous diet 1 consumers plus those that acquire the trait anew by social learning, or $\phi(1-e-\varepsilon)$; and the increase in the proportion of diet 1 consumers due to selection, or $s[\varepsilon + \phi(1-e-\varepsilon)]$. The effect of social learning on the frequency of individuals consuming the advantageous diet can be seen by observing that positive values of c will increase f_1 provided $e + \varepsilon < 1$.

We are also in a position to explore how this form of social learning affects the mean fitness of a population. Rogers (1988) suggests that social learning does not increase mean fitness. Boyd and Richerson (1995) show that this result is robust, but nonetheless find some exceptions. Both of these analyses assume that the capacity for individual and social learning are inversely related. In contrast, here social learners acquire foraging information from conspecifics with no decrement in their ability to test, or adjust, this information on the basis of their own foraging experience. The mean fitness of the population, $\overline{W}$, is given by

$$\overline{W} = 1+s(x_1''+x_3'')$$

which can be reexpressed as

$$\overline{W} \approx 1 + s[\varepsilon + (1+x_4c)(x_1+x_3)(1-e-\varepsilon)]. \quad (5)$$

In contrast to Rogers, here it can be seen that positive values of c will increase the mean fitness provided $e + \varepsilon < 1$.

We consider the invasion of the S genotype into a population of exclusively L individuals. Prior to invasion, the population reaches an equilibrium of L_1 and L_0 individuals, where the frequency of L_1 is given by

$$x_1 = \frac{s(1 - e - 2\varepsilon) - e - \varepsilon + \sqrt{(s(1 - e - 2\varepsilon) - (e + \varepsilon))^2 + 4\varepsilon s(1 + s)(1 - e - \varepsilon)}}{s(1 - e - \varepsilon)} \tag{6}$$

Figure 2 shows the region of parameter space for which the culture genotype will not invade (individual learners only), the region for which it will fix (social learners only), and a polymorphic region (learners and social learners). In this model social learners can attend to social cues with no decrement in their ability to learn from their own experience. This means that populations composed exclusively of social learners contain individuals relying on both individually and socially acquired information, and not individuals relying exclusively on the latter. Hence such populations are capable of effectively tracking environmental variation.

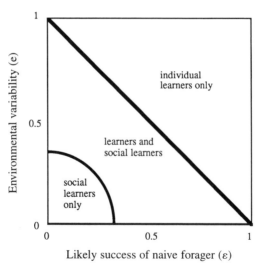

Fig. 2. The equilibrium frequency of social (S_1 and S_0) and individual (L_1 and L_0) learners, given the level of environmental variability (e), and the likely success of an individual forager (ε).

The results make intuitive sense. When the probability is high that individual foraging will allow a naive animal to locate the nutritional diet, and when the environment changes very rapidly, social learning cannot invade, since it will not on average enhance foraging success. Where the likely success of a naive forager (ε) is low, and when the rate of change in the environment is not too rapid, all individuals in the population should be social learners, since all individuals are more likely to locate diet 1 if they rely on socially transmitted information. There is an intermediary region of the parameter space where populations should be polymorphic for individual and social learners. Note, that populations fixed for individual or social learners will typically still exhibit variability in diet consumption.

A diploid model gives similar results. The affect of dominance is to change the frequency of diet 1 consumers, and the mean fitness of the population, without changing the conditions for invasion or the conditions under which social learning increases mean fitness.

The results are consistent with Boyd and Richerson (1985, 1988), but also suggest:

(i). Social learning can evolve in a rapidly changing environment when the probability of individual foraging success is low. Here the time scale of environmental change is relative to daily or more frequent foraging bouts, hence even low values of e will represent a rapidly changing environment relative to the generation time of the organism. For instance, for a daily forager, a low value of e, say 0.2, represents an environment in which two-thirds $(1-(1-e)^t)$ of the nutritious food patches no longer exist in their original location after 5 days.

(ii). Social species for which individuals can only search a small section of the population range each foraging episode, or that consume food sources that are sparsely dispersed, should rely more on social learning.

(iii). Neophobia decreases the value of ε, and hence may only be adaptive in populations with some social learners.

(iv). There may be many species which are polymorphic for social/individual learning, and which exhibit considerably between individual variability in the probability of social learning.

(v). When environments do not change too rapidly, and the chances of individual foraging success are low, (when $e + \varepsilon < 1$), social learning will increase the number of individuals consuming the nutritional diet, and thereby increase mean fitness.

THE ADAPTIVE CONSEQUENCES OF ANIMAL SOCIAL TRANSMISSION

Theoretical work has also begun to explore the adaptive consequences of animal social transmission. The diffusion of a learned trait through a population may modify some of the physical and social conditions that cultural organisms have to deal with, and affect their subsequent evolution (Bateson 1988; Laland 1992a; Plotkin 1988). Wilson and colleagues have argued that culturally driven changes in niche may generate a "behavioral drive," which accelerates morphological evolution by fostering the fixation of genetic mutations (Wilson 1985; Wyles, Kunkel, & Wilson, 1983). Most analyses have focused on the effects of cultural transmission on alleles which directly influence the probability of expressing the information that is socially transmitted. These theoretical analyses have usually found that culture is likely to slow down genetic change (Boyd & Richerson, 1985; Feldman & Cavalli-Sforza, 1989; Laland 1992a; Maynard-Smith & Warren, 1982). This finding rests on the assumption that, by allowing traits to spread through populations socially, culture shields genetic variants of low fitness from selection. In contrast, Laland (1992a) has investigated the effects of a culturally generated change in selection pressures acting on alleles that simply influence the nature of a biological or behavioral trait exposed to this novel selective environment. He found that animal protocultures would have to be atypically stable to elevate rates of genetic change by this means, and that a slowing down in genetic change is more likely.

In fact, cultural processes probably act to both accelerate and decelerate evolu-.tionary rates. This assertion is not contradictory since theoretical work has established that there are a number of different forms in which the interaction could take, each of which will have different dynamics. Where social transmission acts to enhance fitness differences between genotypes, fix novel mutations, generate sexual selection, or initiate new coevolutionary scenarios, it is likely to accelerate rates (Laland, 1992a,b, 1994). Where social transmission acts to minimize fitness differences between the members of a population, or damp out environmental variability, it is likely to slow down rates of genetic change (Laland 1992a; Maynard-Smith & Warren, 1982). For populations with genetic variants that are not too different from the optimal, social transmission will tend to "correct" the "poorer" genotypes, reducing differences in fitness. However, consider a population that suddenly finds itself in a changed environment, for which it is poorly adapted. The population may survive by adapting through social transmission, but subsequently, in its new niche, the population may be exposed to novel, often strong selection pressures. This hypothesis was first put forward by Baldwin in 1896, but has received comparatively little attention (Bateson, 1988; Plotkin, 1988). In this way social transmission may both protect genetic variation from selection in the short term, and expose more variation to selection in the long term.

Boyd and Richerson (1985, 1990a,b) show that, in principle, social learning can potentiate a form of group selection. Group selection based on genetic variation is thought unlikely because migration will prevent local populations from maintaining genetic differences (Aoki 1982; Wade 1978). When within-group processes are stronger, such as when they rely on cultural mechanisms, it is more likely that local populations can resist the effects of migration. Symbolic marker systems like human language or bird song may generate assortative mating or assortative social learning that resist the effects of migration and lead to finely tuned local adaptations (Boyd & Richerson, 1987; Cavalli-Sforza, Piazza, Menozi, & Mountain, 1988; Nottebohm, 1972).

Theoretical work also illustrates how culturally transmitted mating preferences can generate strong sexual selection for genetically transmitted traits (Kirkpatrick & Dugatkin, 1994; Laland, 1994; Richerson & Boyd, 1989). Studies of the mating behavior of female mammals and birds at leks, and even some fish and crustaceans, suggest that social-learning processes may strongly influence mate choice (Clutton-Brock, Hiraiwa-Hasegawa, & Robertson, 1989; Gibson, Bradbury, & Vehrencamp, 1991; Schuster & Wade, 1991). The evolutionary consequences of this "copying" behavior, in which females choose mates which other females have chosen, have also been explored (Kirkpatrick & Dugatkin, 1994; Wade & Pruett-Jones, 1990). Theoretical analysis led Wade and Pruett-Jones (1990) to conclude that copying will increase the opportunity for sexual selection, while the empirical findings of Schuster and Wade (1991) implied that it could maintain alternative male phenotypes. Laland (1994) found that, since socially transmitted mating preferences can spread more rapidly than the equivalent genetic variation, the sexual selection for trait alleles that is generated can be extremely strong, and may result in rapid genetic responses. It is apparent that social-learning processes may prove to have far reaching evolutionary consequences.

CONCLUSION

Animal social learning suffers from a lack of coherent theory. We have attempted to outline the benefits to developing such a theory, reviewing and extending what theory we do have, and pointing to the areas where further theory is required. Our principal conclusion is that there is a need for the greater integration of empirical and theoretical findings, so that each tradition can inform and stimulate the other. We hope that this chapter will be viewed as a step toward this integration.

REFERENCES

Aoki, K. (1982). A condition for group selection to prevail over counteracting individual selection. *Evolution,* **36,** 832–842.

Aoki, K., & Feldman, M. W. (1987). Toward a theory for the evolution of cultural communication: Coevolution of signal transmission and reception. *Proceeding of the National Academy of Science USA,* **84,** 7164–7168.

Baker, A. J. & Jenkins, P. F. (1987). Founder effect and cultural evolution of songs in an isolated population of chaffinches, *Fingilla coelebs,* in the Chatham Islands. *Animal Behaviour,* **35,** 1793–1803.

Bateson, P. P. G. (1988). The active role of behavior in evolution. In M-W. Ho & S. W. Fox (Eds.), *Evolutionary processes and metaphors.* Chichester: Wiley.

Bond, N. W. (1982). Transferred odor aversions in adult rats. *Behavioral and Neural Biology,* **35,** 417–422.

Boyd, R., & Richerson, P. J. (1976). A simple dual inheritance model of the conflict between social and biological evolution. *Zygon,* **11,** 254–262.

Boyd, R., & Richerson, P. J. (1985). *Culture and evolutionary process.* Chicago: University of Chicago Press.

Boyd, R., & Richerson, P. J. (1987). The evolution of ethnic markers. *Cultural Anthropology,* **2,** 65–79.

Boyd, R., & Richerson, P. J. (1988). An evolutionary model of social learning: The effects of spatial and temporal variation. In T. Zentall & B. G. Galef Jr. (Eds.), *Social learning: Psychological and biological perspectives.* Hillsdale, NJ: Erlbaum.

Boyd, R., & Richerson, P. J. (1990a). Group selection among alternative evolutionary stable strategies. *Journal of Theoretical Biology,* **145,** 331–342.

Boyd, R., & Richerson, P. J. (1990b). Culture and cooperation. In J. Mansbridge (Ed.), *Beyond self-interest.* Chicago: University of Chicago Press.

Boyd, R., & Richerson, P. J. (1995). Why does culture increase human adaptability. *Ethology and Sociobiology,* **16,** 125–143.

Cavalli-Sforza, L. L., & Feldman, M. W. (1981). *Cultural transmission and evolution: A quantitative approach.* Princeton: Princeton University Press.

Cavalli-Sforza, L. L., & Feldman, M. W. (1983). Paradox of the evolution of communication and of social interactivity. *Proceeding of the National Academy of Science USA* **80,** 2017–2021.

Cavalli-Sforza, L. L., Piazza, A., Menozi, P., & Mountain, J. (1988). Reconstruction of human evolution: Bringing together genetic, archaeological, and linguistic data. *Proceeding of the National Academy of Science USA* **85**, 6002–6006.

Chou, L., & Richerson, P. J. (1992). Multiple models in social transmission among Norway rats, *Rattus norvegicus. Animal Behaviour,* **44**, 337–344.

Clutton-Brock, T. H., Hiraiwa-Hasegawa, M., & Robertson, A. (1989). Mate choice on fallow deer leks. *Nature,* **340**, 463–465.

Curio, E., Ernst, U., & Vieth, W. (1978). The adaptive significance of avian mobbing II. Cultural transmission of enemy recognition in blackbirds: Effectiveness and some constraints. *Zeitscrift fur Teirpsychologie,* **48**, 184–202.

Dawkins, R. (1976). *The selfish gene.* Oxford: Oxford University Press.

Feldman, M. W. & Cavalli-Sforza, L. L. (1989). On the theory of evolution under genetic and cultural information with application to the lactose absorption problem. In M. W. Feldman (Ed.), *Mathematical evolutionary theory.* Princeton, NJ: Princeton University Press.

Feldman, M. W., & Zhivotovsky, L. A. (1992). Gene-culture coevolution: Toward a general theory of vertical transmission. *Proceedings of the National Academy of Science USA,* **89**, 11935–11938.

Galef, B. G. Jr. (1985). Direct and indirect behavioral pathways to the social transmission of food avoidance. *Annals of the New York Academy of Science,* **443**, 203–215.

Galef, B. G. Jr. (1986). Social identification of toxic diets by Norway rats (Rattus norvegicus). *Journal of Comparative Psychology,* **100**, 331–334.

Galef, B. G. Jr. (1988). Imitation in animals: History, definitions and interpretation of the data from the psychological laboratory. In T. Zentall & B. G. Galef Jr. (Eds.), *Social learning: Psychological and biological perspectives.* Hillsdale, NJ: Erlbaum.

Galef, B. G. Jr. (1993). Functions of social learning about food: a causal analysis of effects of diet novelty on preference transmission. *Animal Behaviour,* **46**, 257–265.

Galef, B. G. Jr., & Allen, C. (1995). A new model system for studying behavioural traditions in animals. *Animal Behaviour,* **50**, 705–717.

Galef, B. G. Jr., Attenborough, K. S., & Whiskin, E. E. (1990). Responses of observer rats (Rattus norvegicus) to complex, diet-related signals emitted by demonstrator rats. *Journal of Comparative Psychology,* **104**, 11–19.

Galef, B. G. Jr., & Whiskin, E. E. (1994). Passage of time reduces effects of familiarity on social learning: functional implications. *Animal Behaviour,* **48**, 1057–1062.

Gibson, R. M., Bradbury, J. W., & Vehrencamp, S. L. (1991). Mate choice in lekking sage grouse revisited: the roles of vocal display, female site fidelity, and copying. *Behavioral Ecology,* **2**, 165–180.

Giraldeau, L. A., & Lefebvre, L. (1987). Scrounging prevents cultural transmission of food-finding behavior in pigeons. *Animal Behaviour,* **35**, 387–394.

Giraldeau, L. A., Caraco, T., & Valone, T. J. (1994). Social foraging: Individual learning and cultural transmission of innovations. *Behavioral Ecology,* **5**, 35–43.

Heyes, C. M. (1994). Imitation and culture: Longevity, fecundity and fidelity in social transmission. In P. Valsecchi, D. Mainardi, & M. Mainardi (Eds.), *Ontogeny and social transmission of food preferences.* Harcourt.

Kirkpatrick, M., & Dugatkin, L. (1994). Sexual selection and the evolutionary effects of copying mate choice. *Behavioral Ecology and Sociobiology,* **34,** 443–449.

Kroodsma, D. E., & Baylis, J. R. (1982). A world survey for evidence of vocal learning in birds. In D. E. Kroodsma & E. H. Miller (Eds.), *Acoustic Communication in Birds (Vol. 2).* New York: Academic Press.

Laland, K. N. (1992a). A theoretical investigation of the role of social transmission in evolution. *Ethology and Sociobiology,* **13,** 87–113.

Laland, K. N. (1992b). Learning and evolutionary rates: A critical review. *Evolution and Cognition,* **2,** 63–78.

Laland, K. N. (1994). Sexual selection with a culturally transmitted mating preference. *Theoretical Population Biology,* **45,** 1–15.

Laland, K. N., & Plotkin, H. C. (1990). Social learning and social transmission of digging for buried food in Norway rats. *Animal Learning and Behavior,* **18,** 246–251.

Laland, K. N., & Plotkin, H. C. (1991). Excretory deposits surrounding food sites facilitate social learning of food preferences in Norway rats. *Animal Behaviour,* **41,** 997–1005.

Laland, K. N., & Plotkin, H. C. (1992). Further experimental analysis of the social learning and transmission of foraging information amongst Norway rats. *Behavioural Processes,* **27,** 53–64.

Laland, K. N., & Plotkin, H. C. (1993). Social transmission in Norway rats via excretory marking of food sites. *Animal Learning and Behavior,* **21,** 35–41.

Laland, K. N., Richerson, P. J., & Boyd, R. (1993). Animal social learning: Toward a new theoretical approach. *Perspectives in Ethology,* **10,** 249–277.

Landau, M. (1984). Human evolution as narrative. *American Science,* **72,** 262–268.

Lavin, M. J., Freise, B., & Coombes, S. (1980). Transferred flavor aversions in adult rats. *Behavioral and Neural Biology,* **28,** 25–33.

Lefebvre, L. (1986). Cultural diffusion of a novel food finding behavior in urban pigeons: An experimental field test. *Ethology,* **71,** 295–304.

Lefebvre, L. (1995a). Culturally-transmitted feeding behavior in primates: Evidence for accelerating learning rates. *Primates,* **36,** 227–239.

Lefebvre, L. (1995b). The opening of milk bottles by birds: evidence for accelerating learning rates, but against the wave-of-advance model of cultural transmission. *Behavioral Processes,* **34,** 43–53.

Lefebvre, L., & Giraldeau, L. A. (1994). Cultural transmission in pigeons is affected by the number of tutors and bystanders present. *Animal Behaviour,* **47,** 331–337.

Lefebvre, L., & Palameta, B. (1988). Mechanisms, ecology and population diffusion of socially learned food finding behavior in feral pigeons. In T. Zentall & B. G. Galef Jr. (Eds.), *Social learning: Psychological and biological perspectives.* Hillsdale, NJ: Erlbaum.

Mason, J. R. (1988). Direct and observational learning by redwing blackbirds (*Agelaius phoeniceus*): The importance of complex visual stimuli. In T. Zentall & B. G. Galef Jr. (Eds.), *Social learning: Psychological and biological perspectives.* Hillsdale, NJ: Erlbaum.

Maynard-Smith, J., & Warren, N. (1982). Review of C. Lumsden & E. O. Wilson (1981) *Genes, Mind and Culture Evolution,* **36,** 620–627.

Nottebohm, F. (1972). The origins of vocal learning. *Am. Nat.,* **106,** 116–170.

Palameta, B., & Lefebvre, L. (1985). The social transmission of a food finding technique in pigeons: What is learned? *Animal Behavior,* **33,** 892–896.

Plotkin, H. C. (1988). Learning and evolution. In H. C. Plotkin (Ed.), *The role of behavior in evolution.* Cambridge, MA: MIT Press.

Plotkin, H. C., & Odling-Smee, F. J. (1981). A multiple-level model of evolution and its implications for sociobiology. *Behavioral and Brain Sciences,* **4,** 225–268.

Richerson, P. J., & Boyd, R. (1978). A dual inheritance model of the human evolutionary process. I: basic postulates and a simple model. *Journal of the Society for Biological Structure,* **1,** 127–154.

Richerson, P. J., & Boyd, R. (1989). The role of evolved predispositions in cultural evolution: or, Human sociobiology meets Pascal's Wager. *Ethology and Sociobiology,* **10,** 195–219.

Rogers, E. M. (1983). *Diffusion of innovations, 3rd Ed.* New York: Free Press.

Rogers, A. (1988). Does biology constrain culture? *American Anthropologist,* **90,** 819–831.

Shuster, S. M., & Wade, M. J. (1991). Female copying and sexual selection in a marine isopod crustacean, Paracerceis sculpta. *Animal Behaviour,* **41,** 1071–1078.

Sugita, Y. (1980). Imitative choice behaviour in guppies. *Japan Psychological Research,* **22,** 7–12.

Wade, M. J. (1978). A critical review of the models of group selection. *Quarterly Review of Biology,* **53,** 101–114.

Wade, M. J., & Pruett-Jones, S. G. (1990). Female copying increases the variance in male mating success. *Proceedings of the National Academy of Science USA,* **87,** 5749–5753.

Wilson, A. C. (1985). The molecular basis of evolution. *Scientific American,* **253,** 148–157.

Wyles, J. S., Kunkel, J. G., & Wilson, A. C. (1983). Birds, behavior, and anatomical evolution. *Proceedings of the National Academy of Science USA,* **80,** 4394–4397.

Social Learning: Synergy and Songbirds

Meredith J. West

Andrew P. King

Department of Psychology
Indiana University
Bloomington, Indiana 47405

ANIMATE AND INANIMATE PUZZLES

When Thorndike placed animals in puzzle boxes, we suspect that he watched the actions of the animals more than he watched the actions of the boxes (Thorndike, 1911/1965). After all, he had fashioned the boxes to do certain things, so that he could measure objectively animals' capacities to solve problems. Similarly, one attraction of Skinner boxes is presumably that their behavior is programmed, allowing investigators to focus on the animal. This is not to say that the animals' interactions with such devices are always as predictable as one might hope, but that the animal's degrees of freedom are constrained while it is interacting within the apparatus (Breland & Breland, 1961).

Those studying social learning face a dilemma: they typically watch at least two or more animate "puzzle boxes," neither of which they have designed. Such investigators may go to considerable lengths to achieve some proximate control of one or both subjects, by manipulating motivation, controlling rearing, limiting physical access (e.g., one-way mirrors), or changing levels of consciousness (e.g., anesthetizing one "puzzle"). But ultimately, the success of studies of social learning depends on an investigator's ability to decipher synergy, i.e., to detect and measure outcomes from more than one perspective.

In the accounts of our research provided here, we have attempted to focus on

Social Learning in Animals: The Roots of Culture
Copyright © 1996 by Academic Press, Inc.
All rights of reproduction in any form reserved.

155

the use of multiple contexts to provide multiple chances to capture synergistic effects. After providing information about these contexts and the resulting outcomes, we return to questions of organizing principles relating to the study of social learning, principles Thorndike also considered in his simplified settings. Our aim is not to argue against studies in the laboratory, but to reflect on some of the problems one faces when the learning that one studies is social.

Although the nature of the animal's social surroundings has been of paramount concern to those studying primates and those investigating anomalous communication in species such as dolphins or parrots (Pepperberg, 1986), the topic of social environments and their consequences has received less attention from those studying songbirds, our particular area of research. As in other areas of ethology, in the field of birdsong, the question of understanding learning has often been approached by creating a special scientific strain—Kaspar Hausers, young animals raised alone or in impoverished social circumstances [named for a German boy supposedly subjected to analogous deprivation, (Thorpe, 1958)]. If such animals behave differently than do normally reared counterparts, then one can begin to search for learning processes that may account for the differences observed in the behavior of the two groups. That songbirds reared in social isolation often differ from their wild-reared counterparts is relatively easy to establish. But what is not so easy to identify is the processes responsible for the differences. Even in more social conditions, i.e., when birds are given tutors, it is often difficult to provide descriptions of the actions of the tutor and pupil, beyond outcome measures of who copied whom. Part of the problem is purely practical—when to watch and what to watch; but part is purely conceptual—what is in the environment and what details of the environment are most important? The idea supporting the Kaspar Hauser work is that the possibility of certain experience has been removed, an assumption that is potentially testable. However, the possibility that the potential for other experiences has been given to Kaspar Hausers is less often entertained. In the examples that follow, we hope to illustrate why we find puzzling the failure to consider the added experiences of animals raised in isolation.

One reason that birdsong researchers may have been less concerned about the details of social housing is that many laboratory studies are preceded or accompanied by investigations in the animal's natural habitat, affording a biologically pleasing view of behavior in both natural and artificial contexts. With time and technological advances, however, the focus of studies devoted to vocal learning in songbirds has shifted from the birds themselves to their songs, regardless of the setting in which the song is performed. The ability to see vocal behavior in considerable acoustic detail in sonagrams has made looking at the singer seem less pressing. While we recognize the benefits resulting from advances in acoustic technology, we worry about the cost to science of losing sight of the subjects themselves, especially within the confines of a laboratory. The cost seems even higher as it becomes clear how social song learning actually is. Thus, we emphasize social effects here to suggest that the nature of proposed learning mechanisms may change when social effects on song production are considered as important as vocal effects on song production. We recognize that many of the points we will make about social learning might strike those studying primates or other mammals as quite elementary. They are. We think the reason for the differences in conceptual levels rests partially in the differences in our perceived obligations to different species of animals. We doubt, for example, that investigators studying primates' social cognition would put their animals in the conditions of social isolation or restriction still routinely used with many songbirds. And, although the direction of influence we advocate is for those studying songbirds to imitate those studying primates, we do wish to point to an advantage of songbirds' size and social nature. It is possible to give captive birds considerable space to regulate their social behavior and still impose experimental conditions. This task has proved harder with many mammals. Thus, we invite those studying other taxa to consider the social ways of birds to enrich their understanding of the processes associated with social learning and imitation.

LEARNING BY ASSOCIATING: A BIRD'S EYE VIEW

Some of our convictions about the need to attend to context when studying vocal learning came from lessons learned in an unorthodox laboratory, our home. Like others before us, we have found that the opportunity to see an animal's behavior in familiar contexts over many weeks or years rivals any human soap opera in its addictive properties (Lorenz, 1957; Nice, 1937). And, although the practice is rewarding in and of itself, some years ago, it actually got us an invitation to receive an award from the American Psychological Association. Our focus was on a male starling (*Sturnus vulgaris*) named Rex (West, Stroud, & King, 1983b). Rex and his cage mate, a male cowbird (*Molothrus ater*), had lived with us for several years. It was Rex's voice on the tape that won over the judges. His mimicry of human speech was clear enough that he could have given his own oddball acceptance speech. He could have walked, or maybe flown, to the podium, uttered "Hi," cocked his head, and whistled an off-key version of "The Star Spangled Banner," interspersed with a few notes of "Dixie." And if properly motivated, he might even have tossed out one of his pet sounds "Basic research, basic research!"

Rex had learned to mimic human sounds while in our care, but we cannot say that Rex had learned as a result of our tutelage. Until Rex uttered something that sounded like "Good morning," we were completely unaware of starlings' ability to mimic human speech. We had been too busy studying Rex's companion, the cowbird. We had paid attention to Rex, but we had come to see him as a companion. (Technically speaking, he was also a necessary part of a "control condition" providing generic social stimulation in an experiment with acoustically naive cowbirds.) But it was easy to forget that role for Rex's scientific role as he shared our morning coffee, vocalized to music, chattered at lab meetings, and patrolled the house on foot, energetically inspecting all surfaces for something to probe. That Rex had acquired a repertoire of human-derived sounds came as a surprise—as it has to the many other families who accidentally acquired starlings (West & King, 1990).

Once we had uncovered starlings' skills, however, we did some experimenting. As part of our efforts to understand conditions favoring mimicry of human sounds, we tried to tutor home-reared starlings with audiotapes of human speech and whistling. Would they, for example, favor the voices of their caregivers? After extensive tutoring, we looked for evidence of taped voices in the starlings' repertoires. We found none. Did this mean that starlings do not learn by inanimate tutoring as many songbirds do? No, they can (Chaiken, Bohner, & Marler, 1993), but whether they do depends on what else the environment offers. Our starlings found the recorded voices far less salient than other events such as the sight and sound of the reels of a tape recorder turning and the presence of tape hiss. The hiss was mimicked. Our experiences with starlings illustrate our synergistic theme: what animals do in an environment depends not only on what we offer them, but also on what we recognize as being there. Until we heard the sound of tape hiss out of the mouth of a starling, we had not really noticed the sound on the tape. To us, as experimenters, it was not something to be learned. Had we not been sensitized to starlings' ways, we might have concluded they could not mimic taped sounds. If we had not, in some sense, shared the environment in which our starlings lived, we would have misgauged their abilities.

The starlings' mimicry suggests that the motivating conditions for vocal learning in songbirds are not simple ones. Sounds heard most frequently by the starlings were not present in their mimicry (such as "No!" or "Here's your food"); sounds experienced only once seemed overrepresented ("Does Hammacher Schlemmer have a toll-free number?") And sounds we cherished and reinforced could be changed so that "basic research" was reduced to "sic(k) research, sic(k) research."

More important, we knew that not all home-reared starlings mimicked human behaviors. We had studied other starlings in superficially quite similar human environments: these birds had lived in homes (with cowbird companions) and were cared for by humans but afforded no opportunities for extended social contact (no lab meetings, no coffee or tea, no sharing a shower). They remained in their cages in acoustically busy parts of homes. These birds did mimic, but no human-derived mimicry was found; they mimicked their avian companion and sounds from outside their cage or outdoors such as doors closing or dogs barking. Although they heard speech routinely, the speech was not directed to them and they did not imitate it (West et al., 1983b).

Such observations reinforced our belief that other songbirds' imitation was more socially entwined with context than first imagined by Thorpe, who initiated the formal study of birdsong learning. He had reasoned that vocal imitation was cognitively less complex than visual imitation, and thus more plausible in birds, because birds could hear, but not see themselves. Birds, indeed all animals, had a self-contained device to compare the sounds they produced to those produced by another, but no similar means to compare visual movements with those of another (Thorpe, 1961). Our observations of starlings suggested that they could see the consequences of their sounds, i.e., that certain sounds led to predictable verbal and behavioral actions. As a result, the imitated sounds became something akin to social sonar. It is impossible to ignore a bird that vocalizes "Hi" as one walks by its cage—it seems, in fact, almost rude to do so given the precision and intonation a starling can achieve. Nine times out of ten, a passerby says "Hi" back to the vocalizing bird. Thus, starlings can learn that sounds result in actions, as well as sounds. One of our starlings, when seeing a caregiver, frequently mimicked the sounds of a microwave oven beeping. It became a signal to the human to thaw and offer some peas to the bird. At one point, telephone companies in the United States encouraged use of the classified section of telephone books (listing services and businesses) by saying "let your fingers do the walking." The starlings' mimicry essentially allowed the birds to let their voices do the walking: they could explore the properties of their social world without leaving their perch.

Thus, what we saw in starling learning contained elements of traditional social shaping, except that the nature of the reinforcer was often hard to articulate. None of the birds had been trained using explicit food or model-rival techniques such as those used with parrots (Moore, this volume; Pepperberg, 1986; Todt, 1975). When we analyzed the starlings' vocal repertoires, what we saw in their mimicry were mixtures of common household events, greetings, farewells, conversational fragments, and sequences related to interactions involving caregiver and bird. The recipe for the mixture had several parts: it was the compound product of access, salience, and consequence of experiencing and experimenting with certain sights and sounds. Similarly complex processes have been described to explain selectivity in some primates' imitative acts (Russon & Galdikas, 1995). Our experiences with starlings indicated that the interwoven processes of experiencing and experimenting with sound is an essential building block for social/imitative learning in the other species that we study, cowbirds.

COWBIRDS: MAINTAINING A BIRD'S EYE (AND EAR) VIEW

One of the appealing features of hand-reared starlings, their potentially intense sociability toward humans, is a potential scientific liability: as investigator/caregiver/competitor for the same armchair or cup of tea, you often feel more like a parent than a scientist, perpetually uncertain if you are part of the solution or part of the problem. Thus, the majority of our scientific efforts to understand social learning have come from studies of cowbirds' social experience under conditions more mundane, but methodologically more tractable that those we had used with starlings. Although the housing conditions we have used with cowbirds seem superficially quite like those used for many laboratory subjects, we have made socially oriented modifications. In some cases, the modifications seemed like minor ones, but, as with the starlings, what seemed minor to us sometimes produced major effects (West & King, 1994).

One of our first experiments demonstrated that female cowbirds could be used to test the stimulating properties of male song (King & West, 1977). Female cowbirds, deprived of male companions when in breeding condition, respond to the acoustic playback of conspecific song with copulatory postures. One could not ask for a more unambiguous response; females either responded to the recorded song with full copulatory postures or paid so little attention to the song that you would not know from the female's behavior that a song had been played. Females did not respond to the songs of other species and did not respond equally often to the songs of all male cowbirds—songs of some of the males elicited far more responding, providing a means to correlate song properties with other properties of singers.

We exploited this female bioassay to ask many questions about the acoustic nature of species identification in cowbirds and met with considerable success (West, King, Eastzer, & Staddon, 1979). We also used converging methods to understand the female-bioassay procedure from a more ecological perspective. Did females housed in sound-attenuating chambers react similarly to the sounds of males compared to females living in more spacious and socially complex captive colonies? In other words, did the songs responded to most by male-deprived females also appeal to females housed with males and actively being courted by males?

To answer this question, we conducted studies of mate choice in large aviaries. Under such circumstances, male and female cowbirds did not choose mates randomly (Eastzer, King, & West, 1985; West, King, & Eastzer, 1981; West, King, & Harrocks, 1983a). Males we determined were most dominant in the winter and were the ones most successful in courting females in the spring. To understand more about the playback bioassay, we recorded the songs of successful and unsuccessful males living in resident aviaries and played them back to females who had never met the males. The females' only exposure to these males was acoustic playback of the males' songs. We found significant positive correlations between the potency of a male's song and his courtship success, giving us further confidence that playback studies could complement other assessments of male behavior.

The aviary studies addressed a methodological worry, but they also introduced new social issues. As part of these studies, we had found evidence to suggest that females played an important role in stimulating males to sing certain song types, even though the females themselves could not sing. For example, we found that the most efficacious way to get an adult male cowbird to change his song repertoire in a subsequent year was to house him with females from a distant geographic area, while housing him just with males from the distant area had no effect (West & King, 1985; West et al., 1983a). It seemed that the males needed a social incentive, provided by the females, to learn new songs. The nature of their audience, not the nature of auditory stimulation, mattered. Returning to the starling example, the male cowbirds needed access (hearing the sound), salience (the right context), and consequence (seeing females react).

We went on to explore the capacity of female cowbirds to influence male vocal development. We determined whether acoustically naive males, with no vocally appropriate models, would show evidence of social influence from female conspecifics. It is important to note that female influence on song development should not have occurred if Thorpe's original reasoning, or that proposed later by others (e.g., Konishi, 1985; Marler, 1976) was right. Song learning had consistently been described as the product of vocal imitation. If imitation were the mechanism guiding song development, the songs of the males housed with females from different populations should not have differed, because their opportunities to imitate themselves or rehearse from inherited "templates" were equivalent (all the males in our studies had been captured at the same time from the same area (King & West, 1988).

Some of the most striking data came from studies of hand-reared male cowbirds collected in North Carolina (NC) and housed individually with Texas (TX) females (who preferred song variants never sung by wild NC males) or housed with NC females or heterospecifics (King & West, 1983a). The three groups of males differed reliably from one another in measures of vocal outcome. Males with TX females produced song variants distinguishable by acoustic measurement and playback from those of males housed either with NC females or with heterospecifics. We also set up conditions in which other hand-reared males were tutored (and thus had the "right" vocal models) but were housed with females from different areas. They too learned the songs resembling their female companions' native preference. The results from both studies indicated that female cowbirds could communicate their song preferences, even if they could not sing them.

How were the females exerting their effects? We gathered several lines of evidence to begin to answer this question. First, we investigated the ecological setting we suspected was the site of male/female interactions affecting song production, the large winter flocks formed by cowbirds with other blackbird species. Did young males have access to females during the time of year that their songs seem most vulnerable to modification? The answer was yes. Juvenile males seemed to associate quite closely with females, much more closely than they associated with adult males (King & West, 1988).

We then returned to the laboratory to videotape interactions between male and female cowbirds during the time period in the spring when males and females are on prospective breeding grounds. We found that males sang a lot and that females appeared to do very little, but when females did react to males, their reaction was conspicuous, at least to the males. We were, in fact, alerted to the behavior of females by the behavior of males as every once in a while, when a male sang, he would change the pace of his singing as well as suddenly move toward the female. When we retraced his steps on videotape, we found that such changes in males' behavior were preceded by wing movements by females (movements we call wing strokes) (West & King, 1988). But what did wing strokes signify? To find out, we employed playback procedures during the breeding season. Would females show copulatory postures in response to the same songs that had elicited wing strokes in the winter? As it turned out, the songs that elicited wing strokes in the winter were highly effective releasers of females' copulatory postures in the spring, leading us to believe that wing strokes represented positive feedback for males.

We now know that playback responsiveness and wing stroking, events that take place at different times of year, are related. Females that are very responsive in the winter, as evidenced by wing stroking, tend to be most responsive to song in the breeding season. Moreover, males housed with such females develop song repertoires judged more potent by an independent group of females listening to play-backs (King & West, 1989; West, King, & Freeberg, 1994). Thus, we conclude that wing stroking serves to shape males' singing toward geographically appropriate signals. As with the starlings, experience and experimentation seemed part of the learning process. By virtue of access to females, males come to associate certain songs with particular consequences. Certainly imitation normally plays a role in providing material for males, but our results indicated that exclusive focus on mechanisms to explain imitation would be insufficient to explain observed out-comes. Thus, studies tracing only the course of imitation of tutors by pupils, and studies using only acoustic assays of transmission, seem likely to miss social mecha-nisms altogether.

MULTIPLE CONTEXTS, MULTIPLE MEANINGS

Our data on wing strokes indicated some future directions we knew we had to take to look more carefully at social interactions early in development. They also af-fected statements we had made before we realized the effects of social housing on song development. In particular, we were led to reexamine one of our early play-back experiments in which we had tested the effects of songs of acoustically naive cowbirds on female cowbirds (King & West, 1977). At the time, the results we obtained surprised us. The females responded with as many copulatory postures to the atypical songs of naive males as to the typical songs of wild males. Although we could fashion an argument as to why this was beneficial for a brood parasitic species like cowbirds, it still seemed odd that depriving an animal of something normally available appeared to have beneficial effects.

Our subsequent work, uncovering female influences, allowed us to see those early efforts in a new light. The males we had used in our early experiments had been isolated from song, but not from females. Thus, they had been exposed to potentially influential social tutors. Moreover, these males had been given social tutors without the added complication of the presence of other males. Although adult male cowbirds are essential to song learning in that they can serve as tutors, such lessons are not without cost to the pupils. In our aviary studies, we had found that males appeared to have to "earn" the right to sing certain song types by assuming a dominant rank and by negotiating access of females by means of prior interactions with males. Thus, males often engaged in extended bouts of counter-singing. Moreover, males would sometimes be attacked by another male after vocalizing. By means of playback studies, we determined that males with potent songs were more likely to be attacked. The social dynamics were multifaceted, and probably facilitated in their natural habitat by the physical nature of the birds' surroundings. In aviaries, males could hear other males singing to females. In nature, males could find themselves in situations where females were close by and males far away and could therefore appeal to one sex without alerting the other.

If we apply these observations to the original studies of males in sound-attenuating chambers, our method of song deprivation looks quite different than it had before. We had allowed males' close access to females with no risk of attack. We had also created a situation in which males did not have to balance the costs of singing to females against the risks of aggression from males. We had thus created something of an ivory tower for our isolated males. Consequently, the finding of effective song from males in deprived settings no longer seemed paradoxical. Instead, it seemed a paradigmatic example of the need to assess individuals developmentally by viewing them in multiple contexts. Indeed, the finding of potent male song from acoustically naive males highlighted the potential danger of overly simplified contexts as bases for developmental conclusions.

We proposed earlier that social learning was harder to study than individual learning because, in the former case, the experimenters had less control over the puzzles before them. But it is just as important to recognize what kinds of controls we do have. For example, in the studies of acoustically naive males with females, we arranged the surroundings so that males or females would be hard pressed to avoid interaction. Would males have sung to females without the close confinement? Would females have responded in more natural surroundings? So too, in playback studies, we, as experimenters, usurped more than the male's voice. We not only chose the song to be played, but also the amplitude, the time of onset, and the time since the female had heard a previous song from him or another male. How many of these steps represented part of the process of learning to be an effective singer? Said another way, how much of the developmental scaffolding that we provided was naturally learned by males? Did males not only have to learn what to sing, but how? We investigated these questions as part of a recent set of studies of males housed either with female conspecifics or with canaries. As in previous studies, we viewed the canaries as "controls" for general avian stimulation. Canaries are social and are responsive when housed with cowbirds. And although cowbirds would not normally encounter canaries in the wild, cowbirds do spend a lot of time in the presence of other species: they hatch into the nests of over 200 different species and subspecies and during winter in mixed-species flocks (Friedmann, 1929; Friedmann, Kiff, & Rothstein, 1977). The question for us was whether "generic" social housing was sufficient to teach males to play the roles we had been playing—to seek out, attend to, and attempt to attract females.

SOCIAL LEARNING: THE PATHWAYS TO COMMUNICATING

We explored the question of the pragmatics of communicating as part of a study of cowbirds from an ancestral part of the species range, the Black Hills of South Dakota (SD) (Freeberg, King, & West, in press; West, King, & Freeberg, in press a; West, King, & Freeberg, in press b). We had asked these SD cowbirds the same questions we had asked of other populations: how does vocal development proceed in juvenile males deprived of extensive experience with adult males? Thus, the SD males were individually housed either with female cowbirds or with canaries but with no adult males. We also chose to expand our measures and the context in which we observed the birds to explore the male's ability to use his song repertoire to court females. A reasonable analogy for our efforts might be some early studies of primates in which investigators asked what social capacities socially naive monkeys possessed when placed into more complex social settings (Harlow, 1974; Harlow & Harlow, 1962; Mason, 1960). Thus, we wondered if young male cowbirds would be at all like Harlow's original socially isolated monkeys and display deviant social behavior when placed in new contexts? We suspected that the social behavior of our birds would not be seriously disturbed for two reasons: (1) the brood parasitic habit of cowbirds has led many to suspect that the species must have a "safety net" to compensate for the species unusual upbringing, and (2) our SD males were neither hand reared nor completely socially naive. They had been wild caught in flocks of cowbirds in SD and had some 50+ postfledging days of experience with conspecifics.

From late August until May, we housed one group of SD males with female cowbirds from South Dakota (the "FH" males) and the other group of SD males with canaries (the "CH" males). In May, we recorded their song repertoires and then placed the males first in large flight cages and then in large aviaries with unfamiliar females and unfamiliar canaries. We used a two-step procedure because we had no idea how the males would react when they had to deal not only with more birds, but also with both sexes and new surroundings.

The observations in these new settings suggested that social skills could not be assumed. The level of "skillessness" was especially surprising in the SD males housed with canaries. In the cages and then in the aviaries, the CH males sang and pursued *canaries,* although female cowbirds were present and very attentive. The CH males' persistence was remarkable. Even when the canaries flew away or ignored the CH males and even when female cowbirds approached and solicited copulations from them, the males were not deterred. Their behavior did not change even after they witnessed successful courtship of females by experienced SD males. The CH males showed few signs of species identification. The "error," absence of species recognition, seems to be at so fundamental a level as to undermine any concept of a "fail-safe" mechanism in a species many proposed required genetic protection from environmental variability (Lehrman, 1974; Mayr, 1974; Todd & Miller, 1993).

Based on the behavior of the CH males with canaries, we presumed that the FH males would be strongly attached to female cowbirds. The FH males paid no attention to canaries in either the flight cage or the aviary, but only in the first setting, when confined in a cage with females, did the FH males show signs of attentiveness to females, i.e., singing, chasing, and copulating. Once in the aviaries, they allocated most of their singing to one another, directing a majority of all vocalizations to other female-housed males, even though, at all times, there were twice as many females in the aviaries as males. The behavior of the FH males ruled out a simple imprinting-like explanation, i.e., if the CH birds preferred canaries, the FH birds should prefer female cowbirds. Although we are still exploring possible means to explain the behaviors we witnessed, we find ourselves returning to asking about access, salience, and consequence, as a way to begin the process. During the social transition we imposed on the males as they went from sound chamber to cage to aviary, the social environment may have changed less for CH males. At each point, CH males had access to canaries, albeit unfamiliar ones. Moreover, the canaries' behavior seemed similar in all three contexts: they tolerated the cowbirds' attention. Thus, the CH males remained oriented toward canaries because no other individuals acted to change their orientation: the female cowbirds generally ignored them and the FH males, by virtue of their lack of interest in canaries, tended to locate themselves in different parts of the aviary from CH males, who went where the canaries went (often to the aviary's indoor section). We had placed some unfamiliar starlings in the aviary to give all the male cowbirds a "new" species as a control for familiarity. The CH males paid little attention to them, no more so than did other cowbirds in the aviary. Taken as a whole, the acquired preferences of the CH males resulted in continued attention to canaries and the social dynamics of the setting may have operated to maintain the CH males' social preferences.

The transitions for the FH males involved more opportunities for changes in social dynamics and these changes were apparent in the changes in the FH males' behavior across contexts. We know from other studies that male cowbirds prefer to copy effective songs of other males (West & King, 1986). We also know that males singing very effective songs are physically attacked more often than males with less effective songs (West et al., 1981). We know too that females tend not to mate with males unless males court them persistently over several days. This set of conditions may explain why FH males, once able to hear one another, found each other's presence more salient than the presence of females. The consequences of interacting with males meant, by definition, that the FH males could allocate less time to courting females. FH males possibly also reduced the female's interest in them because they showed little persistence in following and singing to the same female. That the females were potentially responsive was made abundantly clear by the success of an unplanned group: wild male cowbirds that courted from outside the aviary wire and provoked several females into solicitation postures.

DEVELOPMENTAL PLASTICITY BEYOND THE FIRST YEAR

The data collected in this first year gave us much to think about in terms of the ontogeny of mate recognition, a new topic for us, but shed little light on our original question, how do males acquire skills in song use? All we had thus far was ample evidence of the consequences of lack of skills: few copulations and complete dominance and usurpation of the females by normal adult SD males when they were introduced in a final attempt to "teach" the young males social skills (Freeberg et al., 1995; West et al., in press a,b).

We thus chose different social conditions for the CH and FH males' second year. We collapsed over the CH and FH males and assigned them to one of two new groups, making sure that each new group had roughly comparable numbers of FH and CH males. The first new group consisted of FH and CH males housed with each and with SD females in a group. The second group consisted of the rest of the FH and CH males, SD females, and older, SD males. Thus, the two groups differed in that the second group could interact with experienced male cowbirds. We asked two questions. Would the younger males' social deficiencies change and would experience with older males facilitate such remediation? The results provided affirmative answers to both questions. We could find no signs of the male's first year housing when we viewed them in a courtship context in their second breeding season. Even when canaries were added to the aviary, the original CH males showed no interest. But the changes in the males' behavior could not be ascribed to maturation as both groups did not achieve an equal level of social skills. The males that had been housed with the older SD males prior to the breeding season were significantly more successful at obtaining pairings and matings than males housed only with each other and females. The difference seemed all the more impressive because the older males were not present in the breeding aviaries; thus, the social differences between the FH and CH males exposed to these conditions derived from accumulated experiences over the winter and spring. Thus, one clear route for the transmission of social skills appears to be interactions between younger and older males, experiences most likely made more salient if females are present.

We view the data from the males' second year as indicative of a fundamental social role for males, and as such, one of the more important outcomes of the study. But we still had to ponder the effects of the canaries in the first year. Why had we not seen the effects of canaries in earlier studies of acoustically naive males? Either we again had failed to look at the right measures or perhaps birds from SD were different from the eastern populations we had studied. To begin to answer these concerns, we repeated the entire, 2-year experiment with young male cowbirds collected in Indiana (IN). These cowbirds represent the same subspecies (*M. a. ater*) as the eastern populations we had studied earlier and a different subspecies than the cowbirds in SD (*M. a. artemisiae*). Thus, we housed some young males with canaries and some with IN female cowbirds. We again looked at social responses in flight cages and then in large aviaries when the birds came into reproductive condition. The results indicated a difference between the two populations in susceptibility to heterospecific influence. Although the IN CH males showed some attention to canaries in the flight cages and in the aviaries, they eventually developed some interest in female cowbirds, although their courtship skills were inadequate: they would sing and chase female cowbirds, but tended not to pursue the same female cowbird across hours or days, a necessary condition for successful courtship. The FH males' behavior in their first year was like that of the SD FH males: they vocalized most to FH males, and showed less interest in persistent courtship of female cowbirds than normally reared, wild-caught IN males.

Because housing with canaries had produced the more extreme outcome in SD cowbirds, we focused only on the CH males from IN for the second year's manipulation. Thus, we housed half of the IN CH males with each other and IN female cowbirds and half with each other, IN females, and IN males. We obtained the same result we had obtained earlier. The IN CH males living with older males showed far superior courtship skills to the CH males housed only with peers (that had had the same CH experience in their first year) and females. Thus, we now know why we had not seen the effects of heterospecifics in our earlier studies: not all cowbird populations are equally malleable with respect to species or mate recognition. But thus far, we have found no populational differences in the role of social exposure to female and male conspecifics as necessary for the development of effective courtship skills.

As part of these studies of SD and IN males, we had also made extensive recordings of their vocalizations. We had carried out playback tests and confirmed that the songs of FH males were more effective than were those of CH males in both populations. However the aviary observation revealed a potential weakness of the playback assessment: playback females responded as much to the songs of FH males as to those of normally reared males and yet only the latter group courted successfully. The reason for the discrepancy is not hard to identify, given what we said earlier. FH males failed to sing as much to females in the aviaries as did the normally reared males and in this social system, females do not mate with males unless they have been courted extensively. Thus, "having" a good song does not insure success. Males must use their songs under the right conditions which means negotiating life with males and females.

At a methodological level, the difference in predictions based on playback versus aviary performance is important. Relying only on playback tests is clearly insufficient to reveal what males have or have not learned from social companions. In the same way, we would have reached different conclusions had we ended the experiment after the cage tests, as we would have concluded that the FH males showed a "normal" level of interest in females. Although some birdsong studies include playback components, the majority do not, and only a few have ever included actual tests of mating (Williams, Kilander, & Sotanski, 1993). Most importantly, had we stopped our experiment where most birdsong studies stop, with analyses of the recordings of the males' crystallized songs, we would never have uncovered new roles for social learning. Moreover, had we stopped after one year, using the males' first summer's breeding efforts as our endpoint, we would have missed what is perhaps our major finding, i.e., that the social behavior observed is always a product of contextual variables. If social behavior were to become independent of context, it would cease to be adaptive, i.e., the dynamic product of attention and responsiveness to current, local circumstances.

We believe that the differences in the kind of "laboratory" used are also relevant. A concern others have expressed about laboratory studies of birdsong concerns interpretation: "a single caveat looms above the bewildering array of data on vocal learning in the laboratory: what is actually happening in nature may not be adequately reflected by results of laboratory studies. The complex physical and social processes occurring in nature cannot be duplicated in the laboratory" (Kroodsma, 1982). While we concur with the caveat, we would counter with a question: how many investigators have tried to duplicate the social and physical processes in laboratory settings? We can well guess some of the reasons that attempts at simulation are rare. The social assays we described above require an elaborate script. We first housed males in small enclosures with females or canaries for many months, then looked at them in cages, then in large aviaries containing female cowbirds from several populations, canaries, starlings, and unfamiliar males. Just at the level of space (a volatile topic to most scientists), the changes in behavior are worth noting. The males went from .62 m^3/bird in the sound-attenuating chambers to 17 m^3/bird in the aviaries, a 27-fold increase in space. At the level of complexity, males went from the possibility of interacting with one or two females to the possibility of interacting with any one or multiple of 10 different females, 4 different same-aged males, 5 adult males, 5 canaries, and 5 starlings. Moreover, we rotated birds in and out of the aviary to simulate some of the kinds of changes that take place on breeding grounds, as males and females move or disappear. From our perspective as investigators, we went from the easy reckoning of a static context when birds live in the same chamber or cage throughout the year to changing patterns of social interaction across many days and contexts, from what wild birds never experience to something closer to what happens in nature. We make no claim that our conditions were natural, but our efforts at semisimulation suggest to us that the possibilities for ecological manipulation are far greater than most researchers suppose.

Cowbirds offer a great advantage in that their breeding biology is simplified, no nests, no parental duties. However, juncos (*Junco hyemalis*) living in the same aviaries with our cowbirds did mate and build nests and raise young as part of an investigation of social learning in that species (Titus, 1995, unpublished data). Moreover, data on the male junco's social skills after different social experiences, revealed differences quite like those found in cowbirds. An animal's actual habitat(s) will always remain the standard against which to test knowledge gained in other contexts. But what we hope we have conveyed is that many of the obstacles to learning about social and cognitive capacities in songbirds may have been overly limited by scientific tradition as to what constitutes a laboratory setting and what roles are ascribed to the subjects versus the experimenters.

FEMALE COWBIRDS: GENDER-BASED TRANSMISSION?

We have now also seen how similar conditions affect another part of communication in cowbirds, in a set of studies aimed at probing how far one can manipulate dimensions of mate recognition and preference (Freeberg, in preparation). Among the many discoveries is that social learning may be easier to facilitate in larger social contexts. We had focused in the past primarily on malleability in male cowbirds' social skills, as the origins of our interest were in vocal learning. Thus, one might conclude that it is the female's lack of social learning, her "hard wiring," that makes the whole system work. We had attempted to address this question by housing females with males either in cages or sound-attenuating chambers during their first fall, winter, and spring. We found little or no evidence of female modifiability (King & West, 1983b, 1987; King, West, & Eastzer, 1986). Freeberg took another approach, using socially more complex circumstances to attempt to manipulate female preference. In Freeberg's study, juvenile female and male cowbirds, collected in South Dakota, were housed from the onset of the study in large indoor–outdoor aviaries in groups containing both male and female adults (numbering around 32 individuals in each condition). For two cohorts, the older generation (the potential "demonstrators" in Galef's lexicon (Galef, 1975; Galef & Wigmore, 1983) were IN males and females and for two cohorts, the demonstrators were SD adults. The aim was to see if juvenile birds, all from the same geographic area, would show different social preferences for potential mates, based on the winter experience with IN or SD demonstrators.

The results from the bird's first two breeding season suggest that females' pairing preferences are clearly influenced by social housing: the SD females paired more often with SD males whose experiences matched their own, i.e., males that had been exposed to the same kind of demonstrators, although not the same individuals, thus controlling for familiarity. Moreover, at no point did any of the younger birds, the pupils, actually see adults mating, ruling out local copying as the mechanism.

Thus, we are inclined to believe that past failures to find malleability in females may have been because our housing contexts worked against us. Females may need to witness the reactions of other young females, as well as older females, to stimulate their interest in a particular class of males. And, in keeping with the theme of synergy, these experiences may have to occur at certain times of year when females' responses to song are less tied to immediate needs to mate and lay eggs. The major point, though, is that less traditional designs may bring forth new capacities. More "animal friendly" instead of "experimenter friendly" conditions may thus be needed. In a study of pygmy marmosets (*Cebuella pygmaea*), Elowson and Snowdon (1994) reported vocal changes in the structure of trills when two colonies' social housing was modified. Thus, in primate species as well, changes in experimental design to retain more of the natural dynamics of social transitions may also reveal vocal plasticity in primate groups, an area in which failures vastly outweigh successes.

COPING WITH SYNERGISTIC EFFECTS

In outlining our research, we have attempted to illustrate the organizing power of social interactions in relation to learning. As a result, we come to a different conclusion about one of the basic premises of birdsong learning, i.e., the status of imitation as the guiding mechanism. In our view, the ability to imitate sound may be as reflexive and cognitively uncomplicated as the ability to breathe. It is how imitation affects and is affected by context, by ongoing social behavior, that must be studied before assuming its explanatory power. What we need is a way to manipulate opportunities to see imitative behavior in contexts that have an identifiable relationship to an animal's life history.

We are thus led by the evidence to wonder about the current status of imitation in other social systems. In many overviews, the question is asked as to which animals are "intelligent" enough to meet criteria for imitation. The standard for comparison is often human's imitative capacities. However, from a synergistic perspective, the "act" of imitating may be evaluated quite differently depending on phylogenetic and ecological dimensions. Thus, copying a sound or series of sounds may be fundamentally different from copying a problem-solving technique, e.g., how to open milk bottles or select a diet. Of all the parts of the system of influences we assume to be at work in avian communication, the actual role of imitation seems among the most rudimentary. And, as impressed as we are by starling's ability to scrounge sounds, we are more puzzled by the sounds they do snatch. At times, their mimicry seems impressive for the same reasons that an "idiot savant" attracts attention: the few seconds or minutes of seemingly connected acts stand in contrast to the hours and days of seemingly nonsensical but apparently obligatory repetition. One of our starlings repeated his cowbird companion's song 279 times in one-half hour. No human encouraged a repeat performance. Another uttered the string of sounds, "OK, I think I guess it's true," and never was heard to repeat it a second time, despite many invitations. The starling's winning combinations are no harder to explain than the losers—the bird repeats, rephrases, recombines, repeats, and so on.

So too, with cowbirds, the act of imitating seems the least of their challenges. Even hand-reared birds housed with heterospecifics and tutored only with recordings of cowbird songs produces highly precise copying (King & West, 1988). Housing male cowbirds with females and providing the same tutoring also produces rapid and accurate copying, followed by extensive improvisation, i.e., undoing of the copying. Why change the copied material? Because other social skills may be more important. The ability to produce acoustic photographs of sound, while an extraordinary motor skill (Greenewalt, 1968), seems to reveal little about social or cognitive consolidation of communicative behavior. Thus, demonstrating that an animal can imitate (which still seems devilishly hard to do to everyone's satisfaction in anything but a songbird) may say little a priori about the behavior's adaptive role for the species in question and may say different things in different species about underlying mechanisms.

These thoughts about the explanatory status of imitation also naturally lead us back to Thorndike and some of his more general thoughts about the human nature of scientists, qualities leading them to focus on only one side of many questions: to prefer the marvelous to the mundane, to remember the one dog or cat who found their way home but not the many who failed. In his words, such a bias leads to the creation of an "abnormal or supernormal psychology" of animals (Thorndike, 1911). Perhaps, some instances of imitation fall into the same category. Evidence, for example, that human neonates imitate facial expressions seems on the marvelous end of the continuum. Recent experiments suggest, however, that the key may be in the more mundane coincident operations of the effects of an interesting visual display on the tendency for infants to explore interesting phenomena by tongue protrusion, prior to the onset of reaching (Jones, in press; Meltzoff & Moore, 1983).

Does Jones's explanation take anything away from the infant's achievement? We think not. As Jones's work reveals, rather than possessing an episodic ability to match an expression, babies physically track moment-to-moment changes in their social world and show evidence of modulating their level of interest from a very early age. And what do adults do when a baby shows signs of modulating attention—just about anything as far as we can tell from scrunching their face to wiggling their ears to talking in baby talk so silly they cannot do it outside the presence of the "helpless" infant. Like the starlings and their caregivers, like the male and female cowbirds, the synergistic outcome, the particular patterns of interaction, is inherent in neither organism, but in the emergent process of mutual attention.

Perhaps, the greatest obstacle to analyzing synergistic operations is that it vastly increases the burden of description on investigators. While we have experienced that burden, we hope to have communicated that the wait is worth it. Indeed, having glimpsed some of the possibly social means by which behaviors are acquired, it is hard to imagine any other course. Partly what attracts us is the possibility of seeing how behaviors from two or more animals connect in ways that are actually quite objective and measurable; part of what attracts us is just seeing the behaviors! Songbirds are small, they fly, and in the field, they make good use of vegetation when possible. Thus, like someone looking through a microscope at a familiar object, for example a feather, the magnified view of what birds actually look like, how they move, and the many ways they can manipulate objects, social and nonsocial, seem to put their sounds into proper perspective, as but part of a whole.

Our rereading of Thorndike also reminded us of his recognition of behavioral parts and the processes by which they become connected (Thorndike, 1913). There is striking commonalty between the "facts" as he called them and the goals and words of those who have turned to nonlinear systems to account for complex connections. Thorndike sought to avoid invoking an inner, mental world to which he, as a scientist, had no access; those turning to systems theories also resist the invisible for the visible (Cole, 1994; Gleick, 1988; Thelen & Ulrich, 1991; Timberlake, 1994). Thorndike used a different vocabulary but his message was that learning is a system made up of many simple parts that share the propensity to respond to change. Among his terms were "multiple response," "cooperation of the animal's set or attitude with the external situation," the "activity of parts or elements," and "shifting of a response from one situation to another" (Thorndike, 1913). In the lexicon of systems theorists, we find terms such as the "cooperation of many anatomical elements and physiological processes," the idea that behavior is "constructed from the available elements for a specific task," that "movement solutions . . . arise in the process of functional action," and that new behavior comes about "because the elements are free to reassemble" (Thelen, 1993). At the core is the need to find methods disclosing actions as they actually occur.

At the outset, we stated that Thorndike obviously watched his animals more than the boxes in which he observed them. Consider how ludicrous it would have seemed if Thorndike had put the cat inside the box, left the room, and then came back after a time to see where the cat was. Many current paradigms to study song learning seem, to us, to come close to such an impoverished approach. The sounds of songbirds have so dominated the collective attention of researchers that it has taken a long time to see through their vocal charms to the contributions of companions. The challenge now is to create contexts in which neither scientists nor their subjects are deprived of the chance to watch animals connect.

REFERENCES

Breland, K., & Breland, M. (1961). The misbehavior of organisms. *American Psychologist,* **16,** 681–684.

Chaiken, M., Bohner, J., & Marler, P. (1993). Song acquisition in European starlings, *Sturnus vulgaris:* A comparison of the songs of live-tutored, untutored, and wild-caught males. *Animal Behaviour,* **46,** 1079–1090.

Cole, B. J. (1994). Chaos and behavior: The perspective of nonlinear dynamics. In L. A. Real (Eds.), *Behavioral mechanisms in evolutionary ecology* (pp. 423–444). Chicago: University of Chicago Press.

Eastzer, D. H., King, A. P., & West, M. J. (1985). Patterns of courtship between cowbird subspecies: Evidence for positive assortment. *Animal Behaviour,* **33,** 30–39.

Elowson, A. M., & Snowdon, C. T. (1994). Pygmy marmosets, Cebuella pygmaea, modify vocal structures in response to changed social enviroment. *Animal Behaviour,* **47,** 1267–1277.

Freeberg, T. M. (submitted). Experimental evidence for the social induction of pairing patterns in female brown-headed cowbirds.

Freeberg, T. M., King, A. P., & West, M. J. (1995). Social malleability in cowbirds: Species and mate recognition in the first two years of life. *Journal of Comparative Psychology,* **109,** 357–367.

Friedmann, H. (1929). *The Cowbirds: A study in the biology of social parasitism.* Springfield, IL: C. C. Thomas.

Friedmann, H., Kiff, L. F., & Rothstein, S. I. (1977). A further contribution to knowledge of the host relations of the parasitic cowbirds. *Smithsonian Contributions to Zoology,* **235,** 1–75.

Galef, B. G. (1975). Social transmission of acquired behavior: A discussion of tradition and social learning in vertebrates. In E. Tobach, L. R. Aronson, & E. Shaw (Eds.), *Advances in the study of behavior* (pp. 77–97). New York: Academic Press.

Galef, B. G., & Wigmore, S. W. (1983). Transfer of information concerning distant foods: A laboratory investigation of the "information center" hypothesis. *Behaviour,* **31,** 748–758.

Gleick, J. (1988). *Chaos: making a new science.* New York: Penguin books.

Greenewalt, C. (1968). *Bird song: Acoustics and physiology.* Washington, DC: Smithsonian Institution Press.

Harlow, H. F. (1974). Induction and alleviation of depressive states in monkeys. In N. F. White (Ed.), *Ethology and psychiatry* (pp. 197–208). Toronto: University of Toronto Press.

Harlow, H. F., & Harlow, M. K. (1962). Social deprivation in monkeys. *Scientific American,* **207,** 136–146.

Jones, S. S. (in press). Imitation or exploration: Newborn infants' matching of adult oral gestures. *Child Development.*

King, A. P., & West, M. J. (1977). Species identification in the North American cowbird: Appropriate responses to abnormal song. *Science,* **195,** 1002–1004.

King, A. P., & West, M. J. (1983a). Epigenesis of cowbird song: A joint endeavor of males and females. *Nature,* **305,** 704–706.

King, A. P., & West, M. J. (1983b). Female perception of cowbird song: A closed developmental program. *Developmental Psychobiology,* **16,** 335–342.

King, A. P., & West, M. J. (1987). Different outcomes of synergy between song production and song perception in the same subspecies (*Molothrus ater ater*). *Developmental Psychobiology,* **20,** 177–187.

King, A. P., & West, M. J. (1988). Searching for the functional origins of cowbird song in eastern brown-headed cowbirds (*Molothrus ater ater*). *Animal Behaviour,* **36,** 1575–1588.

King, A. P., & West, M. J. (1989). Presence of female cowbirds (*Molothrus ater ater*) affects vocal improvisation in males. *Journal of Comparative Psychology,* **103,** 39–44.

King, A. P., West, M. J., & Eastzer, D. H. (1986). Female cowbird song perception: Evidence for different developmental programs within the same subspecies. *Ethology,* **72,** 89–98.

Konishi, M. (1985). Birdsong: From behavior to neuron. *Annual Review of Neuroscience,* **8,** 125–170.

Lehrman, D. S. (1974). Can psychiatrists use ethology? In N. F. White (Eds.), *Ethology and psychiatry* (pp. 187–196). Toronto: University of Toronto Press.

Lorenz, K. (1957). Companionship in bird life. In C. H. Schiller (Eds.), *Instinctive behavior: The development of a modern concept* (pp. 82–128). New York: International Universities Press.

Marler, P. (1976). Sensory templates in species-specific behavior. In J. Fentress (Ed.), *Simpler networks and behavior* (pp. 314–329). Sunderland, MA: Sinauer Associates.

Mason, W. A. (1960). The effects of social restriction on the behavior of rhesus monkeys: I. Free social behavior. *Journal of Comparative and Physiological Psychology,* **53,** 582–589.

Mayr, E. (1974). Behavior programs and evolutionary strategies. *American Scientist,* **62,** 650–659.

Meltzoff, A. N., & Moore, M. K. (1983). Newborn infants imitate adult facial gestures. *Child Development,* **54,** 702–709.

Nice, M. M. (1937). Studies of the life history of the song sparrow I. *Transactions of the Linnean Society of New York,* **4,** 1–247.

Pepperberg, I. M. (1986). Acquisition of anomalous communicatory systems: Implication for studies on interspecies communication. In R. J. T. Schusterman & F. Wood (Eds.), *Dolphin behavior and cognition: Comparative and ethological aspects* (pp. 289–302). Hillsdale, NJ: Erlbaum.

Russon, A. E., & Galdikas, M. F. (1995). Constraints of great apes' imitation: Model and action selectivity in rehabilitant orangutan (*Pongo pygmaeus*) imitation. *Journal of Comparative Psychology,* **109,** 5–18.

Thelen, E. (1993). Motor aspects of energent speech: A dynamic approach. In L. B. Smith, & E. Thelen (Eds.), *A dynamic systems approach to development: Applications.* (pp. 339–361). Cambridge, MA: MIT Press.

Thelen, E., & Ulrich, B. D. (1991). Hidden skills. *Monographs of the Society for Research in Child Development,* **56**(No. 1), 1–97.

Thorndike, E. L. (1911/1965). *Animal intelligence.* New York: Hafner.

Thorndike, E. L. (1913). *Educational psychology.* New York: Teachers College, Columbia University.

Thorpe, W. H. (1958). The learning of song patterns by birds, with especial reference to the song of the chaffinch, *Fringilla coelebs. Ibis,* **100,** 535–570.

Thorpe, W. H. (1961). *Bird song.* London: Cambridge University Press.

Timberlake, W. D. (1994). Behavior systems, associationism, and Pavlovian conditioning. *Psychonomic Bulletin and Review,* **1,** 405–420.

Todd, P. M., & Miller, G. A. (1993). Parental guidance suggested: How parental imprinting evolves through sexual selection as an adaptive learning mechanism. *Adaptive Behavior,* **2,** 5–47.

Todt, D. (1975). Social learning of vocal patterns and models and their applications in grey parrots. *Zeitscrift fur Tierpsychologie,* **39,** 178–188.

West, M. J., & King, A. P. (1985). Social guidance of vocal learning by female cowbirds: Validating its functional significance. *Ethology,* **70,** 225–235.

West, M. J., & King, A. P. (1986). Song repertoire development in male cowbirds (*Molothrus ater*): Its relation to female assessment of song. *Journal of Comparative Psychology,* **100,** 296–303.

West, M. J., & King, A. P. (1988). Female visual displays affect the development of male song in the cowbird. *Nature,* **334,** 244–246.

West, M. J., & King, A. P. (1990). Mozart's starling. *American Scientist,* **78,** 106–114.

West, M. J., & King, A. P. (1994). Research habits and research habitats: Better design through social chemistry. In E. F. Gibbons Jr., E. J. Wyers, E. Waters, & E. W. Menzel Jr. (Eds.), *Naturalistic environments in captivity for animal behavior research* (pp. 163–178). Albany, NY: SUNY Press.

West, M. J., King, A. P., & Eastzer, D. H. (1981). Validating the female bioassay of cowbird song: Relating differences in song potency to mating success. *Animal Behaviour,* **29,** 490–501.

West, M. J., King, A. P., Eastzer, D. H., & Staddon, J. E. R. (1979). A bioassay of isolate cowbird song. *Journal of Comparative and Physiological Psychology,* **93,** 124–133.

West, M. J., King, A. P., & Freeberg, T. M. (1994). The nature and nurture of neophenotypes. In L. A. Real (Ed.), *Behavioral mechanisms in evolutionary ecology* (pp. 238–257). Chicago: University of Chicago Press.

West, M. J., King, A. P., & Freeberg, T. M. (in press a). Building a social agenda for birdsong. In C. T. Snowdon & M. Hausberger (Eds.), *Social influences on vocal development* Cambridge: Cambridge University Press.

West, M. J., King, A. P., & Freeberg, T. M. (in press b). Social malleability in cowbirds: New measures reveal new evidence of plasticity in the eastern subspecies (*Molothrus ater ater*). *Journal of Comparative Psychology*.

West, M. J., King, A. P., & Harrocks, T. H. (1983a). Cultural transmission of cowbird song: Measuring its development and outcome. *Journal of Comparative Psychology,* **97,** 327–337.

West, M. J., Stroud, A. N., & King, A. P. (1983b). Mimicry of the human voice by European starlings: The role of social interaction. *Wilson Bulletin,* **95,** 635–640.

Williams, H., Kilander, K., & Sotanski, M. L. (1993). Untutored song, reproductive success and song learning. *Animal Behaviour,* **45,** 695–705.

Contagious Yawning and Laughter: Significance for Sensory Feature Detection, Motor Pattern Generation, Imitation, and the Evolution of Social Behavior

ROBERT R. PROVINE

Department of Psychology
University of Maryland Baltimore County
Baltimore, Maryland 21228

ontagious yawning and laughter in humans offer insights into a variety of problems in the neural, behavioral, and social sciences. Contagion is the probable response of "stimulus feature detectors" triggered specifically by yawns in the visual domain and laughs in the auditory domain (Provine, 1986, 1989b, 1992, 1996a). It does not require conscious effort for an observer to imitate a yawning or laughing person. In the language of classical ethology, these neurological stimulus detectors would be "innate releasing mechanisms" (IRMs) evolved to detect the "releasing stimuli" of yawns or laughs (Alcock, 1989; Provine, 1986, 1996a). Such stimulus feature detectors are more likely to have evolved to select the simple, stereotyped, species-typical acts of yawning or laughing than more arbitrary and variable behaviors learned during a lifetime of the individual. Because observed yawns or laughs trigger their respective neurological detectors to evoke identical acts,[1] contagious yawning or laughter may be used to assay the activity

---

1. Although any signal is capable of synchronizing the behavior and physiology of a group (i.e., an alarm cry triggers escape and/or fear), only a response in which the *identical* act is replicated in an observer is propagated as a behavioral chain-reaction from individual to individual.

and determine the selectivity of the underlying detection process. Contagious behavior, thus, provides a novel, noninvasive approach to the neural basis of sensory feature detection.

The study of species-typical and stereotyped yawning (Provine, 1989b) offers advantages over other approaches to the detection of faces (a visual feature) that rely on neuropsychological studies of rare clinical conditions (i.e., prosopagnosia) (Meadows, 1974; Whiteley & Warrington, 1977) or the electrophysiological recording of face-specific brain neurons in animal models (Bruce, Desimone, & Gross, 1981; Kendrick & Baldwin, 1987; Perrett, Mistlin, & Chitty, 1987; Perrett, Rolls, & Caan, 1982). In the auditory domain, the search for a detector for structurally simple, stereotypic, and species-typical laughter offers advantages over more complex and culturally varied speech (Provine, 1992, 1993a, 1996a). The simplicity, stereotypy, and species typicality of yawning and laughter offer similar tactical advantages in the search for motor pattern generating circuits (Provine, 1986, 1996a; Provine & Yong, 1991).

In studying contagious yawning and laughter, we move seamlessly from the neural to the social level of analysis. Yawning and laughter offer a rare opportunity to examine the neurological basis of that significant but neglected class of social behavior—contagion (Provine, 1989b; Provine, 1992). Typically, social psychologists focus on behavior learned during the life time of individuals and neglect innate or neurologically mediated social behavior. Social psychologists describe contagious-like behavior in the context of higher level processes such as "social facilitation," "conformity," "peer pressure," or "modeling," and seldom consider the possible biological roots of contagious phenomena.

Because the biologic and genetic determinants of yawning and laughter are stronger than that of many behaviors studied by social scientists, they should not be relegated to the category of "interesting footnotes." We should not segregate them from other, more familiar, social acts shaped more directly by learning and experience. A thoughtful position on such matters is offered by an expert on behavioral contingencies, B. F. Skinner (1984). When asked to respond to the challenges to operant behavior by the discovery of "biological constraints on learning," "feature detectors" in sensory systems, and "pattern generating circuits" in motor systems (Provine, 1984a), central themes of this chapter, Skinner (1984) replied "that a given species is predisposed by its genetic history to see particular stimuli in preference to others or to behave in particular ways in preference to others are facts of the same sort. A different kind of selection has been at work." In other words, the contingencies of natural selection can shape structure and behavior during phylogenesis in a way similar to the process that shapes behavior during the life of the individual.

Contagious yawning and laughter provide insights into imitation, a common topic in this book, and a process of general behavioral significance (Piaget, 1951; Provine, 1989a). Instead of venturing into the semantic quicksands of definition, this chapter has the more modest goal of broadening the range of acts evaluated in imitation studies. Consider, for example, the controversy over the existence and nature of facial imitation by human neonates (Meltzoff & Moore, 1977, 1983). Most researchers of facial imitation suggest the involvement of high-level cognitive processes, and have not considered the precedents of contagious yawning and laughter. Although contagious behavior may not qualify as imitation as commonly defined, it is a ubiquitous, ancient form of social coupling that coexists with modern, consciously controlled social behavior (Provine, 1986, 1989a, b, 1992).

This chapter describes contagious yawning and laughter and shows how these acts can be used to study a variety of issues in the neural, behavioral, and social sciences. The neuroethological approach taken here has a strong descriptive foundation. Consequently, this account begins with the description of the motor acts of yawning and laughter because, in the case of contagious behavior, the motor act is both the stimulus and the response, and defines the nature of the stimulus feature detector supporting contagion.

YAWNING

Yawns as Stereotyped Action Patterns

The word for yawning is derived from the Old English "ganien," meaning to open wide as in gape. Yawns are slow, involuntary, gaping movements of the mouth that begin with a slow inspiration of breath and end with a briefer expiration. Yawning is a behavior of the type called "fixed," "modal," or "stereotyped," by ethologists (Alcock, 1989; Provine, 1986). The term "stereotyped action pattern" will be used here to refer to such acts. Consideration of yawning will begin with a description of its duration, frequency, and intrasubject stability. These descriptions define yawning and provide baseline data necessary for later experiments.

The mean yawn duration was 5.9 ± 1.9 s (SD) for the 34 of 37 subjects who yawned at least once during the observation period[2] (Provine, 1986). The mean

---

2. The formidable problem of getting subjects to yawn in the laboratory was solved by having them "think about yawning" and record their own yawns by pushing a button at the start of a yawn and keeping it depressed until the yawn is complete. This technique avoids inhibitions associated with the well-known social sanctions against public yawning. Unless otherwise noted, this procedure was used to induce and record yawns.

duration of yawns for individual subjects ranged from 3.5 to 11.2 s. Yawning frequency of the 34 yawning subjects ranged from 1 to 76 (X = 27.5 ± 18.4) during the 30-min period of observation, an average rate of about one yawn per min. The periodicity of yawns, the onset-to-onset interyawn interval, for the 31 subjects performing at least two yawns, was 68.3 ± 33.7 s. No significant correlation was detected between yawn duration and interyawn interval, indicating that infrequent yawners did not compensate by performing long yawns and vice versa. (The significance of this finding for respiratory function is considered below.) Whatever their style, individual yawning patterns were stable. The frequency and duration of yawns performed by the same subjects during 10 min sessions separated by 1 to 3 weeks were correlated significantly.

Once a yawn is initiated, it goes to completion with the inevitability of a sneeze. Yawns are hard to stifle. The implications of this experience were examined by having subjects yawn with clenched teeth. This procedure tests for the effects of eliminating or modifying movement-produced feedback associated with the gaping component of the yawn while permitting normal respiration through the clenched teeth (Provine, 1986). The frequency and duration of normal and clenched-teeth yawns as estimated by the respiratory component were similar. Thus, the underlying motor pattern generator for yawning was able to run normally with abnormal sensory feedback. However, subjects reported that such yawns were unpleasant, did not satisfy the urge to yawn, and gave the impression of being "stuck" in midyawn. The gaping of the jaws must be performed to achieve a satisfying yawn; the respiratory component is insufficient. Try a clenched-teeth yawn yourself.

Another yawn variant is informative. Try a "nose yawn" in which your lips remain sealed and you inspire through the nose. Most subjects report being unable to perform nose yawns (Provine, Tate, & Geldmacher, 1987). Unlike normal breathing that can be done with equal facility through either the nose or mouth, yawns require inhalation through the mouth. The difficulties of performing the nose and clenched-teeth yawns suggest that the principal function of a yawn is not respiratory; deep breaths can be taken through either the clenched teeth in the clenched-teeth yawn, or the nose in the nose yawn.

The function of yawning is elusive. However, one of the most common popular explanations of yawning can be rejected. Yawning is not a response to elevated CO_2 or decreased O_2 in the blood or brain. Yawning by laboratory subjects was neither increased by breathing a gas mixture high in CO_2 (3% or 5%), nor inhibited by breathing 100% O_2 (Provine et al., 1987). (The normal composition of air is 20.95% O_2, 79.02% N_2 and inert gases, and .03% CO_2.) However, both the CO_2 and the O_2 conditions increased breathing rate, providing clear evidence that they had physiological effect. A second study found that exercise sufficient to double breathing rate had no effect on yawning. Taken together, the gas inhalation and exercise studies indicate that yawning and breathing are triggered by different internal states and are controlled by separate mechanisms.

In summary, data concerning the exclusivity of breathing route during yawns (mouth priority over nose), lack of yawn frequency/duration interaction, and absence of CO_2, O_2, and exercise effects, all argue against yawning having a principal, respiratory function. It is, however, difficult to eliminate all possible respiratory hypotheses because physiological redox effects can act in so many ways, in so many systems, and at so many levels.

The folklore that bored people yawn a lot is confirmed. The hypothesis that people yawn more while observing uninteresting than while observing interesting stimuli was tested by comparing yawns produced by subjects observing a 30-min rock video with those observing a 30-min video (colored bars) test pattern without an audio track (Provine & Hamernik, 1986). Whatever your opinion about rock videos, they are much more interesting than test patterns. Subjects watching the uninteresting (boring) test pattern produced more and longer yawns than those watching the more interesting rock video.

The folklore that sleepy people yawn a lot also was confirmed. The relationship between sleepiness, yawning, and stretching (an act often associated with yawning) was studied by having subjects record their yawns and sleep times in a log book for 1 week (Provine, Hamernik, & Curchack, 1987). Subjects yawned most during the hour before bedtime and the hour after waking, times when they were presumably sleepy. The temporal proximity of yawning to sleep and waking times and boring situations (reported above) is probably the basis for yawning as a paralinguistic signal for drowsiness or boredom. It is curious that stretching, a behavior often associated with yawning, has not assumed this signal function, and is not contagious. The higher degree of conscious control over stretching relative to yawning would give stretching more flexibility as a social signal.

On waking, we typically yawn and stretch (Provine et al., 1987). Late in the evening, shortly before bedtime, we usually yawn without stretching. There are two peak periods per day for yawning (after waking and before bedtime), but only one for stretching (after waking). A relationship between yawning and stretching is indicated by the involuntary stretching movements of the otherwise paralyzed limbs of hemiplegics during yawns (Walshe, 1923). Pharmacological evidence for a yawn–stretch relationship comes from observations that drugs that produce yawning also produce stretching in a variety of animals (Dourish & Cooper, 1990; Gessa, Pisano, Vargiu, Crabai, & Ferrari, 1967; Yamada & Furukawa, 1980). These lines of evidence suggest that the yawn should be considered a form of stretch and that investigations of yawn function should include the correlates of stretching.[3]

Although not studied systematically, there are several significant differences between yawning and stretching. There is less conscious control over yawning than stretching, yawning is more contagious than stretching, yawning has as respiratory element lacking in stretching, yawning has a greater involvement of neck and head structures than stretching, and yawning has a Valsalva-like (breath holding and "bearing down") maneuver lacking in stretching.

At present, there is much speculation, but little evidence about a function for yawning. However, yawning, like stretching, is a high-amplitude maneuver that probably has numerous consequences through the body. Each of these physiological correlates may be a "function" (i.e., have some plausible benefit). There is no evidence that yawning either increases or decreases alertness. However, yawning is linked with some changes in behavioral states. We yawn during the transition from sleep to wakefulness, from wakefulness to sleep, and when becoming bored. The association between yawning and change in behavioral state was pointed out by fish behaviorist Arthur Myrberg (1972), who noticed that when the damselfish he studied on a reef yawned, they would soon switch from one to another class of activity. Yawning may facilitate such state transitions.

The search for a yawn function should also consider the phylogenetic antiquity of the act. Most vertebrates yawn, a fact that indicates a motor pattern generating process and perhaps at least one physiological correlate (i.e., "function") common to all yawning organisms. Additional motor components and physiological correlates may have evolved from this primal prototype.

The occurrence of yawning during the first trimester of human prenatal development opens the possibility of a role of yawning in embryogenesis. Would it not be surprising if a function of yawning is to ensure the proper articulation of the jaw joint by moving it during development? The sculpting and maintenance of developing joints is an important function of prenatal movement (Provine, 1993b). Yawn functions noted in contemporary humans, such as opening the eustachian tube to equalize pressure in the middle ear and the ambient environment (Laskiewicz, 1953), may be secondary consequences of an act evolved in the service of some other environmental or developmental challenge.

---

3. Yawning may have evolved as the facial component of a generalized stretched response that has an added respiratory element (Provine, Hamernik, & Curchack, 1987). However, given the phylogenetic antiquity of yawning, it is also possible that stretching evolved after and may be an elaboration and caudal extension of a yawn, the primal stretch.

Whatever the physiological and behavioral consequences of normal yawning, yawning is symptomatic of a wide range of pathology, including brain lesions and tumors, hemorrhage, motion sickness, chorea, and encephalitis (Graybiel & Knepton, 1976; Jurko & Andy, 1975; Barbizet, 1958; Heusner, 1946). Psychotics are reported to yawn rarely, except when suffering from organic brain syndrome (Lehmann, 1979). This intriguing observation, considered with the finding that antidopinergic agents often produce yawning, suggests that yawning may provide a metric for the pathogenesis of schizophrenia (associated with elevated dopamine levels) and a useful assay for the titration of antidopaminergic neuroleptic dosages. Lehmann (1979) notes further the old clinical observation that people suffering from acute physical illnesses never yawn when their condition is serious; a return of yawning signals convalescence. In regard to neurotransmitters, yawning is associated with cholinergic and peptidergic excitation and dopaminergic inhibition. Because yawning is stimulated by hormones (i.e., testosterone, oxytocin, ACTH, MSH) and drugs (i.e., apomorphine, piribedil, pilocarpine) with known mechanisms of action, yawning can serve as a useful, noninvasive, behavioral assay of chemical events within the brain. Clinically, yawning is therapeutic in preventing atelectasis, the collapse of alveoli, a frequent postoperative respiratory complication (Cahill, 1978).

To conclude this discussion of yawning as a motor act, it is appropriate to return to the initial ethological theme and review the properties that qualify yawning as a stereotyped action pattern (Provine, 1986).

1. Yawning is species-typical in humans, performed by all members of our species. [We do not show the higher rates of male yawning reported for more dimorphic primates (Schino & Aureli, 1989).] Yawning is not, however, species exclusive; most vertebrates yawn (Baenninger, 1987; Deputte, 1994; Huesner, 1946).
2. Yawning is consistent in duration (average duration ~ 6 s).
3. Yawning occurs periodically (average interyawn interval ~ 68 s).
4. Yawning is under strong genetic control because it is already performed by embryos during the first trimester of prenatal development (DeVries, Visser, & Prechtl, 1982) and is obvious in both normal and anencephalic human newborns (Heusner, 1946; Provine, 1989a).
5. Yawns are unitary, being performed at so-called "typical intensity." Fractional (atypically short) yawns are seldom seen.
6. The amplitude and duration of yawns are independent of the amplitude of the releasing stimulus (if present). Further, once initiated, yawns go to

completion with minimal influence of sensory feedback; everyone is familiar with the difficulty of trying to stifle a yawn (Provine, 1986).

7. Yawns can be "released" by witnessing yawns or yawn-related stimuli (Provine, 1986; 1989b), the basis of the contagious yawn response.

8. Yawns are complex in spatiotemporal organization and have facial, respiratory, and other components, e.g., yawns are not simple reflexes of short duration.

9. The motor components of a yawn occur in only one order and the timing of components is consistent from yawn to yawn. This stability of sequence contributes to the yawns unmistakable appearance, an important property for a releasing stimulus.

10. The finding that yawns are prominent in people who are waiting, or performing monotonous work (Provine & Hamernik, 1986), and of dogs on the threshold of aggression, or participating in an aversive activity, is consistent with the performance of yawns as "displacement acts" (Provine, 1986).

Given these many properties, yawning has been recognized as one of the best examples of stereotyped action pattern and releasing stimulus in humans (Alcock, 1989). Yawns are not reflexes. As traditionally understood, reflexes are simpler acts of short duration, are evoked by stimuli, have short response latencies, and have response amplitudes that are correlated with stimulus amplitudes.

Contagious Yawning

The contagiousness of yawning is legendary. Viewing, reading about, and thinking about yawning evokes yawns (Provine, 1986). Although yawning is interesting in its own right, contagious yawning is a means of assessing the yawn-evoking potency of various facial features. Thus used, the search for the ethological releasing stimulus for yawns provides insights into face detection, an issue in perception and neuropsychology (Provine, 1989b). The discovery of a perceptual process activated exclusively by visually observed yawns establishes a precedent for a facial feature and/or expression detector in humans. Similar detectors may exist for facial expressions (actions) other than yawns, and for other complex visual stimuli, but their activity may be more difficult to monitor because they lack a contagious response as a behavioral assay.

A series of studies evaluated the yawn-evoking potency of various features of a yawning face (Provine, 1986, 1989b). The yawn-evoking capacity of variations in a

5-min series of 30 videotaped repetitions of a yawning face (one yawn every 10 s) were compared with each other and with a control condition of a series of 30 videotaped smiles (Provine, 1989b). Single frames of the monochrome video stimuli in midyawn or midsmile are shown in Fig. 1. The 360 subjects, 30 per stimulus condition, were instructed via videotape to observe a video monitor and to record their yawns by pressing a button.

The normal yawning face (Fig. 1a) was an effective stimulus, causing 16 of 30 subjects (Fig. 2, upper) to produce a total of 92 yawns (Fig. 2, lower), significantly more yawns than to the smile. The yawn-detection process was not axially specific; yawns in orientations of 90°, 180°, and 270° were as potent or nearly as potent as normal, upright, 0° yawns. The number of subjects who yawned in response to the high-contrast yawn (Fig. 1b) did not differ significantly from those who yawned in response to normal-halftone yawns (Fig. 1a) or smiles (Fig. 1h). A tonic (still) yawn video frame of a yawner in midyawn (Fig. 1a) produced a number of yawners midway between, and not significantly different from that produced by normal, animate yawns or smiles.

The "no-mouth" yawn (Fig. 1c) was the only stimulus with a deleted feature that produced as many yawning subjects as the complete face and significantly more yawners than did the smile (Fig. 2). This initially counterintuitive and disconcerting result was, however, consistent with other data. Consider, for example, the relative ineffectiveness of the "mouth-only" yawn (Figs. 1e and 2). The gaping mouth, the most obvious candidate for the ethological "sign stimulus" for yawning, is not necessary to evoke contagious yawns. Instead, the yawn detector may be triggered by the overall configuration of the yawning face, perhaps being driven by cues involving the squinting of the eyes, tilting of the head, and movement of the jaw. The importance of the overall configuration and dynamic cues in the discrimination of facial expressions is reinforced by findings of Leonard, Voeller, and Kuldau (1991). In monkeys, a lack of axial and feature specificity in many face-specific neurons suggest a stimulus analysis of the sort described in the present behavioral analyses of human yawns (Bruce et al., 1981; Perrett, Mistlin, & Chitty, 1987; Perrett, Rolls, & Caan, 1982). Monkeys even have neurons specific for yawning faces. These diverse behavioral and neurophysiological results suggest common underlying processes. It is unlikely that complex neural mechanisms for similar perceptual tasks would evolve independently and have radically different principles of operation.

Determination of latencies of the contagious yawn responses provides additional information about the dynamics and nature of the underlying process (Provine, 1986). The stimulus in the latency study was the animate video of the normal

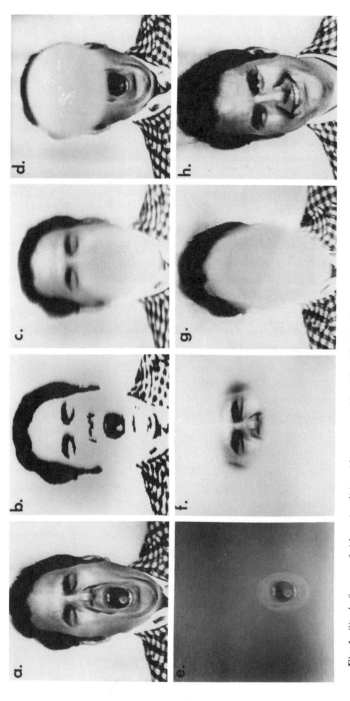

Fig. 1. Single frames of video stimuli in mid-yawn or mid-smile. All stimuli were animate except for the single tonic (still) condition resembling (a). (a) Normal halftone yawn. (b) High-contrast yawn. (c) No-mouth yawn. (d) No-eyes yawn. (e) Mouth-only yawn. (f) Eyes-only yawn. (g) No-face yawn. (h) smile. From Provine (1989).

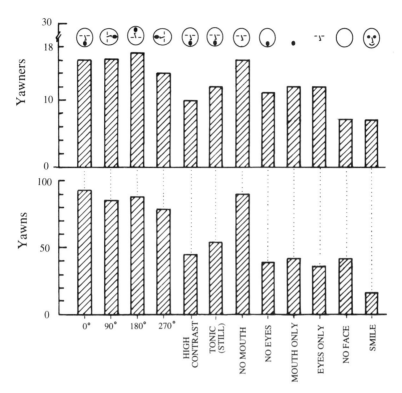

Fig. 2. The number of subjects who yawned (upper graph) and the total yawns (lower graph) evoked by various yawn and smile stimuli. The stimuli are shown in cartoon form at the top of the figure and labeled at the bottom of the figure. From Provine (1989).

yawning face described previously. As in the previous experiment, yawns were potent yawn-inducing stimuli; 23 of 42 subjects (55%) yawned during a 5-min session. Only 5 of 24 subjects (21%) yawned while viewing the control condition of a recurrent series of videotaped smiles. The proportion of subjects yawning while viewing yawning gradually increased during the 5-min session. These data are consistent with the involvement of a complex, higher order perceptual process involving polysynaptic processes; the contagious yawn mechanism is not a reflex having a short and consistent latency.

The complexity of the contagious yawn mechanism is suggested further by the variety of nonvisual stimuli that can evoke it. As readers may have concluded already, simply reading about yawning is sufficient to trigger yawns (Provine, 1986; Carskadon, 1991, 1992). The potency of the text-induced yawning effect was tested by comparing the yawns performed by subjects reading about yawning with yawns performed by subjects reading a control passage about hiccupping. Significantly more subjects either yawned or thought about yawning while reading about yawning than reading about hiccupping.

Most stimuli associated with yawning can evoke yawns. For example, even the sound of yawning, or thinking or reading about yawning, triggers yawns. Given the variety of potential yawn-evoking stimuli, further exploration of the range of yawn-inducing stimuli may yield diminishing returns. Although the contagious yawn is a highly mechanistic social response to yawn-related stimuli, the underlying process does not exclusively involve a detector for a narrowly defined visual stimulus. Does this mean that the "innate releasing mechanism (IRM)" of classical ethology is less selective and more modifiable in humans than in other animals? Or does the human data inform us of the true nature of released behavior as performed throughout the animal kingdom? Our subjective experience of contagious yawns may provide valuable evidence that similar released behavior in nonhumans is not as rigidly determined and selective as is often assumed.

Age of onset of visually evoked contagious yawn responses has not been established (Provine, 1989a). However, spontaneous yawning is already present by the end of the first trimester of prenatal development (DeVries et al., 1982) and is obvious in newborns. In one of the rare developmental references, Piaget (1951) suggested that yawning becomes contagious during the second year of life. Thus, the present tentative evidence suggests that contagious yawning develops after the superficially similar facial-"imitation" response reported in human neonates (Meltzoff & Moore, 1977, 1983). If subsequent research confirms this chronology, the releasing mechanism that triggers contagious yawns develops and becomes active long after the motor pattern generator for yawning.

LAUGHTER

Laughter as a Stereotyped Action Pattern

Laughter is a common, species-typical vocal act and auditory signal that is prominent in social discourse (Provine, 1996a; Provine & Fischer, 1989; Provine & Yong, 1991). Laughter, like smiling and talking, is performed almost exclusively during social encounters; solitary laughter seldom occurs except in response to media, a source of vicarious social stimulation (Provine & Fischer, 1989). Although many aspects of laughter have been studied (Black, 1984; Fry, 1963; Gregory, 1924;

Piddington, 1963; Sully, 1902; Stearns, 1972), including its development (Sroufe & Waters, 1976), social context (Bainum, Loundsbury, & Pollio, 1984; Provine & Fischer, 1989), contagion (Provine, 1992), ethnology (Apte, 1985), evolution (Darwin, 1872; Provine, 1994; van Hoof, 1972), physiological correlates (Averill, 1969; Fry & Rader, 1977; Fry & Savin, 1988; Fry & Stoft, 1971), potential health benefits (Cogan, Cogan, Waltz, & McCue, 1987; Cousins, 1976), pathology (Black, 1982), relation to humor (Chapman & Foot, 1976, 1977; Durant & Miller, 1988; McGhee & Goldstein, 1983a,b), play (Aldis, 1975), and tickling (Fridlund & Loftis, 1990), we know less about the structure of laughter than we do about calls and songs of many nonhuman species. Most ethologists neglect human behavior, and language-oriented analyses of human speaking (Levelt, 1989) and listening (Handel, 1989) do not consider the nearly ubiquitous paralinguistic signal of laughter. However, laughter's stereotypy, simple structure, species-wide distribution, and presumed strong genetic basis, make it ideal for studies ranging from the neurobehavioral mechanisms of vocal production and perception, to the origins of human communication (Provine, 1996a).

Research shows human laughter to have a common underlying structure consisting of "notes" (i.e., "ha," "ho," "he") that have similar durations and recur at regular "internote intervals" (i.e., "ha-ha-ha") (Fig. 3) (Provine & Yong, 1991). The mean note duration is about 75 ms and the mean onset-to-onset internote interval is about 210 ms. Contrary to some popular opinion, everyone does not laugh in a different way. If they did, how would we know they were laughing? The stereotypy of laughter is due in part to strong constraints on the range of laugh-like sounds that our vocal apparatus can produce. It is difficult to laugh in other than the normal manner. It is informative to try to laugh with longer or shorter than normal notes or internote intervals.

Laugh notes begin with a voiceless aspirant (hissing sound not produced by vibration of the vocal cords) of about 200 ms duration similar to that of the English /h/ in "ha" (Provine & Yong, 1991). Although this aspiration is most obvious in the first note of a laugh sequence, it is present before and after each note. The aspirant is probably not important in laugh perception because laughter with the initial and internote aspiration edited out sounds normal. The notes of laughter have temporal symmetry, a fact indicated by their passing the "reversibility test," the ability of recorded laughter played backward on a tape recorder to still sound laugh-like (Provine & Yong, 1991). Additional evidence of laugh note symmetry is the arbitrariness of laugh notation found in various languages. In Italian operas, for example, laughter is notated as "ah" in contrast to the "ha" of the parallel English translation. However, laughter is sung identically in both languages.

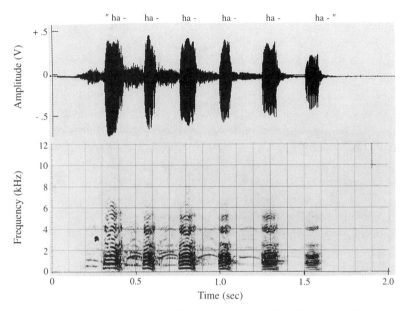

Fig. 3. Waveform of a 6-note laugh from a 46-year-old male (upper). Frequency spectrum of the same laugh (lower). In the frequency spectrum, the evenly spaced horizontal bands of the voiced laughter are harmonics of the notes' fundamental frequency. Both the waveform and the frequency spectrum show the unvoiced aspiration that precedes and follows each voiced laugh note. From Provine and Yong (1991).

The voiceless /h/ that initiates a laugh note is followed by a forcefully voiced vowel-like sound, /a/ being the most common variant (Provine & Yong, 1991). A specific vowel sound does not define laughter because vocalizations composed of notes having one of a variety of vowel-like sounds (i.e., "ha-ha," "ho-ho," "he-he") are at least occasional laugh forms of some people. However, the same vowel is usually used in all notes of a given laugh episode. Someone may preform "ha-ha" and "ho-ho" laughs at different times, but never "ha-ho-ha-ho" laughs. Such a vocal maneuver is difficult. Try simulating a "ha-ho-ha-ho" laugh. The vowel-like structure of most laugh notes has a marked harmonic structure characterized by multiples of the note's fundamental frequency, a feature illustrated by the evenly-spaced stacks of frequency bands of the sound spectrum of Fig. 3. The explosively voiced laugh notes have less formant structure and more harmonic structure than is usually associated with vowels. (Formants are frequency bands whose amplitudes are modulated by resonances in the vocal tract.) As in speech, female laughter is higher in pitch (fundamental frequency of 502 Hz) than male laughter (fundamental frequency of 276 Hz).

The duration of laugh notes of males and females is nearly identical, averaging 75 ms (Provine & Yong, 1991). Notes are generally consistent in duration regardless of their position in a sequence. Exceptions are the first and last notes of a multinote laugh that are sometimes longer in duration than those in the middle of a note sequence (i.e., "haa-ha-ha," "ha-ha-haa"). The first and last notes are also the position where variations in vowel and/or consonant sounds are most likely to appear (i.e., "he-ha-ha," "ha-ha-ho," "cha-ha-ha").

One of the most striking features of multinote laughs is the regular intervals between notes (Provine & Yong, 1991). The onset-to-onset intervals between adjacent notes average 210 ms with no differences between males and females.

In contrast to the relative temporal stability of note duration and internote interval, note amplitude decreases dramatically during laughter. This decrescendo probably results from the depleted air supply available for notes late in a sequence. The decrescendo is one aspect of laughter that definitely does not pass the reversal test; recorded laughter played backward produces a bizarre sounding crescendo.

Laughter in the Speech Stream

Laughter is a vocal anachronism, a kind of behavioral fossil, that coexists with modern speech. The stereotypy and simplicity of laughter is more suggestive of an animal call than complex and flexible speech. Although laughter is a vocalization in the universal human library, we do not speak laughter.

Consider the low degree of conscious control over laughter.[4] Try to laugh voluntarily, or ask someone else to do so. Typically, the request is met by the comment "I can't laugh by just trying to do it." If voluntary laughter is produced, it usually sounds strained and abnormal. Clearly, laughing is not the act of saying "ha-ha-ha," a task that is easily performed. Yet, laughter is sprinkled through the stream of modern, consciously controlled speech. Much can be learned by examining the relationship between these very different vocalizations.

Given the neglect of laughter in research of human speech production (Levelt, 1989) and perception (Handel, 1989), it is hardly surprising that little is known

---

4. Although laughter (a vocal stimulus) and smiling (a visual stimulus) both signal positive affect, we have much greater voluntary control over smiling than laughter. Subtle differences between "false" and "felt" smiles aside (Ekman & Friesen, 1982), it is much easier to produce a smile than a laugh on command.

about the placement of laughter in speech (Provine, 1993). However, this is a topic of considerable scientific and clinical significance. Although speech and laughter may have different evolutionary histories, neural mechanisms, acoustic structure, and may convey different classes of messages, they share the same final common pathway, the organ of vocalization. Thus, the temporal segregation (time sharing) of speech and laughter on the single vocalization channel reveals the presence or absence of an underlying organizational principle. A nonrandom time sequence of laughter and speech would indicate a lawful organizational process and its rules, and whether laughter or speech has channel priority (Provine, 1993a).

To study the placement of laughter in speech, 1200 cases of naturally occurring laughter were observed in anonymous adults, mostly college students, in public places (Provine, 1993a). Observers sought out groups of laughing people and recorded who laughed (speaker or audience), gender of speaker and audience and what was said immediately before laughter occurred.

During conversation, laughter by speaker or audience almost exclusively followed complete statements or questions (Provine, 1993a). Laughter by either speaker or audience seldom (8 in 1200 cases) interrupted the phrase structure of speech. Although people may recall instances when they or others "break up" in laughter and have difficulty finishing sentences during humorous stories, such instances are rare. Because laughter by speaker or audience occurred immediately after complete phrases, laughter can be said to "punctuate" speech. This *punctuation effect* is so strong and prominent that it may be confirmed by cursory observations of everyday social conversations.

The finding that laughter seldom interrupts speech indicated that there is a lawful and probably neurologically programmed process responsible for this temporal organization (Provine, 1993a). The near absence of speech interruptions by laughter indicates further that speech has priority over laughter in gaining access to the single vocalization channel. Laughter's subservience to speech is similar to the situation with breathing and coughing, other airway maneuvers that occur almost exclusively during speech pauses. The placement of signed laughter in the stream of signed speech of congenetically deaf signers would be of considerable interest because their speech is not constrained by respiratory patterns or a common organ of vocalization (Provine, 1993a).

Analyses of what was said immediately before laughter indicates that most laughter is not a consequence of structured attempts at humor such as joke or story telling (Provine, 1993a). Only 10–20% of prelaugh comments were even mildly humorous. Although the determination of what is funny is highly subjective and context dependent, few would consider "Oh yeah"? or "I wouldn't say so"

TABLE 1

Distribution of Speaker (S) and Audience (A)
Laughter of Males (M) and Females (F)
in Same Sex and Different Sex Dyads[a]

Dyad	N	Speaker (%)	Laugher audience (%)
$S_M A_M$	275	208 (75.6%)	165 (60.0%)
$S_F A_F$	502	430 (86.0%)	250 (49.8%)
$S_M A_F$	238	328 (66.0%)	169 (71.0%)
$S_F A_M$	185	163 (88.1%)	72 (38.9%)
SA	1200	958 (79.8%)	656 (54.7%)

[a]From Provine (1993).

promising material for a comedian. Yet, such is the stuff of most prelaugh comments.

Our database of 1200 cases of naturally occurring laughter provided material for another unexpected finding. Speakers, especially females, laugh more than their audience (Provine, 1993a). Table 1 shows the distribution of speaker and audience laughter. Overall, speakers laughed 46% more than their audience. The difference was especially great when female speakers were conversing with a male audience, a condition producing 127% more speaker than audience laughter.

Our observations of naturally occurring laughter reveal several previously unappreciated facts about laughter, all of which require consideration in future studies (Provine, 1993a).

1. Laughter is a social vocalization that is almost extinguished by placing people in solitary situations (Provine & Fischer, 1989). Attempts to study individual subjects in a laboratory setting are unlikely to produce much laughter and may not be ecologically valid. Even comedy videos, a form of vicarious social stimulation, will seem funnier and evoke more laughs if they are observed in a group setting.

2. Most laughter is not a response to jokes or other formal attempts at humor (Provine, 1993a). Thus, research based on audience response to jokes or cartoons, although valid in its own domain, is of limited relevance to most

everyday laughter. Exclusive focus on humor deflects consideration of broader and deeper roots of laughter in human vocal communication and social interaction.

3. Speaker and audience laughter must be differentiated. The finding that speakers laugh more than their audience indicates the limits of research reporting only audience behavior, a common approach of humor research.

4. The pattern of speaker and audience laughter depends on gender. For example, neither males nor females laugh as much to female as to male speakers, and female speakers are especially likely to laugh when conversing with a male audience (Provine, 1993a).

The importance of considering the differential contribution and gender of speaker and audience to group laughter is emphasized by contrasting the present findings with those of Grammer and Eibl-Eibesfeldt (1990). Despite some methodological differences, both studies found more laughter from females than males in mixed-sex groups, and neither study detected a difference between the *overall* frequency of same-sex laughter at the level of the *group*. However, because Grammer and Eibl-Eibesfeldt did not distinguish between speaker and audience laughter, they failed to detect the large gender difference in same-sex laughter (from Provine 1993a): $S_m = 75.6\%$, $A_m = 60.0\%$; $S_f = 86.0\%$, $A_f = 49.8\%$). Averaging the laughter of female speaker and audiences cancels out the high level of speaker and low level of audience laughter.

Laughter in Chimpanzees and Humans: A Comparison

The opinion often repeated in the popular media that laughter is unique to humans is unfounded. From at least the time of Darwin (1862), there has been an awareness that chimpanzees (*Pan troglodytes*) and other great apes emit a laugh-like vocalization when tickled and during play (Berntson, Boysen, Bauer, & Torello, 1989; Fossey, 1972; Marler & Tenaza, 1977; Van Lawick-Goodall, 1968; Yerkes, 1943). However, until recently, too little was known about either human or chimpanzee laughter to permit rigorous comparison. Using procedures developed to study human laughter, we were able to contrast the acoustic and social properties of laughter in chimpanzees and humans. Breathy, sometimes grunt-like chimpanzee laughter was found to have some acoustic and social characteristics of human laughter, but to be generated by a different pattern of neuromuscular activity (Provine & Bard, 1994, 1995). In contrast to previous, largely unsuccessful attempts to teach chimpanzees to speak, this work contrasts the ability of chimpanzees and humans to produce laughter, a vocalization already in the vocabularies of both species (Provine, 1996a).

Laughter of young chimpanzees was evoked by the tickling administered by their human caregiver during playful, face-to-face encounters. Such tickle is a typical part of chimpanzee play and mother–infant interactions. The tickle-evoked chimpanzee laughter was contrasted with the sample of human conversational laughter described previously.

Laughter of both humans and chimpanzees occurred almost exclusively in social contexts, the primary difference being the greater physical contact during the play bouts of the chimpanzees (Provine & Bard, 1994). Chimpanzee laughter occurred mostly during rough and tumble play and chasing games in which physical contact, its invitation or threat, such as tickle, was a part. Such physical contact contrasts with the conversational context of most human laughter in which individuals have minimal physical contact.[5]

As noted previously, human laughter is composed of one or more discrete, vowel-like notes ("ha," "ho," etc.) that parse and are performed exclusively during expiration. In contrast, chimpanzee laughter resembles panting or grunting, with a single breathy vocalization being produced during each expiration and inspiration. Both chimpanzee and human laughter have a highly periodic structure, a property revealed in the regularity of intervals between laugh notes.

The most significant difference between chimpanzee and human laughter is in the relationship between breathing and vocalizing (Provine & Bard, 1994, 1995). Humans laugh by parsing an expiration, with a single exhalation having the capacity to produce a series of laugh notes ("ha-ha-ha"). Laughter, like speech, is produced by modulating the expiratory air flow. In contrast, chimpanzees can produce only a single pant-like sound per exhalation or inhalation. Their vocalization is more closely tied to the respiratory cycle. If this coupling of vocalization to breathing is a general property of vocal control, it suggests an important and unappreciated constraint on the evolution of vocal speech in chimpanzees and perhaps other great apes. The decoupling of breathing and vocalization in chimpanzees and humans occurred after the two species diverged from a common ancestor approximately 6 million years ago.

5. Tickle-evoked laughter, however, is probably the ancestral form in both humans and chimpanzees and it continues to play a significant role in the social behavior of both species (Provine, 1996a,b). Like other laughter (Provine & Fischer, 1989), that evoked by tickle virtually disappears in solitary individuals (i.e., you cannot tickle yourself). Tickle is a form of communication, not a tactile reflex.

Contagious Laughter

One of laughter's most curious and informative properties is its contagion, the tendency of laughter to spread through a group in a chain reaction (Provine, 1992). Contagious laughter is not a fragile product of the laboratory. The power of contagious laughter as a social-coupling process is suggested by a persistent epidemic of laughter that began among 12- to 18-year-old girls in a boarding school in Tanganyika and spread throughout a district requiring the closing of schools (Rankin & Philip, 1963). The potency of the contagion effect has long been recognized by the entertainment and broadcast industries which have developed the technology of "canned" laughter. The contagious laugh effect contributes to the laugh-producing property of "laugh boxes," "laugh records" such as that produced by *OKeh* (Demento, 1985), one of the most popular novelty recordings of all time, and the notorious "laugh tracks" of broadcast comedy shows. These are instances of a technology and commercial product emerging before the scientific implications of the underlying phenomenon were appreciated.

Most research relevant to contagious laughter concerns the associated topic of responses to humorous material (McGhee & Goldstein, 1983a,b). The amount of laughter is correlated positively with group or audience size (Andrus, 1946; Levy & Fenley, 1979; Morrison, 1940; Young & Frye, 1966), and whether others are laughing (Brown, Brown, & Ramos, 1981; Brown, Dixon, & Hudson, 1982; Chapman & Chapman, 1974; Chapman & Wright, 1976; Freedman & Perlick, 1979), a variable that is often investigated using prerecorded "canned" laughter (Chapman, 1973a,b; Fuller & Sheehy-Skeffington, 1974; Leventhal & Cupchik, 1975; Leventhal & Mace, 1970; Nosanchuk & Lightstone, 1974; Smyth & Fuller, 1972).

Despite the wealth of research on laughter in the context of humor and its social dynamics, it has gone unnoticed that *laughter* may trigger laughter and mediate the phenomenon of contagion (Provine, 1992). The hypothesis that laughter is a stimulus for laughter and/or smiles was investigated by observing the responses of subjects in three undergraduate psychology classes to a sample of laughter provided by a "laugh box," a small battery operated record player obtained in a novelty store. The laugh stimulus lasted 18 s and was repeated at the beginning of each of 10, 1-min intervals.

A majority of subjects in two of the three classes laughed in response to the laugh stimulus on the first of the 10 trials (Fig. 4, left). The laugh-evoking potency of the stimulus laughter declined over trials until only 3 of the 128 total subjects laughed on trial 10. On all trials, more subjects smiled than laughed in response to the laugh stimulus. (Smiles should not be viewed as low-amplitude laughs. Smiles

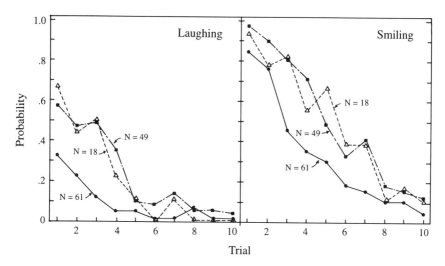

Fig. 4. The probability of self-reported laughing (left) and/or smiling (right) in response to an 18-s sample of "canned" laughter presented during 10 consecutive 1-min trials. From Provine (1992).

are emitted more frequently, and at lower thresholds than laughter, in response to humorous material or to signal positive affect.) Most subjects smiled in response to stimulus laughter on the first trial (Fig. 4, right), but as was the case with laughter-evoked laughter, laughter-evoked smiling declined over trials, until few subjects smiled on trial 10.

The polarity of subjects' responses to the laugh stimulus shifted over trials. Although the question was not asked during earlier trials, after trial 10, most subjects answered "yes" when asked if they found the stimulus "obnoxious." The dual (positive/negative) nature of laughter is supported by everyday experience. To "laugh with" friends consolidates the social bonds between group members, but to "laugh at" someone is to jeer or mock them, to exclude them from the "in group." In this context, it is noteworthy that mobbing, a synchronized group response by some birds and mammals to drive often larger invaders from their territories, is functionally similar to human jeering and involves a simple, staccato, repetitive cry similar to laughter (Eibl-Eibesfeldt, 1989).

In this research, laughter itself was sufficient to evoke laughter in audiences. You can throw away the joke and maintain much of the laughter. A laughing audience is sufficient to evoke laughter in its members. The identification of laughter as a sufficient stimulus for laughter is relevant in a variety of issues in speech science. Humans have evolved the species-typical vocalization of laughter and the neural mechanism dedicated to its detection. Indeed, laughter may qualify as an ethological-releasing stimulus for the stereotyped vocalization of laughter (Eibl-Eibesfeldt, 1989; Provine, 1992). Future psychophysical studies must determine which parameters of laughter (i.e., note structure, note duration, internote interval) are necessary for the perception of laughter and the activation of the hypothetical-laugh detector.

The characteristics of laughter described here are relevant to several central issues in speech science. Contagious laughter involves the replication in a perceiver of a motor pattern that originally generated vocalization in a sender. This intimacy between laugh production and detection suggests a highly specialized functional correlation and/or coevolution of vocalization and perception, a conclusion relevant to motor theories of speech perception and associated issues of modularization of function (Mattingly & Studdert-Kennedy, 1991; Provine, 1996a).

FUTURE DIRECTIONS

Because research on yawning, laughter, and their contagion is at an early stage, it is appropriate to conclude with suggestions for future work. Much remains to be done.

The best next step in the study of yawning and laughter is to provide detailed descriptions of these motor acts. Descriptions using a variety of high-resolution procedures such as electromyography and frame-by-frame cinematography define yawning and laughter in a manner meaningful to both neurophysiologists searching for neural mechanisms and behavioral investigators interested in the production and perception of social signals. Topographic accounts of what yawns look like, or what laughter sounds like, are not adequate for this task. Consider the comparison between human and chimpanzee laughter. The most informative difference between species was in the modulation of breathing to produce vocalization, a distinction beyond the reach of purely acoustic studies. Reduced to their essentials, the signals of yawning and laughter, although under less conscious control and serving other functions, are simply movements like walking. The contagion of yawning or laughter is the contagion of movement. Once the production and control of these movements are described adequately, we can move on to problems of development and evolution. Although studies of motor behavior, development, and evolution are each important lines of inquiry, they converge in significant and unanticipated ways. Motor systems are the critical site of action of the ontogenetic contingencies that shape behavioral evolution (Provine, 1984b, 1994).

Developmental studies of yawning and laughter are a priority. Although yawning develops prenatally, and laughter develops later between 3 and 4 months after birth, little is known about the emergence of contagiousness of either behavior, other than it occurs postnatally. After establishing the developmental trajectory of each contagious behavior, perceptual studies should examine the sensory vectors that mediate the contagious response. A start toward defining the sensory trigger for visually mediated yawning in adults was described in this chapter. Comparable research in the auditory domain should define the vector for contagious laughter, probably some combination of laugh-note duration, note frequency, note-harmonic structure, and internote interval. Developmental studies of the social contexts of yawning and laughing would provide further information about their social roles and how they change over time.

Further comparative research will provide insights into the phylogenesis of yawning and laughter. For example, the present comparison between human and chimpanzee laughter indicates that the evolution of human laughter involved the emancipation of vocalization from chimpanzee-like panting in which the vocalization was tied closely to the cycle of breathing. The limited comparative evidence suggests that other great apes (the orangutan and gorilla) have laughter resembling that of the chimpanzee. Now that the parameters of human laughter are better understood, more rigorous comparative analyses can be undertaken. Future comparative studies should examine both context specific (i.e., evoked by tickle and/or emitted during rough and tumble play) and acoustically similar vocalizations of a wide variety of species, including those not previously considered to laugh. The quantitative analysis of such vocalizations may detect cross-species similarities overlooked by more qualitative, homocentric approaches that focus on the most human-like species and sounds. The search for contagious laughter in nonhumans is complicated by the absence of subjective reports from animals; evidence of contagion must be based on objective records of their actions. Behavioral evidence of contagion would be the synchronization of laughter among group members, perhaps triggered by recordings of conspecific laughter. In chimpanzees, the replay of recorded conspecific laughter evokes a specific vocalization resembling threat barks (but not contagious laughter) and cardiac acceleration, in contrast to cardiac deceleration and lack of vocalizations in response to conspecific screams (Berntson et al., 1989). Although the case concerning contagious laughter is still open, other contagious vocalizations are relatively common among animals.

The comparative analysis of yawning is further underway than that of laughter; most vertebrates are thought to yawn. The study of nonyawning animals or animals that yawn reliably in specific circumstances may suggest yawn functions. Among primates, there may be two classes of yawn-like gestures: one may be the true, normal, spontaneous yawn, and the other may be a voluntary threat gesture that only superficially resembles a true yawn (Deputte, 1994). (A threatening aspect of yawning may be the flashing of formidable canine teeth during a gape.) A microanalysis of true and threat yawns using electromyography and/or cinematography may distinguish between these motor acts (Deputte & Fontenelle, 1978). A superficial similarity between mouth-gaping movements is understandable given the limited freedom of movement of the hinged jaw.

The social dynamics of contagious behavior needs to be examined. Although the fact of behavioral contagion is well established, at least in humans, little is known about its kinetics. What, for example, is the rate at which contagious yawning spreads through a group; what damps or controls the spread of yawning; what is the function of group size on the probability of an individual yawning; is there a "critical mass" effect; and does just having yawned prime an individual and make him/her more susceptible to future observed yawns? These data would provide the basis for mathematical models of behavioral chain reactions having substantial descriptive and theoretical value. We could learn, for example, if behaviors described as contagious differ in rate of spread and relative infectiousness.

Data collected in the naturalistic tradition of ethology offer considerable potential for clinical research and diagnosis. Armed with baseline data concerning yawning and laughter reviewed in this chapter, we can search for conditions that vary from the normal. The section on yawning noted its diagnostic significance for pathologies ranging from brain tumors to hemorrhage. The study of laughter in special and clinical populations offers similar promise. For example, the study of hearing impaired children would provide evidence of the relative contribution of genetic and environmental factors to the development of laughter. Although preliminary reports indicate that even congenitally deaf-blind children laugh (Eibl-Eibesfeldt, 1973), little is known about the acoustic properties of this laughter, its social context, or its placement in the speech stream. Does laughter punctuate the signed speech of congenitally deaf signers in whom there is not a shared organ of expression? Baseline data on laughter permit research on a variety of other topics including the nature of laughter in neuropathology and psychopathology, the specification of some aspects of "flat affect," "inappropriate laughter," and other so-called "soft signs" of abnormal behavior, the sparing of laugh-related affect in aphasia, the hemispheric control of vocalization, the search for pathologies exclusive to laughter, and the possible familial nature of unusual laugh patterns (Provine, 1993, 1996a).

This diverse sample of proposed research converges on the common theme of social communication, its neural substrates, development, evolution, pathology, and variability. The investigation of one aspect of this program illuminates the others. Although it is too soon to determine the rewards of the suggested approach, preliminary findings indicate that yawning and laughter offer unique tactical approaches to a variety of deeper problems in the neural and behavioral sciences. Yawning and laughter both ancient modes of prelinguistic and presymbolic communication that coexist with modern gesturing and speech. Their investigation informs us about the historic roots of human communication and provides a bridge between the study of contemporary human and animal social behavior.

REFERENCES

Alcock, J. (1989). *Animal behavior* (4th ed.). (pp. 26–27). Sutherland, MA: Sinauer Associates, Inc.

Aldis, O. (1975). *Play fighting.* New York: Academic Press.

Andrus, T. O. (1946). A study of laugh patterns in the theatre. *Speech Monographs,* **13,** 114.

Apte, M. L. (1985). *Humor and laughter: An anthropological approach.* Ithaca: Cornell University Press.

Averill, J. R. (1969). Autonomic response patterns during sadness and mirth. *Psychophysiology,* **5,** 399–414.

Baenninger, R. (1987). Some comparative aspects of yawning in *Betta splendens, Homo sapiens, Panthera leo,* and *Papio sphinx. Journal of Comparative Psychology,* **101,** 349–354.

Bainum, C. K., Loundsbury, K. R., & Pollio, H. R. (1984). The development of laughing and smiling in nursery school children. *Child Development,* **55,** 1946–1957.

Barbizet, J. (1958). Yawning. *Journal of Neurological and Neurosurgical Psychiatry,* **21,** 203–209.

Berntson, G. G., Boysen, S. T., Bauer, H. R., & Torello, M. S. (1989). Conspecific screams and Laughter: Cardiac and behavioral reactions of infant chimpanzees. *Developmental Psychobiology,* **22,** 771–787.

Black, D. W. (1982). Pathologic laughter: A review of the literature. *Journal of Nervous and Mental Disease,* **170,** 67–71.

Black, D. W. (1984). Laughter. *Journal of the American Medical Association,* **252,** 2995–2998.

Brown, G. E., Dixon, P. A., & Hudson, J. D. (1982). Effect of peer pressure on imitation of humor response in college students. *Psychological Reports,* **51,** 1111–1117.

Brown, G. E., Brown, D., & Ramos, J. (1981). Effects of a laughing versus a nonlaughing model on humor responses in college students. *Psychological Reports,* **48,** 35–40.

Bruce, C., Desimone, R., & Gross, C. G. (1981). Visual properties of neurons in a polysensory area in superior temporal sulcus of the macaque. *Journal of Neurophysiology, 46*, 369–384.

Cahill, A. (1978). Yawn maneuver to prevent atelectasis. *AORN Journal, 27*, 1000–1004.

Carskadon, M. A. (1991). Yawning elicited by reading: Is an open mouth a sufficient stimulus? *Sleep Research, 20*, 116.

Carskadon, M. A. (1992). Yawning elicited by reading: Effects of sleepiness. *Sleep Research, 21*, 101.

Chapman, A. J. (1973a). Funniness of jokes, canned laughter and recall performance. *Sociometry, 36*, 569–578.

Chapman, A. J. (1973b). Social facilitation of laughter in children. *Journal of Experimental Social Psychology, 9*, 528–541.

Chapman, A. J., & Chapman, W. A. (1974). Responsiveness to humor: Its dependency upon a companion's humorous smiling and laughter. *Journal of Psychology, 88*, 245–252.

Chapman, A. J., & Wright, D. S. (1976). Social enhancement of laughter: An experimental analysis of some companion variables. *Journal of Experimental Child Psychology, 21*, 201–218.

Chapman, A. J., & Foot, H. C. (Eds.). (1976). *Humor and laughter: Theory, research and applications.* New York: Wiley.

Chapman, A. J., & Foot, H. C. (Eds.). (1977). *It's a funny thing, humor.* Oxford: Pergamon Press.

Cogan, R., Cogan, D., Waltz, W., & McCue, M. (1987). Effects of laughter and relaxation on discomfort thresholds. *Journal of Behavioral Medicine, 10*, 139–144.

Cousins, N. (1976). Anatomy of an illness (as perceived by the patient). *New England Journal of Medicine, 295*, 1458–1463.

Darwin, C. (1872/1965). *The expression of emotions in man and animals.* Chicago: University of Chicago Press.

Demento, Dr. (1985). The OKeh laughing record. *Dr. Demento presents the greatest novelty records of all time: Vol. 1. The 1940's (and before)* (Rhino Record R4 70820). Santa Monica, CA: Rhino Records.

Deputte, B. L. (1994). Ethological study of yawning in primates. I. Quantitative analysis and study in two species of old world monkeys (*Cercocebus albigena* and *Macaca facicularis*). *Ethology, 98*, 221–245.

Deputte, B., & Fontenelle, A. (1980). Menace et baillement chez *Macaca fascicularis:* interet de l'etude electromyographique comparee. *Biology of Behavior, 5*, 47–54.

DeVries, J. I., Visser, G. H. A., & Prechtl, H. F. R. (1982). The emergence of fetal behavior. I. Qualitative aspects. *Early Human Development, 7*, 301–322.

Dourish, C. T., & Cooper, S. J. (1990). Neural basis of drug-induced yawning. In S. J. Cooper, & C. T. Dourish (Eds.), *Neurobiology of behavioural stereotypy* (pp. 91–116). Oxford: Oxford University Press.

Durant, J., & Miller, J. (Eds.). (1988). *Laughing matters: A serious look at humor.* London: Longman.

Eibl-Eibesfeldt, I. (1973). The expressive behavior of the deaf and blind born. In M. v. Cranach & I. Vine. (Eds.), *Social communication and movement* (pp. 163–194). London: Academic Press.

Eibl-Eibesfeldt, I. (1989). *Human ethology.* New York: Aldine de Gruyter.

Ekman, P., & Friesen, W. V. (1982). Felt, false, and miserable smiles. *Journal of Nonverbal Behavior,* **6,** 238–252.

Fossey, D. (1972). Vocalization of the mountain gorilla (*Gorilla beringei*). *Animal Behavior,* **20,** 36–53.

Freedman, J. L., & Perlick, D. (1979). Crowding, contagion and laughter. *Journal of Experimental Social Psychology,* **15,** 295–303.

Fridlund, A. J., & Loftis, J. M. (1990). Relations between tickling and humorous laughter: Preliminary support for the Darwin-Hecker hypothesis. *Biological Psychology,* **30,** 141–150.

Fry, W. F., Jr. (1963). *Sweet madness: A study of humor.* Palo Alto, CA: Pacific.

Fry, W. F., Jr., & Rader, C. (1977). The respiratory components of mirthful laughter. *Journal of Biological Psychology,* **19,** 39–50.

Fry, W. F., Jr., & Savin, W. M. (1988). Mirthful laughter and blood pressure. *Humor,* **1,** 49–62.

Fry, W. F., Jr., & Stoft, P. E. (1971). Mirth and oxygen saturation levels of peripheral blood. *Psychotherapy and Psychosomatics,* **19,** 76–84.

Fuller, R. G. C., & Sheehy-Skeffington, A. (1974). Effects of group laughter on responses to humorous material: A replication and extension. *Psychological Reports,* **35,** 531–534.

Gessa, G. L., Pisano, M., Vargiu, L., Crabai, F., & Ferrari, W. (1967). Stretching and yawning after intracerebral injections of ACTH. *Review of Canadian Biology,* **26,** 229–236.

Grammer, K. (1990). Strangers meet: Laughter and non-verbal signs of interest in opposite-sex encounters. *Journal of Nonverbal Behavior,* **14,** 209–236.

Grammer, K., & Eibl-Eibesfeldt, I. (1990). The ritualization of laughter. In W. A. Koch (Ed.), *Naturlichkeit der Sprache und der Kultur. Bochumer Beitrage zur Semiotik* (pp. 192–214). Bochum: Brockmeyer.

Graybiel, A., & Knepton, J. (1976). Sopite syndrome: A sometimes side effect of motion sickness. *Aviation, Space, and Environmental Medicine,* **47,** 873–882.

Gregory, J. C. (1924). *The nature of laughter.* New York: Kegan Paul, Trench, Trubner.

Handel, S. (1989). *Listening: An introduction to the perception of auditory events.* Cambridge, MA: MIT Press.

Heusner, A. P. (1946). Yawning and associated phenomena. *Physiological Review,* **25,** 156–168.

Hoof, J. A. R. M. van (1972). A comparative approach to the phylogeny of laughter and smiling: In R. Hinde (Ed.), *Non-Verbal communication* (pp. 209–241). Cambridge: Cambridge University Press.

Jurko, M. F., & Andy, O. J. (1975). Post-lesion yawning and thalamotomy site. *Applied Neurophysiology,* **38,** 73–79.

Kendrick, K. M., & Baldwin, B. A. (1987). Cells in temporal cortex of conscious sheep can respond preferentially to the sight of faces. *Science,* **236,** 448–450.

Laskiewicz, A. (1953). Yawning in regard to the respiratory organs and the ear. *Acta Ontolaryngology,* **43,** 267–270.

Lehmann, H. E. (1979). Yawning, a homeostatic reflex and its psychological significance. *Bulletin of the Menninger Clinic,* **43,** 123–136.

Leonard, C. M., Voeller, K. K. S., & Kuldau, J. M. (1991). When's a smile a smile? Or how to detect a message by digitizing the signal. *Psychological Sciences,* **2,** 166–172.

Levelt, W. J. M. (1989). *Speaking: From intention to articulation.* Cambridge, MA: MIT Press.

Leventhal, H., & Cupchik, G. C. (1975). The informational and facilitative effects of an audience upon expression and the evaluation of humorous stimuli. *Journal of Experimental Social Psychology,* **11,** 363–380.

Leventhal, H., & Mace, W. (1970). The effect of laughter on evaluation of a slapstick movie. *Journal of Personality,* **38,** 16–30.

Levy, S. G., & Fenley, W. F., Jr. (1979). Audience size and likelihood and intensity of response during a humorous movie. *Bulletin of the Psychonomic Society,* **13,** 409–412.

Liberman, A., & Mattingly, I. G. (1989). A specialization for speech perception. *Science,* **243,** 489–494.

Lieberman, P. (1984). *The biology and evolution of language.* Cambridge, MA: Harvard University Press.

Marler, P., & Tenaza, R. (1977). Signalling behavior in apes with special reference to vocalization. In T. A. Seboek (Ed.), *How animals communicate* (pp. 965–1033). Bloomington, IN: Indiana University Press.

Mattingly, I. G., & Studdert-Kennedy, M. (Eds.). (1991). *Modularity and the motor theory of speech perception.* Hillsdale, NJ: Erlbaum.

McGhee, P. E., & Goldstein, J. H. (Eds.). (1983a). *Handbook of humor research: Basic issues (Vol. I).* New York: Springer-Verlag.

McGhee, P. E., & Goldstein, J. H. (Eds.). (1983b). *Handbook of humor research: Applied studies (Vol. II).* New York: Springer-Verlag.

Meadows, J. C. (1974). The anatomical basis of prosopagnosia. *Journal of Neurological and Neurosurgical Psychiatry,* **37,** 489–501.

Meltzoff, A. N., & Moore, M. K. (1977). Imitation of facial and manual gestures by human neonates. *Science,* **198,** 75–78.

Meltzoff, A. N., & Moore, M. K. (1983). Newborn infants imitate adult facial gestures. *Child Development,* **54,** 702–709.

Morrison, J. A. (1940). A note concerning investigations on the constancy of audience laughter. *Sociometry,* **3,** 179–185.

Myrberg, A. A. (1972). Ethology of the bicolor damselfish, *Eupomacentrus partitus* (Pisces: Pomacentridae). A comparative analysis of laboratory and field behavior. *Animal Behavior Monographs,* **5,** 199–283.

Nosanchuk, T. A., & Lightstone, J. (1974). Canned laughter and public and private conformity. *Journal of Personality and Social Psychology,* **29,** 153–156.

Perrett, D. I., Mistlin, A. J., & Chitty, A. J. (1987). Visual neurons responsive to faces. *Trends in Neuroscience,* **10,** 358–364.

Perrett, D. I., Rolls, E. T., & Caan, W. (1982). Visual neurons responsive to faces in the monkey temporal cortex. *Experimental Brain Research,* **47,** 329–342.

Piaget, J. (1951). *Play, dreams and imitation in childhood.* New York: Norton.

Piddington, R. (1963). *The psychology of laughter: A study in social adaptation.* New York: Gamut Press.

Provine, R. R. (1984a). Contingency-governed science. *Behavioral and Brain Sciences,* **7,** 494–495.

Provine, R. R. (1984b). Wing-flapping during development and evolution. *American Scientist,* **72,** 448–455.

Provine, R. R. (1986). Yawning as a stereotyped action pattern and releasing stimulus. *Ethology,* **72,** 109–122.

Provine, R. R. (1989a). Contagious yawning and infant imitation. *Bulletin of the Psychonomic Society,* **27,** 125–126.

Provine, R. R. (1989b). Faces as releasers of contagious yawning: An approach to face detection using normal human subjects. *Bulletin of the Psychonomic Society,* **27,** 211–214.

Provine, R. (1992). Contagious laughter: Laughter is a sufficient stimulus for laughs and smiles. *Bulletin of the Psychonomic Society,* **30,** 1–4.

Provine, R. R. (1993a). Laughter punctuates speech: Linguistic, social and gender contexts of laughter. *Ethology,* **95,** 291–298.

Provine, R. R. (1993b). Prenatal behavior development: Ontogenetic adaptations and non-linear processes. In G. J. P. Savelsbergh (Ed.), *The development of coordination in infancy* (pp. 203–236). Amsterdam: North Holland.

Provine, R. R. (1994). Pre- and postnatal development of wing-flapping and flight in birds: Embryological, comparative and evolutionary perspectives. In M. N. O. Davis & P. R. Green (Eds.), *Perception and motor control in birds* (pp. 135–159). Berlin and Heidelberg: Springer-Verlag.

Provine, R. R. (1996a). Laughter. *American Scientist,* **84,** 38–45.

Provine, R. R. (1996b). Yawns, smiles, tickles, and talking: Naturalistic and laboratory studies of facial action and social communication. In J. A. Russell & J. M. Fernández–Dols (Eds.), *New directions in the study of facial expressions.* Cambridge: Cambridge University Press.

Provine, R. R., & Bard, K. A. (1994). Laughter in chimpanzees and humans: A comparison. *Society for Neuroscience Abstracts,* **20,** 367.

Provine, R. R., & Bard, K. R. (1995). Why chimps can't speak: The laugh probe. *Society for Neuroscience Abstracts,* **21,** 456.

Provine, R. R., & Fischer, K. R. (1989). Laughing, smiling, and talking: Relation to sleeping and social context in humans. *Ethology,* **83,** 295–305.

Provine, R. R., & Hamernik, H. B. (1986). Yawning: Effects of stimulus interest. *Bulletin of the Psychonomic Society,* **24,** 437–438.

Provine, R. R., & Yong, Y. L. (1991). Laughter: A stereotyped human vocalization. *Ethology,* **89,** 115–124.

Provine, R. R., Hamernik, H. B., & Curchack, B. C. (1987). Yawning: Relation to sleeping and stretching in humans. *Ethology,* **76,** 152–160.

Provine, R. R., Tate, B. C., & Geldmacher, L. L. (1987). Yawning: No effect of 3–5% CO_2, 100% O_2, and exercise. *Behavioral and Neural Biology,* **48,** 382–393.

Rankin, A. M., & Philip, P. J. (1963). An epidemic of laughing in the Bukoba District of Tanganyika. *Central African Journal of Medicine,* **9,** 167–170.

Schino, G., & Aureli, F. (1989). Do men yawn more than women? *Ethology and Sociobiology,* **10,** 375–378.

Skinner, B. F. (1984). Response to Provine. *Behavioral and Brain Sciences,* **7,** 507.

Smyth, M. M., & Fuller, R. G. C. (1972). Effects of group laughter on responses to humorous material. *Psychological Reports,* **30,** 132–134.

Sroufe, L. A., & Waters, E. (1976). The ontogenesis of smiling and laughter: A perspective on the organization of development in infancy. *Psychological Review,* **83,** 173–189.

Stearns, F. (1972). *Laughing.* Springfield, IL: C. C. Thomas.

Sully, J. (1902). *An essay on laughter.* London: Longmans, Green.

Van Lawick-Goodall, J. (1968). The behavior of free-living chimpanzees in the Gombe Stream Reserve. *Animal Behavior Monographs,* **1,** 161–311.

Walshe, F. M. R. (1923). On certain tonic or postural reflexes in hemiplegia with special reference to the so-called "associative movements." *Brain,* **46,** 1–37.

Whiteley, A. M., & Warrington, E. K. (1977). Prosopagnosia: A clinical, psychological, and anatomical study in three patients. *Journal of Neurological and Neurosurgical Psychiatry,* **67,** 394–430.

Yamada, K., & Furukawa, T. (1980). Direct evidence for involvement of dopaminergic inhibition and cholinergic activation in yawning. *Psychopharmacology,* **67,** 39–43.

Yerkes, R. M. (1943). *Chimpanzees, a laboratory colony.* New Haven, CT: Yale University Press.

Young, R. D., & Frye, M. (1966). Some are laughing: Some are not—why? *Psychological Reports,* **18,** 747–754.

PART 2

Imitation

10

Introduction: Identifying and Defining Imitation

Cecilia M. Heyes

Department of Psychology
University College London
London WC1E 6BT
United Kingdom

I mitation has long been regarded as a special kind of social learning, unique in both its psychological complexity and potential to support cultural transmission (e.g., Thorndike, 1898; Washburn, 1908; Piaget, 1962; Boyd & Richerson, 1985; Galef, 1988). Consequently, although imitation may appear to be just one of many types of social learning, it is the principal focus of psychological research on social learning in animals, and, correspondingly, the concern of nearly half of the chapters in this collection.

The dual significance of imitation, its perceived importance not only as a sign of complex psychological processing but also as a means of effecting the nongenetic transmission of information, may have contributed to the problems that have been encountered both in defining imitation conceptually and in identifying it empirically. For at least a century, researchers have been preoccupied with the questions of how the term "imitation" should be applied, which hypothetical class of phenomena it should be understood to name, and how instances of this class can be identified in practice, distinguished empirically from other forms of social learning. These issues may have proved especially intractable in the case of imitation because its dual significance entails that they are addressed by people with a variety of disciplinary backgrounds and theoretical purposes, including ethologists, experimental

Social Learning in Animals: The Roots of Culture

psychologists, primatologists, and behavioral biologists. The potential for cross-talk and confusion has been immense.

The chapters that follow show that, in recent years, genuine progress has been made in defining and identifying instances of imitation, so that we can proceed with greater confidence to examine the phylogenetic distribution of imitation, its evolutionary history, adaptive function, ontogeny, and mechanisms of operation. A subtle but important sign of this progress is that the contributors to this volume exchange views. They acknowledge the positions, and address the arguments, of other authors in this collection and elsewhere. This exchange contrasts with the tradition in research on imitation in which each contributor has, typically, reported his or her own demonstration of imitation, or preferred taxonomy of imitative learning, either in splendid isolation or with reference only to like-minded colleagues.

DEFINING IMITATION BY EXCLUSION

Thomas Zentall begins the first substantive chapter on imitation by highlighting intradisciplinary diversity in the way in which imitation has been defined. Some psychologists (including, in this volume, Whiten and Tomasello) define imitation as learning that is mediated by a particular cognitive process, for example, an intentional process like "perspective taking" or "mind reading," while others eschew reference to unobservable psychological processes, and view imitation as a variety of instrumental conditioning. Zentall argues that the latter approach is preferable because it may not be possible to measure intentionality directly, and he pursues a more behavioral analysis through a survey of varieties of nonimitative social learning.

The purpose of Zentall's broad and incisive review of types of nonimitative social learning (which includes contagion, social facilitation, socially mediated aversive conditioning, local enhancement, stimulus enhancement, observational conditioning, matched dependent behavior, copying, and goal emulation) is to show that "true imitation" can be defined by exclusion; using identified control procedures, one can demonstrate that a social learning phenomenon is *not* nonimitative, and therefore, by default, *is* imitation. The control procedure that Zentall favors in this context is the "two-action" method, in which observer animals witness demonstrators performing one of two different actions on the same object or manipulandum. He argues that if, in a two-action test, observers tend to execute the same action as their demonstrators, it is most unlikely that this behavioral matching is due to nonimitative social learning, and suggests that the two-action method has provided evidence of true imitation in budgerigars (Dawson & Foss, 1965; Galef, Manzig, & Field, 1986), pigeons (Zentall, Sutton, & Sherburne, 1995), and rats (Will, Pallaud, Soczka, & Manikowski, 1974; Heyes & Dawson, 1990).

THE EVOLUTION OF IMITATION

The two-action method is one powerful means of distinguishing imitative learning from cases in which observers and demonstrators perform similar actions either independently (without the demonstrator's behavior having any influence on the observer) or as a result of the demonstrator attracting the observer's attention to a particular place or object, i.e., through stimulus enhancement. Bruce Moore, whose approach to the study of imitation is influenced both by the behavioral or associative tradition and by ethology, begins his chapter on the evolution of imitation by summarizing the results of an ingenious alternative method of achieving the same end. In his studies of a Grey parrot, Okichoro, Moore eliminated the possibility of stimulus enhancement by looking for imitative learning of gestures, rather than actions on objects, and "labeled" each nonvocal gesture in order to distinguish imitation from independent or chance production by the parrot of behavior resembling that of the human experimenter. Thus, the experimenter would, for example, wave his hand while saying "ciao," and subsequently the bird would be scored as having imitated the gesture only if it waved a wing while, or soon after, vocal imitation of this word.

Moore believes that there is satisfactory evidence of movement imitation in only two groups of animals: parrots and great apes. He is sceptical about the two-action method studies of budgerigars, rats, and pigeons because each examines imitation on only one behavioral dimension, and it is not clear, in each case, that the imitated behavior was not already in the observers' repertoires. Thus, Moore takes generality and novelty to be definitive of imitative learning, and argues that it is only parrots and great apes that have been shown to meet these criteria.

Using comparative data and deduction, Moore weaves the parrot and great ape data into a rich set of hypotheses about the evolution of imitative learning. He suggests that the capacity for movement imitation evolved independently in the psittaciformes and hominids, deriving in the former case from song learning via vocal and percussive mimicry, and in the latter from conditioning via skill learning and putting through. Thus, Moore's conception of the hominid sequence is consistent with the behavioral or associative analysis adopted by Zentall and, to some extent, Heyes, while his account of the psittaciform line, and his focus on phylogeny more generally, reflects a more ethological orientation.

CULTURAL BEHAVIOR IN JAPANESE MACAQUES

Michael Huffman's chapter on stone-handling behavior in Japanese macaques is the first of four chapters concerning primates, and it is appropriate that Huffman's contribution should occupy this leading position. As Huffman emphasizes, we cannot be certain that stone handling has been transmitted through imitation. It could be that, while observing seasoned stone handlers, juvenile macaques learn about the properties of stones, but not how to handle them, via stimulus enhancement or emulation (Tomasello, this volume). However, Huffman's meticulous observational studies are very much a part of the Japanese primatological tradition, originating in the work of Imanishi, Itani, and Kawai on Koshima Island, which has had a long standing and profound influence on theorizing about the relationship between imitation and culture.

Huffman's studies focus on careful description of stone-handling behavior and its practitioners, identification of the conditions in which stone handling occurs, and documentation of the route of its spread through populations of individually identified monkeys whose matrilineal relationships are known to investigators. The resulting data reveal that stone handling is unique among the socially transmitted, traditional behaviors described in Japanese macaques because it occurs in the absence of any tangible direct benefits to practitioners. Thus, it would appear that, over a period of less than 20 years, stone handling has become a tradition within certain populations of macaques without contributing to their reproductive fitness.

DO APES APE?

With the introduction of the two-action method, new standards of evidence have been established, and the once easy assumption that apes "ape," that chimpanzees and other nonhuman apes are capable of imitative learning, is being reexamined. The chapters by Andrew Whiten and Deborah Custance, and by Michael Tomasello, which reach contrasting conclusions, are the most recent and most rigorous contributions to this process of reexamination.

Whiten and Custance summarize the results of their recent experiments with chimpanzees and human children using a refinement of Hayes and Hayes (1952) "do-as-I-do" procedure and a two-action method involving "artificial-fruit" apparatus. The within-subjects, seriate design of the do-as-I-do experiment resembles that of the procedure used by Moore with the parrot Okichoro; each subject is presented with a series of distinct gestures, A, B, C, etc., and evidence is sought of imitation by the observer of each of these actions. However, unlike Moore's procedure, the do-as-I-do test is preceded by a training phase in which animals are encouraged, through body molding and reward, to imitate modeled gestures on the command "Do this." Furthermore, the do-as-I-do test relies, not on the subjects' vocalizations, but on the timing of behavior relative to the demonstration of each action to distinguish imitation from independent production of behavior resembling that of the experimenter/demonstrator. Two-action tests, including Custance and Whiten's artificial-fruit procedure, differ from both of these within-subject, seriate procedures, not only in their focus on object-directed actions, but also in controlling for independent production, or chance resemblance, by making an explicit comparison of the frequency with which observers execute action A after observing A and after observing B.

Whiten and Custance conclude from their studies that both chimpanzees and 2- to 4-year-old children are capable of imitation, but that chimpanzees imitate with less fidelity. In the process, they raise the question of what it means for a behavior to be "novel," and, complementing Zentall's survey of varieties of non-imitative social learning, discuss putative types of imitative learning, including emulation (Tomasello, this volume) and "program-level imitation" (Byrne, 1994). It is argued that neither of these should be understood to be discrete categories of phenomena distinct from "true" imitation, because, inter alia, the three cannot be distinguished empirically from one another.

Tomasello's chapter is a substantial review and critique of research on tool use and gestural imitation in wild, captive and home raised or "enculturated" nonhuman apes, and includes comprehensive coverage of experiments on imitation in this group.

The review is prefaced with a lucid statement of how Tomasello's "cognitive" approach to the study of imitation, influenced by Kohler, Piaget, and Vygotsky, contrasts with the behavioral or associative tradition. This cognitivism allows Tomasello to accept that imitation can be distinguished from other forms of social learning as that which involves learning through observation about behavior, rather than learning through observation about the environment (Heyes, 1993). However, Tomasello's cognitive approach also leads him to stress that not all observational learning about behavior is imitation, and not all observational learning about the environment is stimulus or local enhancement. Rather, imitative learning should be distinguished from mere "mimicry," with only the former involving some understanding of the demonstrator's goals or purposes, and stimulus enhancement should be distinguished from "emulation," in which the observer learns about the affordances, or dynamic properties, of environmental objects, not merely to attend to their static features.

With these distinctions, between imitation, mimicry, emulation, and stimulus enhancement, in hand, Tomasello makes the case through his empirical review that imitation occurs only in home-raised apes, and that it is a product of the influence of human culture on these animals. Thus, his conclusion is that nonhuman apes do not, spontaneously, ape.

Whiten and Custance's data are apparently at odds with Tomasello's conclusion because the chimpanzees they tested were captive but not home raised. Their exchange on this issue, in which Tomasello suggests that the artificial-fruit experiment provides evidence of emulation rather than imitation, and Whiten and Custance resist the distinction between emulation and imitation, provides an excellent demonstration of recent progress in research on imitation. That progress has facilitated direct, detailed confrontation between theory and data.

IMITATION IN HUMAN INFANTS

In the fourth and final chapter devoted to primates, Andrew Meltzoff reviews the rapid progress that has been made in the last 20 years of research on imitation in human infants. The majority of this research has been informed and inspired by Meltzoff and Moore's (1977) demonstration, using a "cross-target" procedure, that prelinguistic infants can imitate a variety of actions, including facial gestures, that cannot be perceived in the same sensory modality when executed by the self (observer), and another (a demonstrator).

The logic of Meltzoff's cross-target procedure is identical to that of the two-action method first used by Dawson and Foss (1965) to test for imitation in nonhuman animals. In experiments with human infants, this logic is typically applied in a within subjects or repeated measures design, rather than an independent groups design. However, the terms "cross target" and "two-action" apparently represent independent invention, or convergent cultural evolution, of the same, powerful method of testing for imitation in subjects without language.

While the two-action or crosstarget method is just beginning to be used consistently to find out whether various nonhuman animals can imitate, use of this method in developmental research over two decades has revealed much about the conditions in which infants imitate and the psychological processes involved. Meltzoff reviews studies showing that infants imitate vocalizations and object-related acts as well as facial gestures, that they can imitate after a delay of as long as several weeks, even when they did not have the opportunity to imitate during observation, and that infants will imitate peers as well as adults. Furthermore, in recent, elegantly controlled experiments, Meltzoff and his colleagues have shown that in the process of imitation infants "perfect" actions that were unsuccessfully executed by adult models, and that infants know when they are themselves being imitated by adults.

These findings lead Meltzoff to suggest that, at least in humans, imitation plays a major role in both the acquisition of theory of mind and enculturation. In contrast with Tomasello (e.g., this volume), who takes enculturation and the capacity to attribute intention as prerequisites for imitation, Meltzoff argues that imitation provides the foundation of inferences from first person to third person experience, and contributes to enculturation directly by effecting social transmission of a range of behaviors, and indirectly via its role in the acquisition of phonetic and prosodic features of language.

By surveying what has been achieved using a crosstarget or two-action procedure with infants, Meltzoff's discussion both encourages use of the same procedure to investigate imitation in nonhuman animals, and acts as a reminder that comparative research has not yet identified a nonhuman animal that is, like an infant, an "imitative generalist," capable of imitating a broad range of actions without extrinsic motivation. In emphasizing generality and intrinsic motivation, Meltzoff's characterization of imitation is like that of Moore (this volume). However, unlike Moore and Zentall (this volume), and in common with Heyes, Whiten and Custance (this volume), Meltzoff does not regard the reproduction of *novel* behavior as, in any straightforward way, a defining property of imitation.

RATS AND REALISM

In the final chapter, I summarize the results of bidirectional control experiments on imitation in rats, and use them to make some general points about the way in which imitation has been defined in this book and elsewhere.

The bidirectional control procedure is like other two-action tests in that it allows subjects to observe one of two actions, A and B, on a single object, and then compares the proportions of A and B responses made by observers of A and B, respectively. However, the bidirectional control procedure has a distinctive history [it was derived from Grindley's (1932) method of demonstrating instrumental learning], and, unlike other two-action methods, it uses a perspective manipulation to test for imitation. Subjects witness lateral displacement of a joystick from one position relative to the manipulandum and surrounding chamber, and they are tested, allowed to displace the joystick themselves, in another position. Consequently, although a simple dimension of behavior, direction, is recorded in the bidirectional control procedure, like other two-action tests, it apparently allows imitation to be distinguished from independent production of similar behavior, stimulus enhancement, observational conditioning, and a range of other types of nonimitative social learning.

In Chapter 17, I measure the putative, bidirectional-control evidence of imitation in rats against the standards set by various definitions of imitation, including the definition by exclusion discussed by Zentall (this volume), Thorndike's characterization of imitation as learning to do an act from seeing it done, Thorpe's novelty and "no instinctive tendency" criteria of imitation, and definitions that link imitation with ideation, self-consciousness, intentionality, and cultural transmission. The behavior of observer rats in bidirectional control experiments seems to constitute imitation when the latter is defined by exclusion, to involve ideation, and to meet Thorndike's definition of imitation. However, it is not clear whether the rats' behavior meets Thorpe's criteria for imitation, and it does not imply self-consciousness, intentionality, or cultural transmission. More important, I argue that the bidirectional control example illustrates some more general problems: 1. The various definitions of imitation circumscribe different sets of phenomena, and 2. both Thorpe's criteria and definitions that link imitation with self-consciousness and/or intentionality make it impossible to identify examples of imitation with any confidence.

In contrast with Tomasello (this volume), who upholds that intentionality is characteristic of imitation, and with Zentall and Moore (this volume), who seek to overcome problems of definition by interpreting Thorpe's criteria and avoiding reference to psychological processes, I favor a "realist" approach which adheres to Thorndike's operational definition of imitation while acknowledging that the phenomena it circumscribes are of interest precisely because they are likely to be mediated by some complex, as-yet-underspecified psychological processes. This realist approach is consistent with that of Meltzoff (this volume).

IS IMITATION RARE OR ELUSIVE?

"Can animals learn by imitation?" is a deceptively simple question. It has proved remarkably difficult for researchers to agree on a definition of imitation and to decide which nonhuman animals, if any, can imitate motor, or nonvocal, behavior.

The fact that nearly 100 years of research on imitation in animals has produced only a few convincing examples, could indicate that imitation is rare among nonhuman animals (Galef, 1988; Chapter 1, this volume). However, it is at least equally likely that imitation in animals is elusive, rather than rare. A broad range of species may be capable of imitation, and these animals may even imitate frequently under free-living conditions, while we have been prevented from identifying imitation with confidence by conceptual confusion and a lack of effective experimental methods.

If imitation is elusive rather than rare, then the contributions to this book encourage the expectation that many, reliable examples of imitation in animals will become known in the next few years. Conceptual issues are now being recognized and discussed, rather than written off as merely terminological problems, and it is now broadly agreed that two-action/crosstarget/bidirectional control procedures provide effective methods of distinguishing imitation from its many pretenders.

REFERENCES

Body, R., & Richerson, P. (1985). *Culture and the Evolutionary Process.* Chicago: University of Chicago Press.

Byrne, R. W. (1994). The evolution of intelligence. In P. J. B. Slater & T. R. Halliday (Eds.), *Behaviour and Evolution* Cambridge: Cambridge University Press.

Dawson, B. V., & Foss, B. M. (1965). Observational learning in budgerigars. *Animal Behaviour, 13,* 470–474.

Galef, B. G. (1988). Imitation in animals: History, definition and interpretation of data from the psychological laboratory. In T. R. Zentall & B. G. Galef (Eds.), *Social Learning: Psychological and Biological perspectives.* (pp. 3–28). Hillsdale, NJ: Erlbaum.

Galef, B. G., Manzig, L. A., & Field, R. M. (1986). Imitation learning in budgerigars: Dawson and Foss (1965) revisited. *Behavioral Processes, 13,* 191–202.

Grindley, G. C. (1932). The formation of a simple habit in guinea pigs. *British Journal of Psychology, 23,* 127–147.

Hayes, K. J., & Hayes, C. (1952). Imitation in the home-raised chimpanzee. *Journal of Comparative and Physiological Psychology, 45,* 450–459.

Heyes, C. M. (1993). Imitation, culture and cognition. *Animal Behavior, 46,* 999–1010.

Heyes, C. M., & Dawson, G. R. (1990). A demonstration of observational learning using a bidirectional control. *Quarterly Journal of Experimental Psychology, 42B,* 59–71.

Meltzoff, A. N., & Moore, M. K. (1977). Imitation of facial and manual gestures by human neonates. *Science, 198,* 75–78.

Piaget, J. (1962). *Play, dreams and imitation in childhood.* New York: Horton.

Thorndike, E. L. (1898). Animal intelligence. *Psychological Review Monographs* (Vol. 8) **2.**

Washburn, M. F. (1908). *The Animal Mind.* New York: Macmillan.

Will, B., Pallaud, B., Soczka, M., & Manikowski, S. (1974). Imitation of lever pressing "strategies" during the operant conditioning of albino rats. *Animal Behaviour,* **22,** 664–671.

Zentall, T. R., Sutton, J., & Sherburne, L. M. (1995). Imitation of treadle stepping and pecking by pigeons using the two-action method. Paper presented at the conference on Comparative Cognition, Melbourne, FL.

11

An Analysis of Imitative Learning in Animals

Thomas R. Zentall

Department of Psychology
University of Kentucky
Lexington, Kentucky 40506

The term imitation has been used to identify phenomena ranging from morphological similarity, which appears to be under the total control of natural selection (Batesian mimicry, e.g., the palatable viceroy butterfly mimicking the unpalatable monarch butterfly, see Turner, 1984), to complex symbolic modeling requiring intention or purpose in which the imitator exaggerates the actions of the demonstrator for purposes of humor (e.g., caricature, parody, or satire; see Mitchell, 1987).

For some psychologists, imitation is a process by which an observer learns vicariously through the observation of a model's behavior and the consequences of that behavior (e.g., Bandura, 1969). For others, imitation involves the capacity for taking the perspective of another (Piaget, 1945). Perspective taking can be described by the proposition, "If I were in the place of that individual and I made that same response, then I might experience the same consequence." Such a capacity for perspective taking is similar to what Krebs and Dawkins (1984) have referred to as "mind reading" (see also Whiten, 1991), and what Leslie (1987) has referred to as a natural theory of mind. Thus, imitation (sometimes referred to as true or inferential imitation, social learning, copying, or observational learning) has been characterized as learning that is mediated by cognitive processes.

Other psychologists have tried to integrate imitation within a broadened context of conditioning (see e.g., Gewirtz, 1969; Miller & Dollard, 1941). Gewirtz, for example, has argued that imitation is not inherently different from trial and error learning (instrumental conditioning). When a demonstrator (e.g., a parent) engages in a particular behavior, that behavior is accompanied or followed by a wide variety of responses by the observer (e.g., the child). If by chance, there is some correspondence between the two behaviors (e.g., the word "daddy" spoken by the parent is followed by "dada" spoken by the child), the observer's behavior will often be reinforced (perhaps socially, through the parent's excitement and attention). The word "daddy" then comes to serve as a conditioned stimulus that signals the opportunity to obtain reinforcement for emitting the response "dada." Thus, conditioning theory can explain individual cases of response copying, especially when verbal behavior is involved. But surely every imitated word does not go through such a process of reinforcement by successive approximation (i.e., trial and error shaping).

To account for the extensive use of imitative learning by children, Gewirtz (1969) has proposed that copied responding that occurred initially through selective reinforcement, comes to generalize to other behavior, without the need for additional reinforcement. If generalization, as it is used here, is meant to be explanatory, rather than merely descriptive, however, it requires more than a simple learning mechanism. Stimulus generalization theory (Spence, 1937) is based on the mechanism of physical stimulus similarity, in which reinforced responding in the presence of a particular stimulus will generalize to other stimuli, to the extent that those other stimuli are physically similar to the training stimulus. But how does an infant get from copying "dada" to copying "ball," or more importantly, to gestural copying (e.g., hiding one's eyes).

To account for such generalization, a child must have a concept of identity (i.e., things that are the same, see Zentall, Edwards, & Hogan, 1983). Such an analysis may account for the generalization of stimulus matching with verbal behavior for which comparisons between one's own behavior and that of others is relatively easy because one can hear one's own utterances with relative fidelity (see also, the acquisition of birdsong; Hinde, 1969; Marler, 1970; Nottebohm, 1970; Thorpe, 1961; and vocal mimicry; e.g., Pepperberg, 1986; Thorpe, 1967). This analysis may also apply to behavior that produces a clear change in the environment (e.g., in humans, turning up the volume on a radio—when the knob moves to the right, the volume increases). Such a stimulus (or response) matching approach is less applicable, however, to the example of an adult model who says to a young child, "do this," as the model places his hands over his eyes. In this case, imitation requires a comparison of one's own behavior with that of others. But how does the child "know how it looks" to others when it is matching the behavior of the model? For such a match to be made, there must be an existing link between what a response feels like (the proprioceptive stimulus) and what it looks like to others (the visual stimulus; see Mitchell, 1992, 1993).

In the case of children brought up in an industrial culture, one could argue that extensive exposure to mirrors has allowed them to experience the correlation between proprioceptive cues and visual cues seen by others (i.e., as seen by themselves in the mirror). There has been much discussion of the meaning of self-recognition through mirror exposure (Epstein, Lanza, & Skinner, 1981; Gallup, 1970; Heyes, 1994) but little research has dealt with the role of self-recognition in imitative learning.

The extent of imitative behavior across cultures with little exposure to mirrors, however, makes this mirror-exposure account unlikely. Similarly, imitative behavior appears to occur in very young children who have had little opportunity to learn about mirrors (Meltzoff, 1988). Thus, although the mechanism underlying imitation would appear to have an important cognitive basis, the universal appearance of imitation across cultures and developmental stages suggests that it does not involve the combination of such complex processes as perspective taking, together with the learned correspondence between how one's own behavior looks to others and the proprioceptive cues resulting from that behavior (but see Mitchell, 1992).

NONIMITATIVE SOCIAL LEARNING

An alternative approach to the understanding of imitative learning is to ask if it can be distinguished from other socially influenced changes in behavior. To address imitation in this way, by what it is not, may better define the critical characteristics of the behavior being studied.

Early experiments in imitative learning were sometimes cavalier in their choice of a comparison group. In some studies, for example, the rate of acquisition of a bar-press response by rats following observation of a trained bar-pressing rat has been compared with that of a group of rats that were shaped to bar press by the method of successive approximations (Corson, 1967; Jacoby & Dawson, 1969; Powell, 1968; Powell & Burns, 1970; Powell, Saunders, & Thompson, 1969). Such results are difficult to interpret, not only because the shaping procedure is typically not objectively defined, but also because comparison of the two training procedures does not allow for the isolation of any theoretically meaningful variable.

The plan in the present review is to identify some of these theory-relevant variables and either evaluate their role in imitation, or isolate them by elimination or control, from what has been called true imitation. This review was greatly facilitated a number of earlier, well-written analyses (Galef, 1988; Heyes, 1994; Whiten & Ham, 1992).

It should be noted at the outset, that most of the research cited here was conducted with nonprimates—mainly, rats and pigeons. This choice was made because research with nonprimates is relatively easier to analyze in terms of the possible mechanisms underlying reported socially enhanced behavior. Very similar arguments have been proposed, however, in the analysis of socially enhanced behavior in primates (Whiten & Ham, 1992).

Contagion

Contagion refers to a class of matching behavior that is limited to those unlearned responses that are typical of a species. Synchronized courtship, predator evasion (in flocking and herding animals), and coordinated consummatory behavior, for example, are behaviors more readily attributed to genetically mediated mechanisms rather than to learning through observation. Although contagion (also called mimesis; Armstrong, 1951) may play a role in some socially transmitted behavior, one should avoid selecting as a target of imitation, not only behavior that merely serves as a releaser for the same behavior in others (Thorpe, 1963), but also any already acquired response.

Social Facilitation

Because the term *social facilitation* has been used in so many different ways that for some it has lost all semblance of specificity, Galef (1988) prefers the term social enhancement. However, the relative consistency of the use of the term by psychologists, following the appearance of Zajonc's classic (1965) paper, as behavior that is influenced by the mere presence of a conspecific leads me to encourage its continued use.

The motivational mechanisms responsible for social facilitation effects are of interest in their own right, but they should be distinguished from imitative learning which occurs, presumably, through the observation of a target behavior. Although social facilitation cannot directly be separated from imitative learning, one can readily control for it by comparing the rate of task acquisition by a group exposed to the target behavior with that of a group exposed to the mere presence of another animal. The clear isolation of social facilitation has not always been captured in experimental designs, however. Gardner and Engel (1971), for example, exposed observers to demonstrators that were not bar pressing, but were eating from the food dispenser during the observation period. The importance of this added factor is that a significant component of acquisition of the bar-press response involves locating the food dispenser, learning to eat from it, and learning to associate the sound of dispenser operation with the availability of food. Thus, this social facilitation control group was actually exposed to an important component of the to-be-learned task.

The importance of learning to eat from the food dispenser prior to bar-press acquisition in an imitation experiment may also account for faster acquisition by rats that observed a bar-pressing demonstrator and were fed each time the demonstrator was fed, as compared with rats that observed the demonstrator but were not fed (Del Russo, 1971).

When social facilitation has been properly isolated in rats, clear observational learning effects have been found in the rate of response acquisition and extinction (Henning & Zentall, 1981; Sanavio & Savardi, 1979; Zentall & Hogan, 1976; Zentall & Levine, 1972). And contrary to the conclusion of Gardner and Engel (1971), when acquisition of bar pressing by a social facilitation group is compared with that of a group not exposed to a conspecific (i.e., a trial and error control), *retardation* of acquisition has been found (Bankhart, Bankhart, & Burkett, 1974; Zentall & Levine, 1972). On the other hand, as predicted by Zajonc's (1965) theory, the facilitating effects of a conspecific's mere presence on rats' bar pressing can be demonstrated once the target behavior has been well learned (Levine & Zentall, 1974).

Although inclusion of a "mere presence" group may allow control for certain motivational effects on the acquisition of a response, it is possible that the presence of an *active* model, working for reinforcement, provides an additional source of motivation, independently of its demonstration of the to-be-learned response. In other words, observation of the performance of an *irrelevant* response may provide a more complete control for the motivational contribution that the demonstrator might make to the observer's behavior. To my knowledge, this motivational effect has not been addressed in the literature.

Socially Mediated Aversive Conditioning

A special case of social facilitation may be produced when an observer watches a conspecific acquire a response that is reinforced by the absence of an aversive event (e.g., shock avoidance). In this case, a merely present conspecific (social facilitation) control group may not control for the specific motivation associated with observing a conspecific being shocked (i.e., acquired fear). Because fear can be easily conditioned, not only to contextual cues, but also to task-specific cues (see e.g., Curio,

1988), one needs to be particularly careful in interpreting the results of experiments in which imitative learning is studied in contexts involving aversive stimulation.

John, Chesler, Bartlett, and Victor (1968), for example, found that cats trained to make a hurdle jump to a buzzer to avoid foot shock learned slower than cats similarly treated but first permitted to observe another cat learning to avoid foot shock. It may be that being in the presence of a shocked cat is sufficient to sensitize the observer (or increase its fear motivation) and thus facilitate hurdle jumping.

In another study, although rats acquired a discriminative shuttle avoidance response faster following observation of a trained demonstrator than following observation of merely present demonstrator, rats that were simply exposed to the empty shuttle box acquired the response fastest (Sanavio & Savardi, 1979). Thus, it is possible that under certain conditions, observation of a trained demonstrator can actually interfere with the acquisition of an aversively motivated response because the presence of a conspecific can result in a reduction in the fear motivation needed for task acquisition (see Davitz & Mason, 1955; Morrison & Hill, 1967).

One way to avoid problems associated with differential motivational cues encountered with observation of aversively motivated conditioning is to include an observation control group exposed to performing demonstrators but with the observer's view of a critical component of the demonstrator's response blocked. Such a control was included in an experiment by Bunch and Zentall (1980) who used a candle-flame-avoidance task originally developed by Lore, Blanc, and Suedfeld (1971). Bunch and Zentall found that rats learned the candle-flame-avoidance task faster after having seen a demonstrator acquire the task, as compared with (1) a group for which a small barrier was placed in front of the candle such that the observer's view of the rat's contact with the candle was blocked and (2) a social facilitation control group. Thus, although a variety of auditory cues (a potential by-product of the demonstrator's pain), olfactory cues (e.g., potentially produced by singed whiskers, defecation, and urination), and visual cues (e.g., seeing the demonstrator approach and then rapidly withdraw from something directly behind the barrier) associated with the task should have provided comparable motivational cues to these control observers, task acquisition was not facilitated as much as for observers that could also observe the demonstrator's contact with the candle.

Alternatively, one can rule out cues provided by a demonstrator's reaction to an aversive event by using only well-trained demonstrators that do not encounter the aversive event (Del Russo, 1975). Unfortunately, it is possible that in anticipation of the potential aversive event, the demonstrator still generates fear-produced cues that can have general motivational properties for an observer.

One approach that does control for motivational cues provided by a demonstrator exposed to (or in anticipation of) an aversive stimulus has been used by Kohn and Dennis (1972; see also, Kohn, 1976). In this research, rats that observed a demonstrator performing a relevant shock-avoidance discrimination acquired that task faster than controls for which the demonstrator's discrimination was the reverse of the observer's (i.e., what was correct for the demonstrator was incorrect for the observer and vice versa).

Although it may be difficult to control for the social transmission of motivation produced during the acquisition or performance of an avoidance response, it may be that social learning is more likely to occur under conditions of fear motivation because of its greater evolutionary significance.

Local Enhancement

Local enhancement refers to the facilitation of learning that results from drawing attention to a locale or place associated with reinforcement (Roberts, 1941). For example, Lorenz (1935) noted that ducks enclosed in a pen may not react to a hole large enough for them to escape, unless they happen to be near another duck as it is escaping from the pen. The sight of a duck passing through the hole in the pen may simply draw attention to the hole.

Similarly, an *umweg* (barrier circumvention) problem was acquired faster by chicks for which the appropriate response was demonstrated by either a conspecific or a red tennis ball, than by chicks exposed to neither (Reese, 1975). In both cases of demonstration, attention may have been drawn to the path that led around the barrier, and the local enhancement was sufficient to facilitate acquisition (see also, Neuringer & Neuringer, 1974).

Local enhancement has also been implicated in the finding that puncturing the top of milk bottles by great tits spread in a systematic way from one neighborhood to another (Fisher & Hinde, 1949). Although the technique of pecking through the top of the bottle may be learned through observation, it is also likely that attention was drawn to the bottles by the presence of the feeding birds. Once at the bottles, the observers found the reward and consumed it. Learning to identify milk bottles as a source of food, can readily generalize to other open bottles. Finally, drinking from opened bottles can readily generalize to an attempt to drink from a sealed bottle, which in turn can lead to trial-and-error puncturing of the top.

Local enhancement may be studied in its own right. As Denny, Clos, and Bell (1988) have shown, observation by rats of merely the movement and sound of a bar being activated (by the experimenter from outside the chamber) and paired with food presentation can facilitate the acquisition of the bar-press response, relative to various control procedures.

Local enhancement may also account for John et al.'s (1968) finding of socially facilitated acquisition of lever pressing by cats. Cats placed in the same chamber as another cat lever pressing for food, learned to press that lever faster than other cats that observed another cat that was fed periodically without lever pressing. But observation of lever pressing may simply draw attention to the lever. Local enhancement is especially likely in this context, in which observation of the moving lever might encourage lever approach upon removal of the demonstrator.

Similarly, local enhancement may play a role in the faster acquisition of lever pressing by kittens that observed their mothers as demonstrators, than by kittens that observed a strange female demonstrator (Chesler, 1969), because orientation toward the mother may be more likely than toward a strange cat.

Local enhancement may also be involved in John et al.'s (1968) finding of facilitated acquisition of an aversively motivated hurdle-jump response. The distinction between true imitation and local enhancement may be a subtle one in this case, but observation of the demonstrator simply may draw the observer's attention to the top of the hurdle. In other words, seeing a ball bounce over the hurdle, or even placing a flashing light at the top of the hurdle might be enough to facilitate the hurdle-jumping response.

A combination of local enhancement and social facilitation may also account for Groesbeck and Duerfeldt's (1971) finding of facilitated acquisition of a visual discrimination after rats observed a demonstrator performing that discrimination on an elevated Y maze (knocking down the correct cue card to obtain a water reinforcer). Observers acquired the same discrimination faster than other observers that (a) watched the experimenter knock down the correct cue card and tap on the water bottle, or (b) watched a demonstrator drink from the water bottle. Neither the presence of a demonstrator (social facilitation) nor drawing attention to the correct cue card (local enhancement) appeared to facilitate acquisition, but the two in combination may have been sufficient, without the need for vicarious learning about the relations among the identity of the correct cue card, the response of knocking it down, and reinforcement.

In general, whenever the performance observed involves an object (i.e., a manipulandum) to which the observer must later respond, local enhancement may play a role (Corson, 1967; Denny, Clos, & Bell, 1988; Herbert & Harsh, 1944; Jacoby & Dawson, 1969; Oldfield-Box, 1970).

In other cases, it may be possible to control for local enhancement effects by including proper controls. Lefebvre and Palameta (1988), for example, found that pigeons that observed a model pierce the paper cover on a food well to obtain hidden grain, later acquired that response on their own, whereas those that observed that same response, but with no grain in the well (the model performed in extinction), failed to acquire the response.

Stimulus Enhancement

In the case of local enhancement, the attention of an observer is drawn to a particular *place* by the activity of the demonstrator. The term stimulus enhancement is used when the activity of the demonstrator draws the attention of the observer to a particular *object* (e.g., the manipulandum). Quite often in the study of imitative learning, the object in question is at a fixed location so the two mechanisms are indistinguishable. In the duplicate-chamber procedure (see Warden & Jackson, 1935; Gardner & Engel, 1971), however, a manipulandum (e.g., a lever) is present in both the demonstration chamber and in the observation chamber. Under these conditions, drawing attention to the demonstrator's lever should not facilitate acquisition of lever pressing by the observer. In fact, one could argue that it should retard acquisition of lever pressing by an observer because it should draw the observer's attention away from its own lever. In the case of stimulus enhancement, however, the similarity between the demonstrator's lever and that of the observer may make it more likely that the observer notices its own lever after having its attention drawn to the demonstrator's lever. Thus, stimulus enhancement refers to the combination of a perceptual, attention-getting process resulting from the activity of the demonstrator in the presence of the lever, and stimulus generalization between the demonstrator's and observer's levers. Because it subsumes local enhancement, the term stimulus enhancement may be a preferable term (Galef, 1988).

Stimulus enhancement may also be involved in the facilitated acquisition of an observed discrimination. If the demonstrator is required to make contact with the positive stimulus, that stimulus is likely to attract the observers attention and to facilitate responding to it (Edwards, Hogan, & Zentall, 1980; Kohn, 1976; Kohn & Dennis, 1972; Fiorito & Scotto, 1992; Vanayan, Robertson, & Biederman, 1985).

Observational Conditioning

The observation of a performing demonstrator may not merely draw attention to the object being manipulated (e.g., the lever), but because the observer's orientation to the manipulandum is often followed immediately by presentation of food to the demonstrator, a Pavlovian association may be established. This form of conditioning has been called observational conditioning (Whiten & Ham, 1992), valence transforma-tion (Hogan, 1988), or emulation (not to be confused with goal emulation, which will be discussed later, Tomasello, 1990). Although such conditioning would have to take the form of higher-order conditioning (because the observer would not actually experience the unconditional stimulus), there is evidence that such higher-order conditioning can occur, in the absence of a demonstrator, following the simple pairing of keylight with inaccessible grain in an autoshaping paradigm with pigeons (Zentall & Hogan, 1975). The presence of a demonstrator drawing additional attention to the manipulandum (by pecking) and to the reinforcer (by eating) may further enhance associative processes in the absence of true imitation.

With regard to the nature of the conditioning process, it is of interest that when reinforcement of the demonstrator's response cannot be observed (or the response–reinforcer association is difficult to make) acquisition may be impaired (Heyes, Jaldow, & Dawson, 1994; but see also Groesbeck & Duerfeldt, 1971). Fur-thermore, rats appear to acquire a bar-pressing response faster following observa-tion of a bar-pressing demonstrator if they are fed at the same time as the perform-ing demonstrator (Del Russo, 1971). One interpretation of this finding is that feeding the observer following the demonstrator's response may result in simple rather than higher-order conditioning (i.e., attention to the bar followed by rein-forcement).

Observational conditioning may also play a role in an experiment in which observation of experienced demonstrators facilitated the opening of hickory nuts by red squirrels, relative to trial-and-error learning (Weigle & Hanson, 1980). Differ-ential local enhancement can be ruled out, in this case, because animals in both groups quickly approached and handled the nuts, and the observers actually han-dled the nuts less than controls (because observers were more efficient at opening them). However, observers alone got to see the open nuts and could associate open nuts with eating by the demonstrator.

Socially transmitted food preferences (e.g., Galef, 1988; Strupp & Levitsky, 1984) represent a special case of observational conditioning. Although food prefer-ence may appear to fall into the category of unlearned behavior subject to elicitation through contagion, consuming a food with a *novel* taste should be thought of as an acquired behavior. It would appear, however, that the mechanisms responsible for socially acquired food preferences have strong, simple associative learning compo-nents (e.g., learned safety or the habituation of neophobia to the novel taste), for which the presence of a conspecific may serve as a catalyst. Furthermore, these specialized mechanisms may be unique to foraging and feeding systems.

One of the best examples of observational conditioning is in the acquisition of fear of snakes by laboratory-reared monkeys exposed to a wild-born conspecific in the presence of a snake (Mineka & Cook, 1988). Presumably, the fearful conspecific serves as the unconditioned stimulus, and the snake serves as the conditioned stimulus. It appears that exposure to a fearful conspecific or to a snake alone is insufficient to produce fear of snakes in the observer. For an excellent discussion of the various forms of observational conditioning see Heyes (1994).

Following or Matched-Dependent Behavior

There is considerable evidence that rats can learn to follow a trained conspecific to food in a T maze in the absence of any other discriminative stimulus (Bayroff & Lard, 1944; Church, 1957; Haruki & Tsuzuki, 1967). Although the leader rat in these experiments is clearly a social stimulus, the data are more parsimoniously interpreted in terms of simple discriminative learning. If, for example, the demonstrator were replaced with a block of wood pulled along by a string, or even an arrow at the choice point directing the rat to turn left or right, it is clear that one would identify the cue (i.e., the demonstrator, the block of wood, and the arrow) as a simple discriminative stimulus. Even if following a demonstrator led to faster learning than following a passive signal, it might merely indicate that the social cue was more salient than either a static or even a moving nonliving cue (see Stimbert, 1970). For matched-dependent behavior to be analogous to imitation, the untrained animal would have to follow the demonstrator on the first trial. Even then, however, the motivation to affiliate could account for following behavior.

TRUE IMITATION

True imitation has been defined as "the copying of a novel or otherwise improbable act or utterance, or some act for which there is clearly no instinctive tendency" (Thorpe, 1963). We may now be in a better position to be more specific. First, for true imitation to be demonstrated, the target behavior should not already be part of the observing animal's repertoire—whether improbable or not (Clayton, 1978). Second, one should control for motivational effects on the observer, produced either by the mere presence of the demonstrator or by the mere consequences of the behavior of the demonstrator. And finally, one should control for the possibility that the demonstrator's manipulation of an object merely draws the observer's attention to that object (or one like it), thus making the observer's manipulation of the object more probable.

Imitation of an Alternative Response: The Two-Action Method

One way to deal with the problem of nonimitative learning is to compare the performance of one imitation group with another. Dawson and Foss (1965), found that budgerigars acquired a lid-removal task (by trial and error) in one of three different ways: Pushing the lid off with the beak, twisting it off with the beak, or grasping it off with the foot. Observers were then exposed to these performing birds, and when they were then given the opportunity to perform themselves, each observer removed the lid in same manner as its demonstrator (see also Galef, Manzig, & Field, 1986).

In a variation on this procedure, Zentall, Sutton, and Sherburne (in press) trained pigeons to either step on a treadle or peck a treadle for reward. Observers showed a significant tendency to respond to the treadle with the same part of the body as their respective demonstrator. Similar results have been obtained with Japanese quail (Akins & Zentall, in press).

Will, Pallaud, Soczka, and Manikowski (1974) noted a related effect in a study in which rats observed either a trained demonstrator performing a successive discrimination or an experimentally naive demonstrator. They found that the trained demonstrators typically responded with one of three distinctive patterns when the discriminative stimulus was available, and that the observers learned not only to respond in the presence of one stimulus and not in the presence of the other, but they also learned the pattern of responding of their demonstrator (e.g., alternating a bar press with eating, or making a burst of bar presses followed by eating the accumulated pellets).

Heyes and Dawson (1990) have recently reported similar results by rats that observed demonstrators expressly trained to respond in one of two different ways. After observing demonstrators push an overhead bar either to the left or to the right, Heyes and Dawson found that observers given access to the bar tended to push the bar in the same direction as their demonstrator. Remarkably, the observers matched the demonstrators' behavior in spite of the fact that, because the observers faced the demonstrators during the period of observation, the direction of bar motion (relative to the observer's body) during observation was opposite that of the bar's motion when the observers performed.

Copying

Heyes (1994) has made a distinction between true imitation and copying. According to Heyes, in the case of true imitation, response matching results from observation of "a positive relationship between a demonstrator's response and appetitive reinforcement," whereas copying only requires observation of the demonstrator's response. Vocal imitation (e.g., birdsong) is often given as an example of copying. The importance of reinforcement in obtaining response matching (see Heyes, Jaldow, & Dawson, 1994) suggests that observational conditioning may be implicated. After all, in a Pavlovian conditioning procedure, it would certainly be the case that omission of the unconditioned stimulus (i.e., the reinforcement) would impair acquisition. Thus, it may be that observational conditioning plays an important role in imitation. But the conditioning is not simply the association between movement of the manipulandum and reward. Rather, it is likely to be the association between the demonstrator's *behavior* and its consequence.

Immediate versus Delayed Imitation

One distinction, claimed to be of some importance (Bandura, 1969; Piaget, 1945), is whether the observer has the opportunity to perform the target response at the time of observation (generally the case in the duplicate-chamber arrangement) or the observer must wait to be tested in the absence of the demonstrator (always the case when there is a single chamber, or when there is an observation chamber and a second chamber in which performance is assessed). According to this view, immediate imitation is a less advanced or a more automatic process than delayed imitation (sometimes called observational learning, Bandura, 1969; or deferred imitation, Piaget, 1945). Delayed imitation, according to Bandura, is a cognitive process that requires an understanding of the relation between response and outcome (i.e., what Bandura calls the integration of the observed behavior).

At a less cognitive level, one can imagine two important differences between immediate and delayed imitation. Most obviously, delayed imitation requires the animal to hold in memory the behavior it has observed. Thus, in delayed imitation there is more opportunity for forgetting. On the other hand, in the case of delayed imitation, the opportunity to perform occurs in the absence of the demonstrator, and the social facilitation literature suggests that the presence of a conspecific may actually retard the acquisition of a new response (Zajonc, 1965). Thus, according to this noncognitive view, it is not obvious what the effect on task acquisition should be of delaying the opportunity to perform following observation of a demonstrator.

Huang, Koski, and DeQuardo (1983) compared bar-press acquisition in rats in imitation and social facilitation groups under immediate and delayed conditions. Interestingly, although in both conditions they found significant imitation effects relative to a trial-and-error control, relative to the social facilitation group, facilitated acquisition was found only in the delayed condition.

Whether the mere presence of a conspecific retards the *learning* of a new response or just its *performance* is not clear. If it affects learning, then testing in the absence of the demonstrator should make little difference. If it affects performance, however, removing the demonstrator prior to test may actually facilitate performance of the new response.

Proficiency of the Demonstrator

One might also expect proficiency of the model to affect the rate of acquisition of the behavior by observing pigeons (Vanayan, Robertson, & Biederman, 1985). Contrary to expectation, however, Vanayan et al. found faster acquisition of a successive discrimination by observers when less proficient models were observed. It may be that observation of the consequences of incorrect (nonreinforced) responding is as important (or, in the case of aversively motivated learning perhaps, even more important) than observation of the consequences of correct (reinforced) responding. As mentioned earlier, however, observation of a discrimination being performed may result in stimulus enhancement and the demonstrator-proficiency effects found may result from differential observational conditioning.

Generalized Imitation

Imitation of a particular response can be thought of as one example of a broad class of imitative behavior. One can then ask if an animal can learn to match any behavior of another "on cue" (i.e., can an animal learn the general concept of imitation and then apply it when asked to in a "do-as-I-do" test). Hayes and Hayes (1952) found that a chimpanzee (Viki) learned to respond correctly to the command "Do this"! over a broad class of behavior. The establishment of such a concept not only verifies that chimpanzees can imitate, but it also demonstrates that they are capable of forming a generalized behavioral-matching concept.

Goal Emulation

Under certain conditions an observer may attempt to reproduce the results that the model's behavior achieves, rather than reproduce the behavior itself. Tomasello (1990) has used the term goal emulation to describe this phenomenon. The logic behind this "nonmatching" form of imitation is that an observer may understand that a particular observed behavior has certain consequences, but it may also recognize that the goal could be achieved by any one of a class of behaviors.

The possibility that true imitation could be present in the absence of a match between the behavior of the demonstrator and that of the observer raises problems for the assessment of imitative learning beyond those already mentioned. The procedure reported by Dawson and Foss (1965) provides a useful example. If a budgerigar observes another removing the lid of a food container with its foot but "decides" that it could accomplish the same result (perhaps more easily) with its beak, the observer's behavior would be scored as nonimitative. Thus, the possibility of goal emulation raises problems even when the two-action method is used. On the other hand, the potential ambiguity in interpretation of findings resulting from use of the two-action method would be a problem only if one *failed* to find evidence for matching behavior.

Furthermore, alternatives to the two-action method are even less appealing because they tend to err on the side of failing to rule out simpler mechanisms. Thus, the two-action method remains a useful, albeit relatively conservative, test of true imitative behavior.

Intentionality

Interest in imitation research can be traced, at least in part, to the assumption that true imitation involves some degree of intentionality. This is certainly the case in many of the higher-order forms of imitation, such as the human dancer who repeats the movements of the teacher. Unfortunately, intentionality, because of its indirect nature, can only be inferred, and evidence for it appears most often in the form of anecdote rather than experiment. Ball (1938), for example, noted the case of a young rhesus monkey that while kept with a kitten was observed to lap its water in the same way as a cat. Ball noted further that lapping is extremely rare in rhesus monkeys.

Similarly, Mitchell (1987), in an analysis of various levels of imitation, provides a number of examples of imitation at these higher levels. For example, he discusses the young female rhesus monkey who seeing her mother carrying a sibling, walks around carrying a coconut shell at a same location on her own body (Breuggeman, 1973).

Such anecdotes, by their very nature, are selected and difficult to verify. If there were some way to bring these examples of intentional imitation under experimental control, it would greatly increase their credibility.

Analytic versus Synthetic Approaches to Social Learning

The experimental approach to the study of imitation in animals involves the analysis of factors that affect behavior occurring in a social context. A critical assumption of this approach is that in any context one can define and isolate the various factors that contribute to behavior change. This analytical approach involves assessment (by a process of subtraction) of the necessary and sufficient conditions for producing a particular behavior. For example, the effects of social facilitation can be separated from those of imitation by comparing the performance of an imitation group with that of a social facilitation control group. In principle, this is a powerful approach because it allows for "strong inference" about the role played by any factor that can be experimentally isolated.

The problem with this analytic approach, when applied to imitation, is that it focuses on the isolation of critical variables, often without sufficiently considering the conditions under which imitation is most likely to occur. Ironically, in so doing one may design an experiment that *conceptually* isolates true imitation, while establishing conditions that do not encourage its production. Thus, one may fail to find evidence for imitation (or fail to appreciate the richness of the capacity for imitative learning) for reasons unrelated to the capacity of the animal to imitate.

Related to this concern about experimental conditions that favor imitation, is the issue of whether, for a particular species, social learning has functional value. It is often assumed that learning from others should always be beneficial because it reduces the need to experience consequences that, under some conditions, may be life threatening. It is further assumed that some species (e.g., primates, with their large brains) have the capacity to imitate, whereas other species may not. Although the view of imitation as a higher-order form of learning (see Bandura, 1969) seems intuitive, Boyd and Richerson (1988) have argued that imitation can also be viewed as an evolutionary compromise between fixed action patterns, which are relatively inflexible but are also inexpensive (in terms of the cost of possibly-aversive consequences) and trial-and-error learning, which is relatively flexible but costly (see chapters by Galef, and Heyes, this volume).

In our attempt to understand imitation, we should view such behavior not only as a psychological phenomenon but also as a process that has evolved because of its biological utility. In this context, imitation can be viewed as one mechanism for behavioral change, the evolution of which depends more on the needs of the particular species than on its intellectual capacity.

To this point, research has been focused primarily on the *demonstration* of imitation in a small number of species as an indication of their cognitive capacities but now, a broader, more ecological perspective may be appropriate (see Davis, 1973). The importance of imitation may depend on the sociability or social structure of a species or on the particular class of behavior studied (e.g., foraging, predator avoidance, maternal behavior).

Without abandoning the analytic approach, we should consider including an alternative approach based on synthesis. An assumption underlying this approach is that whether a behavior is learned or species typical, appropriate environmental contexts may serve as releasers. Thus, the *relevance* of an environment may play an important role in determining whether a behavior (e.g., imitation) will occur. This more naturalistic approach focuses on the function of behavior rather than the underlying mechanisms. This approach defines a class of behavior that results in a common outcome rather than defining behavior in terms of its specific topography (see Howard & Keenan, 1993). The advantage of this approach is that it may help to identify the conditions under which imitative behavior evolved and thus, it may provide converging evidence as to its underlying mechanisms.

The main criticism of this approach is that it does not allow for proper *control* over the conditions under which imitation occurs. But these problems can be overcome. First, we should consider converging evidence from laboratory and field research. Second, we should develop theories to predict the conditions under which imitation will occur (see e.g., Boyd & Richerson, 1988). Third, we should identify which species are most likely to show imitative behavior, as a function of expected evolutionary significance of that behavior. And finally, we should test our theories by manipulating environmental conditions thought to affect imitation.

CONCLUSIONS

For a variety of reasons, animal imitation research has not always been well received in the field of animal learning and behavior. First, the broad range of phenomena to which it has been applied (including, e.g., Batesian mimicry) has given the impression that it may not be a useful psychological concept. Second, the large number of alternative accounts of learning through observation (e.g., contagion, social facilitation, stimulus enhancement), as well as the inconsistent use of terminology, makes isolation of true imitation from simpler processes appear impossible. Third, the assumption that true imitation involves intentionality, a phenomenon that can not be directly measured, suggests that it is an intractable concept.

The response to the first concern is to develop a classification system to separate purely morphological processes such as mimicry, and largely biologically programmed processes such as the acquisition of birdsong, from the more flexible and arbitrary forms of imitation more often studied by psychologists (see, e.g., Mitchell, 1987). The response to the second concern is to identify and use procedures, such as the two-action method, that can preclude explanation in terms of simpler perceptual and motivational mechanisms. With regard to the third concern, because it may not be possible to provide direct measures of intentionality, one should perhaps look for evidence for true imitation without regard for the underlying cognitive mechanisms.

ACKNOWLEDGMENTS

Preparation of this chapter was supported by a grant from the National Institute of Mental Health (MH 45979) and by grants from the National Science Foundation (BNS-9019080 and IBN-9414589). I thank Celia Heyes for her helpful comments on an earlier draft.

REFERENCES

Akins, C. K., & Zentall, T. R. (in press). Imitative learning in male Japanese quail (*Coturnix japonica*) using the two-action method. *Journal of Comparative Psychology*.

Armstrong, E. A. (1951). The nature and function of animal mimesis. *Bulletin of Animal Behaviour*, **9**, 46–48.

Ball, J. (1938). A case of apparent imitation in a monkey. *Journal of Genetic Psychology*, **52**, 439–442.

Bandura, A. (1969). Social learning theory of identificatory processes. In D. A. Goslin (Ed.), *Handbook of socialization theory and research* (pp. 213–262). Chicago: Rand-McNally.

Bankhart, P. C., Bankhart, B. M., & Burkett, M. (1974). Social factors in acquisition of bar pressing by rats. *Psychological Reports*, **34**, 1051–1054.

Bayroff, A. G., & Lard, K. E. (1941). Experimental social behavior of animals. III. Imitational learning of white rats. *Journal of Comparative and Physiological Psychology*, **37**, 165–171.

Boyd, R., & Richerson, P. J. (1988). An evolutionary model of social learning: The effect of spatial and temporal variation. In T. R. Zentall & B. G. Galef, Jr. (Eds.), *Social learning: Psychological and biological perspectives* (pp. 29–48). Hillsdale, NJ: Erlbaum.

Breuggeman, J. A. (1973). Parental care in a group of free-ranging rhesus monkeys. *Folia Primatologica*, **20**, 178–210.

Bunch, G. B., & Zentall, T. R. (1980). Imitation of a passive avoidance response in the rat. *Bulletin of the Psychonomic Society*, **15**, 73–75.

Chesler, P. (1969). Maternal influence in learning by observation in kittens. *Science,* **166,** 901–903.

Church, R. M. (1957). Two procedures for the establishment of imitative behavior. *Journal of Comparative and Physiological Psychology,* **50,** 315–318.

Clayton, D. A. (1978). Socially facilitated behavior. *Quarterly Review of Biology,* **53,** 373–391.

Corson, J. A. (1967). Observational learning of a lever pressing response. *Psychonomic Science,* **7,** 197–198.

Curio, E. (1988). Cultural transmission of enemy recognition by birds. In T. R. Zentall, & B. G. Galef, Jr. (Eds.), *Social learning: Psychological and biological perspectives* (pp. 75–97). Hillsdale, NJ: Erlbaum.

Davis, J. M. (1973). Imitation: A review and critique. In P. P. G. Bateson, & P. H. Klopfer (Eds.), *Perspectives in ethology* (Vol. 1, pp. 43–72). New York: Plenum.

Davitz, J. R., & Mason, D. J. (1955). Socially facilitated reduction of a fear response in rats. *Journal of Comparative and Physiological Psychology,* **48,** 149–151.

Dawson, B. V., & Foss, B. M. (1965). Observational learning in budgerigars. *Animal Behaviour,* **13,** 470–474.

Del Russo, J. E. (1971). Observational learning in hooded rats. *Psychonomic Science,* **24,** 37–45.

Del Russo, J. E. (1975). Observational learning of discriminative avoidance in hooded rats. *Animal Learning and Behavior,* **3,** 76–80.

Denny, M. R., Clos, C. F., & Bell, R. C. (1988). Learning in the rat of a choice response by observation of S-S contingencies. In T. R. Zentall, & B. G. Galef, Jr. (Eds.), *Social learning: Psychological and biological perspectives* (pp. 207–223). Hillsdale, NJ: Erlbaum.

Edwards, C. A., Hogan, D. E., & Zentall, T. R. (1980). Imitation of an appetitive discriminatory task by pigeons. *Bird Behaviour,* **2,** 87–91.

Epstein, R., Lanza, R. P., & Skinner, B. F. (1981). "Self-awareness" in the pigeon. *Science,* **212,** 695–696.

Fiorito, G., & Scotto, P. (1992). Observational learning in *Octopus vulgaris. Science,* **256,** 545–546.

Fisher, J., & Hinde, R. A. (1949). Further observations on the opening of milk bottles by birds. *British Birds,* **42,** 347–357.

Galef, B. J., Jr. (1988). Imitation in animals: History, definition, and interpretation of data from the psychological laboratory. In T. R. Zentall, & B. G. Galef, Jr. (Eds.), *Social learning: Psychological and biological perspectives* (pp. 3–28). Hillsdale, NJ: Erlbaum.

Galef, B. J., Jr., Manzig, L. A., & Field, R. M. (1986). Imitation learning in budgerigars: Dawson and Foss (1965) revisited. *Behavioral Processes,* **13,** 191–202.

Gallup, G. G., Jr. (1970). Chimpanzees: Self-recognition. *Science,* **167,** 86–87.

Gardner, E. L., & Engel, D. R. (1971). Imitational and social facilitatory aspects of observational learning in the laboratory rat. *Psychonomic Science,* **25,** 5–6.

Gewirtz, J. L. (1969). Mechanisms of social learning: Some roles of stimulation and behavior in early human development. In D. A. Goslin (Ed.), *Handbook of socialization theory and research* (pp. 57–211). Chicago: Rand-McNally.

Giraldeau, L.-A., & Lefebvre, L. (1987). Scrounging prevents cultural transmission of food-finding behavior in pigeons. *Animal Behaviour, 35,* 387–394.

Groesbeck, R. W., & Duerfeldt, P. H. (1971). Some relevant variables in observational learning of the rat. *Psychonomic Science, 22,* 41–43.

Haruki, Y., & Tsuzuki, T. (1967). Learning of imitation and learning through imitation in the white rat. *Annual of Animal Psychology, 17,* 57–63.

Hayes, K. J., & Hayes, C. (1952). Imitation in a home-reared chimpanzee. *Journal of Comparative and Physiological psychology, 45,* 450–459.

Henning, J. M., & Zentall, T. R. (1981). Imitation, social facilitation, and the effects of ACTH 4-10 on rats' barpress behavior. *American Journal of Psychology, 94,* 125–134.

Herbert, M. J., & Harsh, C. M. (1944). Observational learning by cats. *Journal of Comparative Psychology, 37,* 81–95.

Heyes, C. M. (1994). Reflections on self-recognition in primates. *Animal Behaviour, 47,* 909–919.

Heyes, C. M., & Dawson, G. R. (1990). A demonstration of observational learning in rats using a bidirectional control. *Quarterly Journal of Experimental Psychology, 42B,* 59–71.

Heyes, C. M., Jaldow, E., & Dawson, G. R. (1994). Imitation in rats: Conditions of occurrence in a bidirectional control procedure. *Learning and Motivation, 25,* 276–287.

Hinde, R. A. (Ed.). (1969). *Bird vocalizations.* Cambridge: Cambridge University Press.

Hogan, D. E. (1988). Learned imitation by pigeons. In T. R. Zentall, & B. G. Galef, Jr. (Eds.), *Social learning: Psychological and biological perspectives* (pp. 225–238). Hillsdale, NJ: Erlbaum.

Howard, M. L., & Keenan, M. (1993). Outline for a functional analysis of imitation in animals. *Psychological Record, 43,* 185–204.

Huang, I., Koski, C. A., & DeQuardo, J. R. (1983). Observational learning of a bar-press by rats. *The Journal of General Psychology, 108,* 103–111.

Jacoby, K. E., & Dawson, M. E. (1969). Observation and shaping learning: A comparison using Long-Evans rats. *Psychonomic Science, 16,* 257–258.

John, E. R., Chesler, P., Bartlett, F., & Victor, I. (1968). Observational learning in cats. *Science, 159,* 1489–1491.

Kohn, B. (1976). Observation and discrimination learning in the rat: Effects of stimulus substitution. *Learning and Motivation, 7,* 303–312.

Kohn, B., & Dennis, M. (1972). Observation and discrimination learning in the rat: Specific and nonspecific effects. *Journal of Comparative and Physiological Psychology, 78,* 292–296.

Krebs, J. R., & Dawkins, R. (1984). Animal signals: Mind reading and manipulation. In J. R. Krebs & N. B. Davies (Eds.), *Behavioural ecology: An evolutionary approach* (pp. 380–401). Oxford: Blackwell.

Lefebvre, L. (1994, August). *Ecological correlates of socially learned foraging in columbids.* Paper presented at the Social Learning and Tradition in Animals Conference, Cambridge, England.

Lefebvre, L., & Palameta, B. (1988). Mechanisms, ecology, and population diffusion of socially learned food-finding behavior in feral pigeons. In T. R. Zentall & B. G. Galef, Jr. (Eds.), *Social learning: Psychological and biological perspectives* (pp. 141–164). Hillsdale, NJ: Erlbaum.

Leslie, A. M. (1987). Pretense and representation in infancy: The origins of "theory of mind." *Psychological Review, 94,* 84–106.

Levine, J. M., & Zentall, T. R. (1974). Effect of conspecific's presence on deprived rats performance: Social facilitation vs. distraction/imitation. *Animal Learning and Behavior, 2,* 119–122.

Lore, R., Blanc, A., & Suedfeld, P. (1971). Empathic learning of a passive avoidance response in domesticated Rattus norvegicus. *Animal Behaviour, 19,* 112–114.

Lorenz, K. (1935). Der kumpanin der umvelt des vogels: die artgenosse als ausloesendesmoment socialer verhaltensweisen. *Journal fur Ornithologie, 83,* 137–213, 289–413.

Marler, P. (1970). A comparative approach to vocal learning: Song development in white-crowned sparrows. *Journal of comparative and Physiological Psychology, 71,* 1–25.

Meltzoff, A. N. (1988). The human infant as Homo imitans. In T. R. Zentall & B. G. Galef, Jr. (Eds.), *Social learning: Psychological and biological perspectives* (pp. 319–341). Hillsdale, NJ: Erlbaum.

Miller, N. E., & Dollard, J. (1941). *Social learning and imitation.* New Haven, CT: Yale University Press.

Mineka, S., & Cook, M. (1988). Social learning and the acquisition of snake fear in monkeys. In T. R. Zentall & B. G Galef, Jr. (Eds.), *Social learning: Psychological and biological perspectives* (pp. 51–75). Hillsdale, NJ: Erlbaum.

Mitchell, R. W. (1987). A comparative-developmental approach to understanding imitation. In P. P. G. Bateson & P. H. Klopfer (Eds.), *Perspectives in ethology* (Vol. 7, pp. 183–215). New York: Plenum.

Mitchell, R. W. (1992). Developing concepts in infancy: Animals, self-perception, and two theories of mirror self-recognition. *Psychological Inquiry, 3,* 127–130.

Mitchell, R. W. (1993). Mental models of mirror-self-recognition: Two theories. *New Ideas in Psychology, 11,* 295–325.

Morrison, B. J., & Hill, W. F. (1967). Socially facilitated reduction of the fear response in rats raised in groups or in isolation. *Journal of Comparative and Physiological Psychology, 63,* 71–76.

Neuringer, A., & Neuringer, M. (1974). Learning by following a food source. *Science, 184,* 1005–1008.

Nottebohm, F. (1970). Ontogeny of bird song. *Science, 167,* 950–956.

Oldfield-Box, H. (1970). Comments on two preliminary studies of "observation" learning in the rat. *Journal of Genetic Psychology, 116,* 45–51.

Pepperberg, I. M. (1986). Acquisition of anomalous communicatory systems: Implications for studies on interspecies communication. In R. Schusterman, J. Thomas, & F. Wood, (Eds.), *Dolphin behavior and cognition: Comparative and ethological aspects* (pp. 289–302). Hillsdale, NJ: Erlbaum.

Piaget, J. (1945). *Play, dreams, and imitation in childhood*. New York: Norton.

Powell, R. W. (1968). Observational learning vs shaping: A replication. *Psychonomic Science,* **10,** 263–364.

Powell, R. W., & Burns, R. (1970). Visual factors in observational learning with rats. *Psychonomic Science,* **21,** 47–48.

Powell, R. W., Saunders, D., & Thompson, W. (1969). Shaping, autoshaping, and observational learning with rats. *Psychonomic Science,* **13,** 167–168.

Reese, N. C. (1975). Imprinting as an independent variable in the modeling of a low-probability behavior in chicks. *Bulletin of the Psychonomic Society,* **6,** 28–30.

Roberts, D. (1941). Imitation and suggestion in animals. *Bulletin of Animal Behaviour,* **1,** 11–19.

Sanavio, E., & Savardi, U. (1979). Observational learning in Japanese quail. *Behavioral Processes,* **5,** 355–361.

Sanavio, E., & Savardi, U. (1980). Observational learning of a discriminative shuttlebox avoidance by rats. *Psychological Reports,* **44,** 1151–1154.

Spence, K. W. (1937). The differential response in animals to stimuli varying within a single dimension. *Psychological Review,* **44,** 430–444.

Stimbert, V. E. (1970). A comparison of learning based on social and nonsocial discriminative stimuli. *Psychonomic Science,* **20,** 185–186.

Strupp, B. J., & Levitsky, D. A. (1984). Social transmission of food preferences in adult hooded rats (*Rattus norvegicus*). *Journal of Comparative Psychology,* **98,** 257–266.

Terkel, J., & Aisner, R. (1992). Ontogeny of pine cone opening behaviour in the black rat (*Rattus rattus*). *Animal Behaviour,* **44,** 327–336.

Thorpe, W. H. (1961). *Bird song: The biology of vocal communication and expression in birds.* Cambridge, MA: Harvard University Press.

Thorpe, W. H. (1963). *Learning and instinct in animals* (2nd ed.), Cambridge, MA: Harvard University Press.

Thorpe, W. H. (1967). Vocal imitation and antiphonal song and its implications. In D. W. Snow (Ed.). *Proceedings of the XVI International Ornithological Congress* (pp. 245–263). Oxford: Blackwell.

Tomasello, M. (1990). Cultural transmission in the tool use and communicatory signalling of chimpanzees? In S. Parker & K. Gibson (Eds.), *"Language" and intelligence in monkeys and apes: Comparative developmental perspectives* (pp. 271–311). Cambridge: Cambridge University Press.

Turner, J. R. G. (1984). Mimicry: The palatability spectrum and its consequences. In R. I. Vane-Wright & P. R. Ackery (Eds.), *The biology of butterflies* (pp. 141–161). New York: Academic Press.

Vanayan, M., Robertson, H., & Biederman, G. B. (1985). Observational learning in pigeons: The effects of model proficiency on observer performance. *Journal of General Psychology,* **112,** 349–357.

Warden, C. J., & Jackson, T. A. (1935). Imitative behavior in the rhesus monkey. *Journal of Genetic Psychology,* **46,** 103–125.

Weigle, P. D., & Hanson, E. V. (1980). Observation learning and the feeding behavior of the red squirrel (*Tamiasciurus hudsonicus*): The ontogeny of optimization. *Ecology, 61,* 213–218.

Will, B., Pallaud, B., Soczka, M., & Manikowski, S. (1974). Imitation of lever pressing "strategies" during the operant conditioning of albino rats. *Animal Behaviour, 22,* 664–671.

Whiten, A. (Ed.). (1991). *Natural theories of mind: Evolution, development, and simulation of everyday mind reading.* Oxford: Blackwell.

Whiten, A., & Ham, R. (1992). On the nature and evolution of imitation in the animal kingdom: Reappraisal of a century of research. In P. J. B. Slater, J. S. Rosenblatt, C. Beer, & M. Milinski (Eds.), *Advances in the Study of Behavior* (Vol. 21, pp. 239–283). New York: Academic Press.

Zajonc, R. B. (1965). Social facilitation. *Science, 149,* 269–274.

Zentall, T. R., Edwards, C. A., & Hogan, D. E. (1983). Pigeons' use of identity. In M. L. Commons, R. J. Herrnstein, & A. Wagner (Eds.), *The quantitative analyses of behavior: Vol. 4. Discrimination Processes* (pp. 273–293). Cambridge, MA: Ballinger.

Zentall, T. R., & Hogan, D. E. (1975). Key pecking in pigeons produced by pairing key light with inaccessible grain. *Journal of the Experimental Analysis of Behavior, 23,* 199–206.

Zentall, T. R., & Hogan, D. E. (1976). Imitation and social facilitation in the pigeon. *Animal Learning and Behavior, 4,* 427–430.

Zentall, T. R., & Levine, J. M. (1972). Observational learning and social facilitation in the rat. *Science, 178,* 1220–1221.

Zentall, T. R., Sutton, J., & Sherburne, L. M. (in press). True imitative learning in pigeons. *Psychological Science.*

12

The Evolution of Imitative Learning

BRUCE R. MOORE

Department of Psychology
Dalhousie University
Halifax, Nova Scotia
Canada B3H 4J1

Movement imitation can be defined as the copying of novel, noninstinctive responses in the absence of explicit reinforcement, and in situations where simpler explanations (Galef, 1988; Moore, 1992) are untenable. This chapter describes three attempts to demonstrate such imitation, studies involving three very different creatures. It then considers the paths along which avian and mammalian imitation may have evolved and summarizes evidence that these processes are not homologous. It ends by proposing an evolutionary tree that shows the possible origins of these and other forms of learning and conditioning.

AVIAN IMITATION

The Grey Parrot, Okíchoro

Moore (1992) reported spontaneous movement imitation in a bird, a male Grey parrot, *Psittacus erithacus*. The animal was about one year old when the study began. It bonded to the experimenter and predictably copied his speech. But it also copied his movements. Each modeled movement had been accompanied by an identifying word or phrase. The words and movements modeled together were later mimicked together—each movement with its verbal label. Thus, by design, the parrot *announced* what it was going to imitate. Whenever it said "ciao," for example, it was meant to wave goodbye.

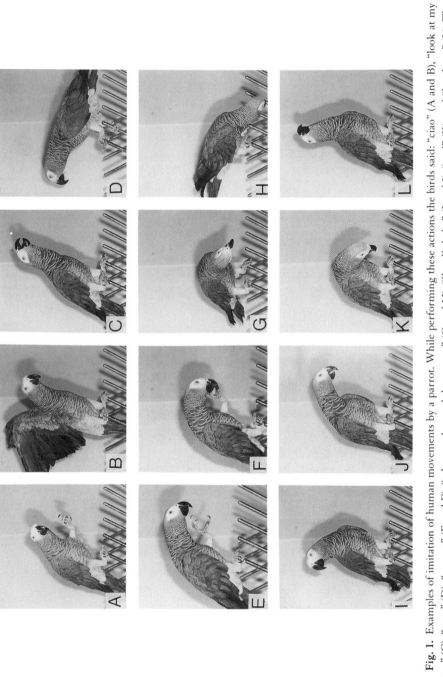

Fig. 1. Examples of imitation of human movements by a parrot. While performing these actions the birds said: "ciao" (A and B), "look at my tongue" (C), "turn" (D), "peanut" (E and F), "whoops, dropped the peanut" (G and H), "[head] shake" (I and J), "nod" (K), and "heads up" (L). The nibbling response (F) occurred only as a sequel to (E), and may have been self-imitative. Reprinted from Moore (1992) with permission of *Behavior*.

The bird did learn to say "ciao" and to wave goodbye with its feet or its wings, as shown in Figs. 1A and 1B. It also said "Look at my tongue," then opened its beak and showed it, (Fig. 1C), and learned to turn in place while saying "turn," as seen in Fig. 1D.

The parrot also copied the way that treats were offered it, as in "peanut" (Fig. 1E). Having done that, it often seemed to nibble an imaginary peanut (Fig. 1F), perhaps an act of self-imitation. When it refused a nut, the experimenter sometimes tossed the offering aside. That too was imitated (not shown), though never with an actual peanut; the bird simply reproduced the movement while repeating the labeling phrase, "Forget it!" Whenever the bird dropped a nut, the experimenter bent down to retrieve it. The bird later copied that (Figs. 1G and 1H) while repeating, "Whoops, dropped the peanut." When the bird was aggressive, the experimenter sometimes chased it with a lateral arm movement, or sudden toss of the head. These reactions, too, were imitated (not shown), while the bird said, "Get back, you," or "Back in your tree."

The parrot copied five different horizontal or vertical head rotations. There were slow unilateral rotations ("ready," not shown), fast bilateral movements ("shake," Figs. 1I and 1J), nodding movements ("nod," Fig. 1K), a slight upward rotation of the head ("microphone," not shown), and a larger rotation ("heads up," Fig. 1L). There was also a slight raising of the shoulder ("jump," not shown). And the bird learned to say, "Remember Lloyd Morgan, don't forget," with the last two words accompanied by emphatic finger (claw) strokes.

In several cases, movements and associated nonverbal sounds were very closely synchronized. For example, the experimenter sometimes knocked on a door within sight of the parrot. The bird copied that by making knocking movements (in the air) with its beak or foot, and synchronized those movements with vocal mimicry of knocking sounds. Most amusingly, the parrot mimicked the sounds of the experimenter's footsteps—and synchronized them with its own (natural) footsteps while walking, and also with imitative footsteps (walking in place).

Novel Responses?

Our definition of imitation follows Thorpe's (1963) in requiring novel, noninstinctive responses. To what extent did the parrot's learning satisfy this requirement? There was a broad spectrum of reactions. At the genetic extreme, the wing-waving ("ciao") response began as species typical—almost *class*-typical—wing folding. Over time, it changed somewhat from that original form, becoming looser and more casual—more like a human wave. But the action remained bilateral, reflecting its genetic origins. The "peanut" response, also, was not novel: It was the reaction that parrots typically use to hold food and small objects.

At the other extreme, the "tongue" response appeared novel, as did the "Lloyd Morgan" reaction, with two strokes of the foot, and the "shake" response, exuberant left-right head flagging. The nodding response, which differed from species-typical threat nodding, was also novel, as were the stylized piston movements of the feet while walking in place, and the way the bird sometimes switched feet while waving goodbye (as shown in Moore, 1992).

Many other responses were analogous to most human imitation. Within obvious limits, humans have the ability to move their hands in an infinite number of directions, distances, and speeds, to move their heads left or right, up or down, at various speeds, and to imitate any such response performed by another. And so it seemed with the parrot. The bird's feet and head were used in at least 13 different reactions. The foot was moved up and down smoothly for "ciao"; moved up and held there for "head scratch"; up and slightly forward, then back and forth, for "Lloyd Morgan"; back and forth in very short, rapid movements for knocking; and up and down in short, bilateral, alternating movements for walking in place. The head was rotated slightly upward for "microphone"; further up for tongue showing; much further still for "heads up"; down and back repeatedly for "nod"; rapidly up and down for "get off"; far to one side and back, slowly, for "ready," or rapidly for "get back in your tree"; and rapidly from side to side for "shake." In short, it appeared that any hand displacement or head movement that the experimenter could manage, the parrot was able to imitate. This ability to copy particular reactions from a near-infinite repertoire of possibilities is quite analogous to human imitation.

Most reactions were never performed when the experimenter was present. They were therefore never reinforced, not even accidentally. Only two exceptions occurred, and neither was allowed to compromise the study. That is, no further data were collected for the "ciao" and "whoops" responses once they had occurred in human presence. Altogether, the parrot imitated a great variety of movements a total of more than one thousand times. Since many of the reactions were novel, and none received explicit reinforcement, it can be claimed that the animal exhibited true movement imitation. We shall later consider how this remarkable capacity might have evolved.

STUDIES OF MAMMALIAN IMITATION

By the definition used here, imitation has been shown in just two groups of mammals: certain primates (humans and great apes), and cetaceans (specifically, dolphins).

Humans clearly imitate, at least after they approach one year of age. Claims of earlier imitation (Meltzoff & Moore, 1977, 1983) have been seriously questioned (Anisfeld, 1979, 1991; Hayes & Watson, 1981; Jacobson, 1979; Masters, 1979; Moore, 1992).

Chimpanzees also imitate very well by our definition, as shown by a dozen people including Fouts and Rigby (1977), Gardner and Gardner (1969), Kohler (1926), Kohts (1928), Lintz (1942), Sheak (1917), de Waal (1982), and, especially, Hayes and Hayes (1952). See Moore (1992) for review, and for discussion of criteria used in weighing anecdotal data.

The other great apes have been studied less frequently than chimpanzees, but also seem to imitate. The physiologist Flourens (1845) described a tame orangutan that played with children and sought to imitate whatever they did (*"cherchait à imiter tout ce qu'on faisait devant lui"*). It also seized the cane of an elderly visitor and used it while mimicking his gait and posture. Less anecdotally, Russon and Galdikas (1993) have provided many other examples of imitation in orangutans. Imitation in gorillas has been claimed by Chevalier-Skolnikoff (1977), Hoyt (1941), and Lintz (1942).

The Common Gibbon, Boubou

The lesser apes have not been extensively studied. But one of the author's students, Eva Rogerson, did conduct a month-long imitation study with a 2½-year-old female common gibbon (*Hylobates lar*). The animal had been handled by humans since birth, and did copy certain human responses, such as mouth opening and tongue protrusion (coincidentally, the same responses studied in human neonates by Meltzoff & Moore, 1983). But these are common reactions in gibbons, as they are in human infants (Jacobson, 1979; Masters, 1979), and therefore do not qualify as imitation in either case.

Rogerson's study looked for imitation of a series of gestures that would have been novel for gibbons: put finger to lips, clap hands, nod head, pat head, shut eyes, cover face, wave, cover ears, join hands behind back, pat thigh, put hands on hips, and fold arms across chest. These are responses that gibbons could easily make, but normally do not. Each week, three of these actions were modeled in exaggerated form, six times per session, in each of 14 sessions. The tests were conducted in a familiar 1.5 × 3 m cage, 2.5 m high, constructed of plywood and chain-link fencing. The model was the person to whom the gibbon was most closely bonded.

In the course of the study, an occasional probe was made using Todt's (1975) "rivalry" procedure. That is, a second human was present, a rival for the teacher's attention. Whenever a response was modeled, the rival quickly copied it, and was patted and rewarded for doing so.

No imitation was seen in any phase of the experiment. None of the 12 responses was ever imitated, and the rivalry probes were less than successful. Whenever the rival received a reward, the gibbon attacked her.

In retrospect, we may have modeled the wrong sorts of things. The author was too much influenced by the parrot work, which was still under way, and parrots imitate gestures. But what is now very obvious is that imitation in great apes is quite different. It almost always involves skill learning or tool use (cf. Beck, 1974): e.g., the use of hammers, saws, screwdrivers, shovels, brooms, brushes, keys, needles, lipstick, spray bottles, canes, outboard motors, or canoes (see Moore, 1992; Russon & Galdikas, 1993). So if we could repeat the experiment, which we cannot, we would model tool use.

Research with monkeys has fared no better. After almost 100 years of efforts beginning with Thorndike (1901), and Watson (1908), there is still no compelling proof of imitation in any old- or new-world monkey (see Moore, 1992, for review). The question is not yet closed. Indeed, several anecdotes involving baboons suggest the need for more research with that group. But at the time of this writing, imitation has been shown in only two groups of primates: humans and great apes (Moore, 1992). Similar conclusions have been reached by Beck (1974), Fragasy and Visalberghi (1989), and Whiten (1989).

Imitation (and vocal mimicry) have also been shown in cetaceans. Tayler and Saaymen (1973) reported that captive dolphins (*Tursiops aduncus*) used makeshift scrapers and sponges to clean the bottom and windows of their tank in clear imitation of the acts of human divers.

The Harbor Seal, Chimo

Vocal mimicry, but not movement imitation, has been reported in one other marine mammal: a Harbor seal (*Phoca vitulina*). The seal *Hoover* was raised by humans and acquired a repertoire of about a dozen English words. They appear to have been produced primarily by combining various vowel-like sounds natural to seals, but the effect was sometimes striking. Audiences are usually able to understand, without prompting, the author's tape recordings of the animal's vocalizations. And few have failed to recognize Hoover's flawless human *belly laugh*.[1]

1. The author once played a tape of Hoover's vocalizations for his students. When the seal laughed, we laughed so loudly that the parrot in the next room heard us. It then joined in, and therefore, for perhaps the first time ever, three very different species laughed together.

A humorous, but carefully researched, account of Hoover's life was published by Hiss (1983), and clear sonograms of several of his vocalizations were published by Rawls, Fiorelli, and Gish (1985).

Now, if seals can copy human sounds, can they also copy movements? To explore this question, a male Harbor seal was taken from Sable Island, Nova Scotia, while still a neonate (placenta still attached), and flown to the mainland, where it was raised by Kelly Stanhope, Edith Foy, and the author.

The animal was tube-fed condensed milk, pureed sardines, and fish oil for the first day, then force-fed smelt for several days, after which it sought them eagerly. At 3 weeks of age, it was switched to a permanent diet of (whole) herring, supplemented by vitamins B1 and E. The animal was housed alone in a large (40,000 l) outdoor tank. The tank was fed by filtered ocean water, supplied at ambient temperature and not recirculated.

The seal was named *Chimo,* from an Inuktitut greeting meaning "Are you friendly?"—and quickly became so. By 4 days of age he followed us everywhere, even up stairs (and later, ladders), and cried pitifully if we moved too far from him. He also tried to nurse from us—male and female alike. (Anatomic details were irrelevant, as he preferred to nuzzle our feet.) Thus, the animal showed apparent filial imprinting: separation anxiety, persistent nursing attempts, and heroic follow- ing. At 7 years of age he also showed apparent sexual imprinting, making unam- biguous advances to one young woman on several occasions. The approaches were unprovoked and unrequited.

In the experiment, we modeled just five simple forelimb and head responses: waving, clapping, a beckoning movement, head shaking, and covering the eyes. Each reaction was demonstrated many thousands of times, beginning when the seal was 4 days old, and continuing for 7 years. But never, in that long period, did the animal unambiguously copy a single human movement, sound, or posture.

At 18 days of age he answered our repeated "hellos" with natural vocaliza- tions, in what may have been a circular reaction (Piaget, 1951/1945). At age 5–11 weeks he made what might have been interpreted as three poor approximations of "hello" and "how are you"? About a dozen more such sequences occurred at age 3– 5 years, along with a few dozen vowel sounds (ee, oo, oh, ow, uh). But that was all. And since even wild seals produce a variety of such humanoid sounds, our study clearly failed to show vocal mimicry. Nor was there evidence of movement imita- tion. Only 16 *possible approximations* of two modeled responses (clapping and head shaking) were recorded between the 3rd and 8th months, and virtually none thereafter.

After 7 years, the animal was sent to the Québec Aquarium, which needed a male for breeding. In that he was successful, apparently reimprinting as various animals sometimes do. We nursed a faint hope that, in breeding season, he might whisper sweet nothings, or strike humanoid poses. And our hopes were soon reinforced by a media report (Roy, 1993) asserting that Chimo was "*le plus humaine*" of seals and that his behavior was sometimes not very seal-like ("*parfois pas très phoque*"). But, on inquiring, we were told that these phrases did not refer to imitative behavior. After 3 years in Québec, his handlers reported that Chimo seemed normal in all ways but two. He swam too close to conspecifics at mealtime, never having learned the norms of social distance. And he was unusually tame. But there had been no obvious signs of vocal mimicry or movement imitation.

Other Claims of Imitation

Movement imitation has been attributed to many other birds and mammals, most notably monkeys, otters, cats, rats, budgies, and pigeons. Many such claims were discussed by Moore (1992), and need not be reexamined here, but three deserve mention.

The cat study of John, Chesler, Bartlett, and Victor (1968) has been widely cited. The animals were said to have learned through imitation to jump hurdles on signal, but the study was very seriously flawed. "Trial spacing was badly confounded, there were no controls for audience effects or habituation to apparatus, and the cats may have been pre-conditioned to treat the buzzer as a danger signal. . . ." (Moore, 1992).

Heyes, Dawson, and Nokes' (1992) rat study, by contrast, had no flaws of that sort. The animals acquired through observation a statistically significant tendency to press a rod in a particular direction. But this effect is so different from accepted examples of imitation that one hesitates to treat it as equivalent. With humans (Piaget, 1951/1945), chimpanzees (Hayes & Hayes, 1951), orangutans (Russon & Galdikas, 1993), dolphins (Tayler & Saayman, 1973), and even parrots (Moore, 1992), imitation is so robust that it appears very clearly in the behavior of individual animals. No group studies, or statistics, are required. Also, animals in these studies have typically copied a great variety of responses, demonstrating the effect's generality. Again, their imitation has typically been spontaneous; reinforcement has not been needed for performance during testing. Finally, the imitated responses have been truly novel, as required by most definitions. The behavior of the Heyes et al. (1992) rats differs on all of these dimensions. It is interesting, and may prove important. But until it is better understood, and its generality demonstrated with other and more novel responses, it would be premature to accept that rats can imitate.

The problem with claims of imitation in pigeons can be stated very simply. All of these studies have directly or indirectly involved locomotion or pecking. But these reactions are species typical. Clearly, the birds were not learning *how* to peck. They were learning where or what or when to peck. Thus, they were demonstrating observational conditioning, not movement imitation (see Moore, 1992; Suboski, 1990). Such conditioning is both impressive and biologically important. But it is an entirely different process from movement imitation. The two processes have little in common and should not be confused.

THE EVOLUTION OF IMITATION

Let us turn now to the question of why some species are able to imitate and others not. How is it that parrots, great apes, and dolphins can manage imitation? How might these processes have evolved?

Psittacine Imitation

Biological processes typically evolve from similar, simpler processes, through series of small, adaptive changes. If we apply that principle to parrots and ask what else they do that might be related to movement imitation, the question answers itself. They copy sounds; they are superb *vocal mimics.* Their ability to copy movements is no doubt related to their ability to copy sounds. The obvious possibility is that movement imitation evolved from the simpler and more common trait, vocal mimicry. But these very similar processes appear to differ in two ways, one sensory, one motor. The sensory difference is that vocal mimicry involves copying auditory models, whereas movement imitation involves visual models. The motor difference is that vocal mimicry involves vocal muscles, and movement imitation, nonvocal muscles. These are small differences, but evolution proceeds by small steps. And it proceeds *one step at a time.* So if movement imitation evolved from vocal mimicry, it had to do so in two steps: one sensory, one motor. And we can deduce which of these would have had to come first. Not the sensory step, because copying the new (visual) models with the old (vocal) muscles does not make sense. Parrots can mimic almost anything; but not even they can (actively) sound like visual images.

But the motor change *could* have evolved first, because nonvocal muscles can be used to mimic some auditory models—namely, percussive sounds. The animal can, in principle, accomplish this by striking something noisily. So the evolutionary hypothesis predicts the existence of a new form of learning, intermediate between vocal mimicry and movement imitation, a process in which movements are used to copy percussive sounds. Moore (1992) called the new process *nonvocal* mimicry. But *percussive* mimicry would have been more descriptive, and should be used hereafter.

The process does seem to exist. The parrot gave two examples: it sometimes mimicked the author's knocking sounds by banging things with the top of its head or tapping them with the top of its beak—neither of which the author had done. The older field-work literature suggests similar learning in starlings (*Sturnus vulgaris*) (Allard, 1939; Witchell, 1896). So there may have been a very simple evolutionary progression:

Vocal mimicry → Percussive mimicry → Movement imitation.

The progression can easily be extended. Slater suggested in 1983, and Hindmarsh in 1986, that vocal mimicry had evolved from song learning. Witchell almost said the same in 1896. And the idea is extremely plausible: Song learning is often constrained to certain frequencies, certain rhythms, or certain sensitive periods. The mere relaxation of those constraints would produce a creature that could copy a great variety of sounds at any time—and we would call that vocal mimicry. It is simply a higher level of song learning.

The parrot also showed *cross-modal* movement imitation. In ordinary imitation, the learner can get a visual match between its own behavior and the model's. But the parrot could not see its own imitative head responses (e.g., shake, nod, ready, microphone, and heads up). So they were cases of cross-modal movement imitation (Guillaume, 1971/1926; Piaget, 1951/1945). Thus, Moore (1992) hypothesized the evolutionary progression shown on the left of Table 1.

Now we must ask whether the hypothesized steps are biologically plausible, and whether the progression is consistent with comparative data. These criteria, while scarcely sufficient to prove an hypothesis, are often quite sufficient to disprove one.

In the present case, the steps between processes can be described, in order, as: a relaxation of constraints, a change in muscle groups, a change in sensory models, and a loss of visual feedback. The first three are the sorts of changes that evolution most easily accomplishes. But the fourth is problematic: it is not clear how parrots accomplish imitation without visual feedback. Piaget (1951/1945) explained the analogous step in human infants (see below). But parrots have not had their Piaget.

The agreement with comparative data appears to be perfect: Over 200 species of birds are known to show song learning (Kroodsma, 1982; Kroodsma & Baylis, 1982). The vocal mimics are a subset of that group (Baylis, 1982; Witchell, 1896).

TABLE 1

Hypothesized Evolutionary Paths of Imitative Learning in Psittacine Birds and in Primates

Psittaciformes	Primates
Cross-Modal Imitation	Cross-Modal Imitation
↑	↑
Visual Movement Imitation	Visual Movement Imitation
↑	↑
Percussive Mimicry	Putting Through
↑	↑
Vocal Mimicry	Skill Learning
↑	↑
Song/Call Learning	Operant Conditioning
	↑
	Thorndikian Conditioning

The known percussive mimics, parrots and starlings, are a small subset of that subset, and the movement imitators a subset of them. Thus, the different levels of learning occur in strictly nested subsets of birds. The comparative data support the hypothesis perfectly.

Cetacean Imitation

Mimetic learning in dolphins is surprisingly bird-like in some ways. Cetaceans are song learners (Payne, Tyack, & Payne, 1983), and also vocal mimics (Herman, 1980; Tayler & Saayman, 1973). Further, their movement imitation is sometimes used socially. They often match their aquatic maneuvers; they are spectacular synchronized swimmers (see also Tayler & Saayman, 1973). For these reasons, there might well be similarities between the evolution of psittacine and cetacean imitation (similarities arising from convergence). But we do not know enough to say much more than that.

Anthropoid Imitation

It is quite clear, however, that primate data do not fit the avian pattern. Imitation in great apes clearly did *not* evolve from vocal mimicry or song learning. So what could it have come from? There is a logical, and possibly evolutionary, hierarchy involving six basic forms of learning seen in anthropoids and some of their relatives, as shown on the right side of Table 1. But, again, we must ask if the hypothesized steps are biologically plausible and compatible with comparative data.

At the bottom of the hierarchy, four levels of instrumental learning form a straightforward progression. In the simplest of these four processes, Thorndikian conditioning, reinforcement-appropriate, species-typical reactions are strengthened by explicit reinforcement (Thorndike, 1898). Operant conditioning (Skinner, 1938) is usually treated as synonymous with Thorndikian learning. But Thorndike (1896) stressed the importance of species-typical reactions elicited by environmental stimuli. Thus, when his cats were confined in wooden crates, gaps between the slats elicited attempts to squeeze through, strings elicited paw responses, and so on. Skinner (1938), by contrast, insisted that operants were *never* elicited reactions. And it is widely understood that true operants are of novel ("arbitrary") form. It is therefore evident that Thorndikian and operant conditioning were conceived by their authors to involve quite different response repertoires.

Both Thorndikian and operant conditioning clearly exist. Most instrumental learning, for either positive or negative reinforcement, involves elicited, species-typical reactions (Bolles, 1970; Breland & Breland, 1966; Thorndike, 1898) and therefore fits the Thorndikian model. But operant conditioning in the strict sense also occurs, and is extremely important, in humans, other primates, and some other taxa. We shall therefore treat operant conditioning as a daughter of Thorndikian learning, but more complex because it involves novel, nonelicited responses.

The emergence of novel responses was not a trivial development. Indeed, Sherrington's *Integrative Action of the Nervous System* (1906) concludes by saying that a major function of the cerebral cortex is to allow species-typical responses to be "modified [with] seeming independence [from] external stimuli. . . ." The "controlling centres can pick out from an ancestrally given motor reaction some one part of it, so as to isolate that as a new separate movement" (1906). Thus, Skinner's (1938) nonelicited, novel reactions, his *operants,* reflect what Sherrington had considered a great achievement of the vertebrate nervous system.

Skill is a step up from operant conditioning in that it does not require explicit reinforcement. (It may, instead, require *implicit* reinforcement.) Learning to skate, draw, type, or drive a car involves this process.

A step up from that is the process called *molding* or *putting through,* which is like guided skill learning. It begins with a passive, rather than active, response (Sherrington, 1900). A teacher guides the passive animal through the response, and it learns from that experience.[2] Humans use putting through in a variety of teaching situations. Parents move the hands of their infants to teach them how to play patty cake, and mold the hands of slightly older children to teach them to hold pencils and form letters. Similar methods have been used for centuries by music and dance teachers (L. Mozart, 1951/1756; Schlaich & DuPont, 1993).

Putting through also occurs in great apes (Fouts & Rigby, 1977; Gardner & Gardner, 1969; see also Moore, 1992), and in monkeys. The method is traditionally used in Southeast Asia, where Thai and Malay people train Pig-tailed macaques (*Macaca nemestrina*) to harvest coconuts. The coconut must be twisted to tighten the stem before the monkey can bite through it, and this is taught by putting through (Bertrand, 1967). Also, Westergaard and Fragaszy (1987) observed an incident in which putting through appeared to be initiated by the learner, an 8-month-old Capuchin monkey (*Cebus appella*).

Putting through differs from skill learning in that the number of supporting stimuli is reduced. If a chimp is taught the sign for *drink* by putting through, it sees its hand make the response, and receives proprioceptive feedback, but no longer experiences a third class of cues—those kinesthetic stimuli uniquely associated with *producing* the response.

The next transition, from putting through to visual imitation (which can also be seen as the transition from instrumental to imitative learning) is a surprisingly small step, in that putting through amounts to *self-imitation*. In putting through the animal repeats what it has seen and felt its own body do, whereas in movement imitation it repeats what it has seen another's body do. It still *sees* a hand make the response, but no longer *feels* it; proprioception is lost. Thus, putting through can be seen both as one of the most complex forms of instrumental learning and as the simplest form of true movement imitation.

The processes hypothesized to have evolved from skill learning now often complement and set the stage for it. Thus, Sherrington long ago (1906) observed that it was impossible to "learn skating or racquets by simple . . . visual observation." The same could be said of putting through. These processes show the learner what to do, but not always how to do it well. Skill learning then shapes the details.

---

2. *Pseudo* putting through, in which the learner's body is moved or stimulated in such a way as to trigger a *reflexive* reaction which is then reinforced, occurs in many mammals. For example, one can train an elephant to raise its foot by tapping the foot with a whip or sharp object. The slight pain elicits the desired reaction, which is then reinforced. But this is simply an efficient variation on Thorndikian learning. True putting through involves passive, nonelicited responses which are not reinforced. Unfortunately, it is sometimes difficult to distinguish between these two very different processes (see Konorski, 1969).

Thus, imitation (or putting through) can teach only the rudiments of such complex behavior; they direct and set the stage for subsequent skill learning. The acquisition of termite "fishing" by chimpanzees (Goodall, 1971; Teleki, 1974) may involve this same combination of imitative and skill learning.

Skill learning, putting through, and normal movement imitation correspond to three levels of stimulus support. The most fundamental of these processes, skill learning, may use any of three classes of stimuli: (i) kinesthetic cues uniquely associated with response production, (ii) other proprioceptive stimuli, and (iii) visual cues. The second process, putting through, can use only classes (ii) and (iii), and movement imitation can use only class (iii).

Altogether, these three processes (skill learning, putting through, and visual imitation) are linked in many ways: their possible controlling stimuli are nested as just described; both putting through and imitation incorporate and set the stage for skill learning. Putting through is like self-imitation. And all three processes involve novel responses and possibly implicit reinforcement.

The suggestion that cross-modal imitation evolved from ordinary visual imitation was first advanced by Piaget (1951/1945). The hypothesis is plausible since the processes are very similar, but cross-modal imitation is somewhat more complex, and occurs later in ontogeny (Piaget, 1951/1945). But how can the mimic match corresponding features of the model's body and its own, in cross-modal cases where it cannot see the latter?

The answer was suggested by Piaget's (1951/1945) very careful observations. Briefly, the human infant uses touch to find corresponding parts on the mother's face and its own. The process begins at about 13–15 months of age when the infant repeatedly touches its parent's facial and cranial features (eyes, ears, nose, mouth, teeth, hair), then locates tactually the corresponding features of its own face (1951). Having found these features, it appears to use them as landmarks for locating others. The mother's forehead is found below her hair, the infant then touches its own hair, and slides its hand down. Similarly, the chin is found beneath the mouth, the tongue between the lips, and so on (1951). Having located the various unseen structures, the child can then *touch* them in cross-modal imitation.

It may also use touch to learn to make facial *movements* that correspond to hers: opening the mouth, sticking out the tongue, closing the eyes, and so forth. Thus, cross-modal imitation, which had seemed to many to imply extraordinary mental powers, appears to be an extension of visual imitation, emerging after a period of developmentally programmed tactile exploration.

Altogether, the six processes discussed here (Table 1, right) do form a plausible progression. The factors required to transform each form of learning to that just above it are all straightforward. They can be described, in order, as: an expansion of repertoire, the dropping of explicit (for implicit?) reinforcement, a loss of kinesthetic feedback, a loss of other proprioceptive feedback, and a product of programmed tactile exploration.

But does the hypothesized progression agree with comparative data? The best short answer is: "probably." The author is not aware any species that shows any of these processes without also showing those lower on the list. But the mammalian data base is too fragmentary to allow a really confident verdict.

It is clear that Thorndikian learning occurs in most or all mammals. Operant conditioning, in the original sense (involving nonelicited, novel responses) certainly occurs in humans, other primates, and some other mammals, but is distinctly less common than Thorndikian learning (Bolles, 1970; Breland & Breland, 1966). Skill learning is most conspicuous in primates, but clearly occurs in a few other groups—e.g., raccoons. True putting through occurs in monkeys and all "higher" primates, and movement imitation in great apes. So, again, the processes appear to occur in nested subsets of species. But in this case, the phylogenetic boundaries are, for the most part, approximate because the data base has too many deficiencies.

There has not been enough work with either skill learning or putting through to really define their phyletic limits. Another serious gap involves the simpler processes. On the avian side, the song-learning literature offered a superb data base involving hundreds of species (Kroodsma, 1982; Kroodsma & Baylis, 1982). But with mammals, where we need to distinguish between species-typical and novel responses, we have the literature of operant conditioning. There, instead of hundreds of well-studied species we find primarily rats, rats, and rats (Beach, 1950). Further, most of the data are from impulse counters, which rarely make subtle distinctions. Indeed, operant researchers themselves, with several honorable exceptions, have not been good at recognizing species-typical reactions. Many saw no need to do so, and therefore treated the pigeon's peck as a prototypic novel response (see Moore, 1973). Even the cat's familiar shin-rubbing reaction was not recognized (see Moore & Stuttard, 1979 re Guthrie & Horton, 1946 and Herrnstein, 1966). For all of these reasons, the research on operant and Thorndikian conditioning in mammals is not nearly so helpful as those on song learning and vocal mimicry in birds.

Are Psittacine and Anthropoid Imitation Homologous?

The details of the hypothesized evolutionary sources (Table 1) are obviously subject to revision. Evolutionary theories should themselves evolve as better data or analyses become available. But the clear implication of Table 1 is that avian and mammalian imitation evolved independently—that the processes are not homologous (Moore, 1992).

The best confirming evidence for this inference is that the last common ancestors of great apes and parrots were extremely primitive reptiles, creatures which vanished 300,000,000 years ago, and therefore long before the appearance of dinosaurs. It is hard to imagine true imitation in any reptiles, let alone pre-Jurassic, even pre-Mesozoic reptiles. These were *Pale*ozoic reptiles. But if they had somehow been capable of true imitation, and had passed it down to great apes and parrots, then we would expect to see it also in some modern reptiles, and most, or all, groups of birds and mammals. And that is not the pattern. Rather, we see it in just three very recently evolved forms: great apes, cetaceans, and parrots. I therefore suspect that imitation has evolved independently at least two or three times—along one path in parrots, another in great apes, and doubtless a third in cetaceans (Moore, 1992).

That the process might have evolved three times should not be surprising. We are told that flight has evolved at least three times, hearing at least 10, and photoreception about 40 times (Salvini-Plawen & Mayr, 1977).

Now, if avian and primate imitation did evolve separately, are they different? They appear to differ in three significant ways. First, avian imitation has incubation periods. Todt (1975) pointed out that parrots never copy new sounds in less than 3 days. The author's parrot confirmed that. It learned 260 words and sounds, including several that it had heard only once (in "dramatic" situations). But all took more than 3 days. And the incubation period for new movements was measured in weeks. In primates, of course, there is no incubation period. In fact, immediate imitation is easier than delayed, and comes earlier in development.

The second difference is that in birds vocal mimicry is far more common than movement imitation—whereas great apes copy movements, but rarely if ever copy sounds. Third, as Beck (1974) pointed out, primate imitation frequently involves tool use. But the parrot's imitation never incorporated any sort of object. When it imitated a response that had involved an object, it always did so without using the object. In fact, its imitation seemed to be an extension of display behavior.

Notice that this apparent social function (display), and the predominance of sounds, and the incubation periods, in short, all three distinctive features of avian imitation, are also properties of song learning, their hypothesized source. Similarly, the properties of primate imitation all reflect its hypothesized origins in skill learning. Thus, the pattern of differences is entirely consistent with the inference drawn from Table 1, and from phylogenetic patterns of occurrence. All suggest that psittacine imitation evolved independently from that seen in mammals.

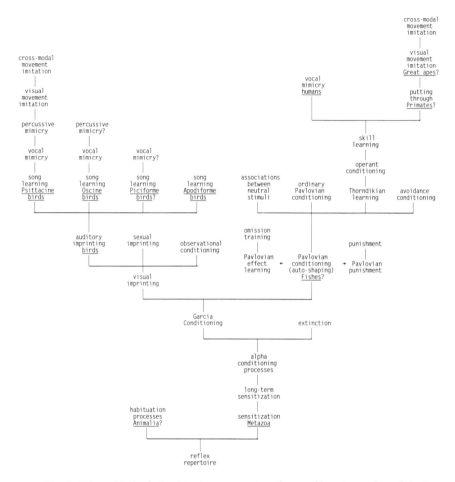

Fig. 2. Hierarchical relationships between various forms of learning and conditioning in vertebrates. In each case, a small modification of the source process would produce the next-higher form of learning. Arrows are used instead of lines where one process is a special case of another. Taxonomic groups, where given, indicate populations in which particular forms of learning may have evolved. From Moore (in preparation).

THE EVOLUTIONARY ROOTS OF LEARNING

Of course, if the preceding conclusion were true, and avian and mammalian imitation were not homologous, that would not imply the absence of any common antecedent. It is certainly possible, perhaps even probable, that all forms of learning are related. Indeed, it should be possible to draw an evolutionary tree (a branching, hierarchical tree) linking them all, and showing the paths through which they came into being.

While available data do not permit us to reach that ultimate goal, they do permit a beginning. The tree shown in Fig. 2 is taken from a work in progress (Moore, in preparation), and based in part on the present work, and on earlier contributions of Kandel (1976), Moore (1992), Piaget (1951/1945), Razran (1971), Slater (1983), and Wells (1968). It is not in any sense offered as final. There are important omissions, and some details are undoubtedly wrong. It is nevertheless a beginning, and offers an evolutionary alternative to the prevailing but anachronistic paradigm in which most forms of learning are still treated as separate creations— much as species were treated by pre-Darwinian biologists.

ACKNOWLEDGMENTS

The author is grateful to B. F. Moore, and Drs. C. M. Heyes, H. James, and M. L. Spetch for many helpful comments on early drafts of the manuscript, and to the Natural Sciences and Research Council of Canada for financial support. Any correspondence should be addressed to the author at Department of Psychology, Dalhousie University, Halifax, N.S., CANADA B3H 4J1, or sent electronically to bmoore@ac.dal.ca.

REFERENCES

Allard, H. A. (1939). Vocal mimicry of the starling and the mocking bird. *Science* **90**, 370–371.
Anisfeld, M. (1979). Interpreting "imitative" responses in early infancy. *Science* **205**, 214–215.
Anisfeld, M. (1991). Neonatal imitation. *Developmental Review* **11**, 60–97.
Baylis, J. R. (1982). Avian vocal mimicry: Its function and evolution. In D. E. Kroodsma, E. H. Miller, & H. Ouellet (Eds.), *Acoustic communication in birds* (pp. 51–83). London: Academic Press.
Beach, F. A. (1950). The snark was a boojum. *American Psychologist* **5**, 115–124.
Beck, B. B. (1974). Baboons, chimpanzees, and tools. *Journal of Human Evolution* **3**, 509–516.
Bertrand, M. (1967). Training without reward: Traditional training of pig-tailed macaques as coconut harvesters. *Science* **155**, 484–486.
Bolles, R. C. (1970). Species-specific defense reactions and avoidance learning. *Psychological Review* **77**, 32–48.
Breland, K., & Breland, M. (1961). The misbehavior of organisms. *American Psychologist* **16**, 681–684.

Chevalier-Skolnikoff, S. (1977). A Piagetian model for describing and comparing socialization in monkey, ape, and human infants. In S. Chevalier-Skolnikoff & F. E. Poirier (Eds.), *Primate biosocial development* (pp. 159–187). New York, Garland Press.

de Waal, F. (1982). *Chimpanzee politics.* New York: Harper.

Flourens, P. (1845). *De l'instinct et de l'intelligence des animaux* [On the instinct and intelligence of animals] (pp. 43–44). Paris, Paulin.

Fouts, R. S., & Rigby, R. L. (1977). Man-chimpanzee communication. In T. A. Sebeok (Ed.), *How animals communicate* (pp. 1034–1054). Bloomington, IN: Indiana University Press.

Fragasy, D. M., & Visalberghi, E. (1989). Social influences on the acquisition of tool-using behaviors in tufted capuchin monkeys (*Cebus apella*). *Journal of Comparative Psychology* **103,** 159–170.

Galef, B. G., Jr. (1988). Imitation in animals: History, definition, and interpretation of data from the psychological laboratory. In T. R. Zentall & B. G. Galef, Jr. (Eds.), *Social learning: Psychological and biological perspectives* (pp. 3–28). Hillsdale, NJ: Erlbaum.

Gardner, R. A., & Gardner, B. T. (1969). Teaching sign language to a chimpanzee. *Science* **165,** 664–672.

Goodall, J. (1971). *In the shadow of man.* London: Wm. Collins.

Guillaume, P. (1971). *Imitation in children* (E. P. Halperin, Trans.). Chicago: University of Chicago Press. (Original work published 1926.)

Guthrie, E. R., & Horton, G. P. (1946). *Cats in a puzzle box.* New York: Rinehart.

Hayes, K. J., & Hayes, C. (1952). Imitation in a home-reared chimpanzee. *Journal of Comparative and Physiological Psychology* **45,** 450–459.

Hayes, L. A., & Watson, J. S. (1981). Neonatal imitation: Fact or artifact? *Developmental Psychology* **17,** 655–660.

Herman, L. M. (1980). Cognitive characteristics of dolphins. In L. M. Herman (Ed.), *Cetacean behavior: Mechanisms and functions* (pp. 363–409). New York: Wiley.

Herrnstein, R. J. (1966). Superstition: A corollary of the principles of operant conditioning. In W. K. Honig (Ed.), *Operant behavior* (pp. 33–51). New York: Appleton.

Heyes, C. M., Dawson, G. R., & Nokes, T. (1992). Imitation in rats: Initial responding and transfer evidence. *Quarterly Journal of Experimental Psychology* **45b,** 81–92.

Hindmarsh, A. M. (1986). The functional significance of vocal mimicry in song. *Behaviour* **90,** 302–324.

Hiss, A. (1983, January 3). Hoover. *New Yorker,* pp. 25–27.

Hoyt, A. M. (1941). *Toto and I.* Philadelphia: J. B. Lippincott.

Jacobson, S. W. (1979). Matching behavior in the young infant. *Child Development* **50,** 425–430.

John, E. R., Chesler, P., Bartlett, F., & Victor, I. (1968). Observation learning in cats. *Science* **159,** 1489–1491.

Kandel, E. R. (1976). *The cellular basis of behavior.* San Francisco: Freeman.

Kohler, W. (1926). Intelligence in apes. In C. Murchison (Ed.), *Psychologies of 1925* (pp. 145–161). Worcester, MA: Clark University Press.

Kohts, N. (1935). [*Infant ape and human child*]. Moscow, State Darwin Museum.

Konorski, J. (1969). Postscript. *Journal of the Experimental Analysis of Behavior* **12,** 189.

Kroodsma, D. E. (1982). Learning and the ontogeny of sound signals in birds. In D. E. Kroodsma, E. H. Miller, & H. Ouellet (Eds.), *Acoustic communication in birds* (pp. 1–23). London: Academic Press.

Kroodsma, D. E., & Baylis, J. R. (1982). Appendix: A world survey of evidence for vocal learning in birds. In D. E. Kroodsma, E. H. Miller, & H. Ouellet (Eds.), *Acoustic communication in birds* (pp. 311–337), London: Academic Press.

Lintz, G. D. (1942). *Animals are my hobby.* New York: Robert M. McBride.

Masters, J. C. (1979). Untitled letter. *Science* **205,** 215.

Meltzoff, A. N., & Moore, M. K. (1977). Imitation of facial and manual gestures by human neonates. *Science* **198,** 75–78.

Meltzoff, A. N., & Moore, M. K. (1983). Newborn infants imitate adult facial gestures. *Child Development* **54,** 702–709.

Moore, B. R. (1973). The role of directed Pavlovian reactions in simple instrumental learning in the pigeon. In R. A. & J. S. Hinde (Eds.), *Constraints on learning* (pp. 159–188). London: Academic Press.

Moore, B. R. (1992). Avian movement imitation and a new form of mimicry: Tracing the evolution of a complex form of learning. *Behaviour* **122,** 231–263.

Moore, B. R. (in preparation). *The evolution of learning: Hierarchical relationships between various forms of learning and conditioning in vertebrates.* Unpublished manuscript.

Moore, B. R., & Stuttard, S. (1979). Dr. Guthrie and *Felis domesticus* or: Tripping over the cat. *Science* **205,** 1031–1033.

Mozart, L. (1951). *A treatise on the fundamental principles of violin playing* (E. Knocker, Trans.). Oxford: Oxford Univ. Press. (Original date 1756.)

Payne, K., Tyack, P., & Payne, R. (1983). Progressive changes in the songs of Humpback whales (*Megaptera novaeangliae*): A detailed analysis of two seasons in Hawaii. In R. Payne (Ed.), *Communication and behavior of whales* (pp. 9–57), Washington, DC: A.A.A.S.

Piaget, J. (1951). *Play, dreams and imitation in childhood* (pp. 57–58). (C. Gategno & F. M. Hodgson, Trans.) London: Heinemann. (Original work published 1945.)

Rawls, K., Fiorelli, P., & Gish, S. (1985). Vocalizations and vocal mimicry in captive Harbor seals, *Phoca vitulina. Canadian Journal of Zoology* **63,** 1050–1056.

Razran, G. (1971). *Mind in evolution.* Boston: Houghton Mifflin.

Rogerson, E. (1985). *Imitative behaviour in the common gibbon (Hylobates lar).* Unpublished manuscript.

Roy, d'Andrée (1993, July 3). Carrière de vedette et vie de chasteté pour deux bébés-filles (Stardom and life of chastity for two baby girls.). *Le Soleil,* Québec City, P. Q., Canada, p. A-2.

Russon, A. E., & Galdikas, B. M. F. (1993). Imitation in free-ranging rehabilitant orangutans. *Journal of Comparative Psychology* **107,** 147–161.

Salvini-Plawen, L. von, & Mayr, E. (1977). On the evolution of photoreceptors and eyes. In M. K. Hecht, W. C. Steere, & B. Wallace (Eds.), *Evolutionary biology* (Vol. 10, pp. 207–263). New York: Plenum.

Schlaich, J., & DuPont, B. (1993). *The art of teaching dance technique*. Reston, VA: National Dance Association.

Sheak, W. H. (1917). Disposition and intelligence of the chimpanzee. *Proceedings of the Indiana Academy of Science,* 301–310.

Sherrington, C. S. (1900). The muscular sense. In E. A. Schäfer (Ed.), *Textbook of physiology* (pp. 1002–1025). London: Pentland.

Sherrington, C. S. (1906). *The integrative action of the nervous system* (pp. 388–392). New Haven, CT: Yale.

Skinner, B. F. (1938). *The behavior of organisms.* New York: Appleton-Century-Crofts.

Slater, P. J. B. (1983). Bird song learning: Theme and variations. In A. H. Bush & G. A. Clark (Eds.), *Perspectives in ornithology* (pp. 475–499). London: Cambridge University Press.

Suboski, M. D. (1990). Releaser-induced recognition learning. *Psychological Review* **97,** 271–284.

Tayler, C. K., & Saayman, G. S. (1973). Imitative behaviour by Indian Ocean bottlenose dolphins (*Tursiops aduncus*) in captivity. *Behaviour* **44,** 286–298.

Teleki, G. (1974). Chimpanzee subsistence technology: Materials and skills. *Journal of Human Evolution* **3,** 575–594.

Thorndike, E. L. (1896). Animal intelligence. *Psychological Monograph Supplements 2* (4, Whole No. 8).

Thorndike, E. L. (1901). Mental life of the monkeys. *Psychological Monograph Supplements 3* (5, Whole No. 15).

Thorpe, W. H. (1963). *Learning and instinct in animals* (2nd ed.), London: Methuen.

Todt, D. (1975). Social learning of vocal patterns and modes of their application in Grey parrots (*Psittacus erithacus*). *Zeitschrift für Tierpsychologie* **39,** 178–188.

Watson, J. B. (1908). Imitation in monkeys. *Psychological Bulletin* **5,** 169–178.

Wells, M. J. (1968). Sensitization and the evolution of associative learning. In J. Salanki (Ed.), *Symposium on the neurobiology of invertebrates* (pp. 391–411). New York: Plenum.

Westergaard, G. C., & Fragaszy, D. M. (1987). The manufacture and use of tools by Capuchin monkeys (*Cebus appella*). *Journal of Comparative Psychology* **101,** 159–168.

Whiten, A. (1989). Transmission mechanisms in primate cultural evolution. *Trends in Ecology and Evolution* **4,** 61–62.

Witchell, C. A. (1896). *The evolution of bird-song* (p. 212). London: Adam and Charles Black.

13

Acquisition of Innovative Cultural Behaviors in Nonhuman Primates: A Case Study of Stone Handling, a Socially Transmitted Behavior in Japanese Macaques

Michael A. Huffman

Department of Zoology
Kyoto University
Kyoto, Japan

THE STUDY OF NONHUMAN PRIMATE CULTURE IN JAPAN

T he question of whether or not animals have culture, and if they do, how does animal culture differ from that of humans has long been a topic of interest and debate (see Halloway, 1969; Kummer, 1971; Dobzhansky, 1972; Mann, 1972; Weiss, 1973; Moore, 1974; Harris, 1979; Galef, 1992). The pioneering studies of Japanese macaques have brought us closer to answering these questions and have played a significant part in bringing to light the importance of social learning in nonhuman primates.

At the time the Kyoto University Primate Research Group, under the leadership of Denzaburo Miyadi and Kinji Imanishi, began investigations of Japanese macaques in 1948, culture was considered to be a uniquely human trait (e.g., Kroeber & Kluckhohn, 1952). Imanishi, often called the father of Japanese primatology, was one of the first to explicitly suggest the presence of culture in animals. In a paper entitled "The evolution of human nature," Imanishi (1952) discussed instinct and culture, emphasizing that unlike instinct, culture could be viewed as the expression of nongenetically inherited acquired behaviors. He reasoned that if one defines culture as learned by offspring from parents, then differences in the way of life of members of the same species belonging to different social groups could be attributed to culture. Imanishi also stipulated that cultural behaviors can be maintained only in species living in perpetual social groups. Imanishi further suggested that if the above requirements are met whether by wasps or monkeys, we should recognize the existence of culture in a species.

While Imanishi was proposing his definition of culture, members of the Primate Research Group were beginning to investigate Japanese macaques

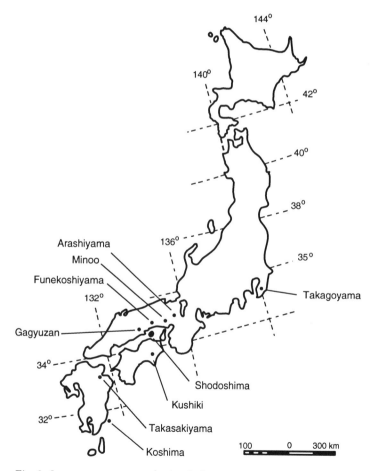

Fig. 1. Japanese macaque study sites in Japan.

throughout Japan (Fig. 1). As time went on, researchers became aware of differences between groups in both social patterns and feeding behavior. For example, while adult males at Takasakiyama regularly displayed paternal care toward 1- to 2-year-olds, such behavior was seldom seen at Arashiyama (Itani, 1959). At Shodoshima the monkeys ate unhusked rice from fields, while monkeys of the Kushiki troop in Tokushima Prefecture did not. Also, monkeys in the Minoo troop in Osaka were observed to dig up and eat the tubers and bulbs of several plant species that monkeys at Takasakiyama would not put into their mouths (Kawamura, 1954, 1965).

Soon thereafter, in the early 1950s, at both Takasakiyama and Koshima, provisioning and individual recognition of all troop members was accomplished, paving the way for long-term research on, and detailed analysis of, social networks, a cornerstone of Japanese primatology. As a result, our early understanding of the acquisition and transmission of new foods and feeding behaviors increased rapidly (e.g., Itani, 1958; Kawamura, 1954, 1959; Itani & Nishimura, 1973).

When provisioning had just begun on the island of Koshima in 1951, the researchers began to spread sweet potatoes along mountain trails and eventually down to the beach in order to lure the monkeys out into the open. By August of 1952, the monkeys, which had previously subsisted in the forest on leaves, nuts, fruits, and insects began to come out onto the beach daily and eat the food provided for them (Itani & Tokuda, 1958). In September 1953, a young female, Imo, began to carry soiled sweet potatoes to a small stream and wash off sand and dirt before eating (Kawai, 1965). The behavior itself, i.e., rubbing and rolling objects between the hands or on the ground, is a typical macaque response both to novel objects and to some common food items. The combining of sweet potatoes and a large open source of water, two previously nonoverlapping elements in macaque life, set the stage for the events that followed.

Sweet potato washing was soon displayed by a play mate of Imo, then Imo's mother and other playmates, eventually spreading to other members of the troop. Next, Imo began to wash her potatoes in sea brine instead of in fresh water and Kawai (1965) proposed that she did this for the salty taste. He classified this behavior as food seasoning.

At about the same time, researchers began to spread wheat grains on the sandy beach. It is undoubtedly a difficult task for monkeys to pick up small grains of wheat when mixed together with grains of sand as there is little necessity for such a technique in the natural foraging substrate of Japanese macaques. Again it was Imo who was first observed to take handfuls of wheat and sand to puddles on the beach or into the ocean and put the mixture into water. In doing so, the grains of wheat and sand were separated making wheat easier to pick up. This activity was named wheat-washing behavior (Kawai, 1965).

These early events have had an important influence on primatology and ethology in general as the foundation for later work on cultural behavior and social learning in animals. Since these early field studies, the criteria for recognizing culture in nonhuman primates have been refined and emphasis put on the processes by which new behaviors are transmitted from one individual to another (e.g., Kummer & Goodall, 1985; Huffman & Quiatt, 1986; Fragaszy & Visalberghi, 1989; McGrew, 1992). Previous assumptions about the way in which novel behaviors are acquired are being challenged and the processes of social facilitation and local enhancement are being examined in the laboratory (e.g., Galef, 1990; Fragaszy & Visalberghi, 1990; Visalberghi, 1993).

COMMON FACTORS OF NEWLY ACQUIRED BEHAVIORS IN JAPANESE MACAQUES

Potato washing and wheat washing are based in part on natural propensities of all members of the species as a whole, but were brought about by the introduction of stimuli new to the particular group in which they were observed. Each new behavior was initiated by a young individual and first transmitted to its mother, sibs and or peers (see Scheurer & Thierry, 1985; Machida, 1990, Higuchi, 1992, for more recent examples). When young females that had learned a behavior matured and had offspring of their own, the behaviors were acquired by their infants (Itani & Nishimura, 1973). It is important to note that many older monkeys do not acquire such new behaviors. In general, adults are quite conservative in this respect suggesting a time window within which many types of new behaviors can be acquired. This plasticity in behavior of subadults underlies the importance of their role in the transmission of new behaviors within a group.

Exceptions to this general rule are observed in the case of the spread of new feeding habits, such as seen in the spread of fish eating among the monkeys of Koshima (Watanabe, 1989). The behavior was started by adult males in the periphery of the troop and was then transmitted to dominant adult females of the group, whereupon it spread more rapidly to other members of the group. The pattern of propagation of fish eating has been suggested to be influenced by a high population density and reduction in provisioned foods on the island (Watanabe, 1989). Indeed, the route of acquisition of new foods should be more flexible than that of other behaviors because of the equal importance for all members to adapt to changing food resource availability.

Many of the behaviors mentioned above came about as the consequence of humans introducing new foods to monkeys. Thus in essence, these acquired behaviors can be considered as a product of acculturation brought about by monkeys' interaction with humans. By introducing new foods into an environment which had previously been little explored by these monkeys, as in the case of potatoes and wheat, conditions were created making it easy for social and behavioral changes to occur.

STONE HANDLING BEHAVIOR

Materials and Methods

Japanese macaques (*Macaca fuscata fuscata*) have been studied since 1954 at the Iwatayama Natural Park, Arashiyama, Kyoto and since 1950 at the Takasakiyama Monkey Park, Oita (see Huffman, 1991a; Baldwin, Koyama, & Teleki, 1980). Both groups are provisioned, During the approximately 15-year period between 1979 and 1994, I have conducted a number of investigations at both sites, focusing mainly on the Arashiyama troops (Huffman, 1991a). One main topic of my investigations has been documentation of the innovation and transmission of a newly acquired behavior called stone play (Huffman, 1984), more recently referred to as stone handling (sic. Huffman & Quiatt, 1986).

In 1986 the 246 members of the main study group at Arashiyama, B troop, underwent fission and two sister troops E and F were formed (Huffman, 1991a). After fission, the dominant E troop became the focal study group of my surveys of stone handling. E troop was selected because its members continued to come to the feeding station and could be more easily followed than could F troop which spent most of its time in the forest.

Between 1979 and 1989 observations of stone handling were recorded mainly with pen and note pad, occasionally supplemented with VHS video recording. *Ad libitum* sampling surveys were conducted with the help of park staff and core-searchers to confirm the identity of all stone handling individuals. From 1989, selected individuals were videotaped, using either a Victor-VHSC or Kyocera-8 mm compact video recorder. From an equal number of males and females of each age, a target individual was selected before each scheduled feeding time and its behavior was continuously recorded on video. The period immediately after feeding is the time in which stone handling is most likely to occur (Huffman, 1984).

Once started, stone handling was sometimes interspersed with other activities such as locomotion, foraging, grooming, etc. On the videotape, time to the tenth of a second was continuously displayed and later used for the exact measurement of the duration of all bouts of behavior. Each video record was transcribed onto a data sheet, noting name, age, and sex of subject, total duration of observation time, duration of contact with stones, and behavioral patterns displayed. This information was then entered into a computer for statistical analysis. The Spearman correlation coefficient and Mann-Whitney U tests, both corrected for ties were used for tests of statistical significance ($p > 0.05$).

Preliminary analysis of a subset of video-recorded focal follows (n = 97) made at Takasakiyama showed that the behavior ceased within 15 min after its initiation. Therefore, the duration of subsequent video sessions were conditionally set at 15 min after the onset of stone-handling activity. An observation was extended in the rare case that the target individual still exhibited stone handling after 15 min.

Analysis of the complete stone-handling video data set (Arashiyama: n = 192; Takasakiyama: n = 71) showed that if stone handling were to be resumed after a bout of other activity then 95% of the time it would be resumed within 120 sec or less. Therefore, to ensure that the data used for the analysis of stone-handling time was complete, only those video records lasting 120 sec or more after stone handling last occurred during the session were used (Arashiyama: n = 169; Takasakiyama: n = 53). Because the average duration of time after the last occurrence of stone handling at the end of a video session was 382.3 sec (S.D. 217.2) at Arashiyama and 535.4 sec (S.D. 238.9) at Takasakiyama, the data are considered to accurately represent complete records of stone-handling activity. The period between the beginning and the end of stone-handling activity during a video session (as defined above) is referred to below as a "stone-handling session."

First Observations

On December 7, 1979, I first observed this peculiar behavior which had never before been observed at Arashiyama. A 3-year-old female, Glance-6476, had several flat stones which she had carried out from the forest onto the open area of the feeding grounds. She first stacked the stones on top of one another in layers of 2–3 stones. She then knocked down the pile and scattered the stones about with the palms of her hands (Fig. 2). When another monkey approached she picked up several stones, moved and sat down again about 5 m away. Although I continued my observations at Arashiyama until September of 1980, this was the first and only time I saw an individual manipulating stones in this way.

By the time I resumed observations at Arashiyama in October of 1983, stone-handling behavior had become a daily occurrence. This behavior was already being transmitted from older to younger individuals.

Fig. 2. First observed case of stone handling at Arashiyama (Glance-6476).

Context in Which Stone Handling Occurs

After further investigation, the circumstances under which the behavior most frequently occurred became clear. As a rule, stone handling is most frequently observed immediately after feeding time, during a period in which individuals usually rest, play, or engage in grooming activities. Individuals seen stone handling always have their cheeks filled with grain and show no signs of distress or other abnormal emotional expressions. Stone handling usually continues until all the grain in the cheek pouches has been ingested.

The stones appropriate for handling can be found almost anywhere in the forest and the brush surrounding the feeding area. Even though there is nothing particularly striking about the stones used in handling to the eye of the human observer, nearby monkeys will approach, and in some cases, snatch the stones away from a handler. It is quite common when another monkey approaches, for an individual handling stones to pick them up and move to another area to resume the activity. Quite frequently, nearby individuals will pick up stones left behind and begin to manipulate them as if they were the only stones available. Attention appears to be drawn to the stones simply because others are manipulating them.

When a stone handler is approached and solicited for example, to play or copulate, he or she will sometimes abandon the stones and join in, but more often than not, the invitee will ignore the solicitor entirely. When approached and groomed by its mother, young (1- to 3-year-olds) will continue to handle stones and stones will remain the handler's focus of attention.

Behavioral Patterns

In 1983, I classified stone-handling behavior into eight basic types: gathering (into a pile in front of ones self), pick up (place in one hand), scatter (about on the ground), roll in hands, rubbing stones (together), clacking (or striking two stones together), carry (from one location to another), and cuddling (Huffman, 1984). Excluding "clacking," all of the above behaviors are commonly exhibited by Japanese macaques when manipulating such objects as twigs and acorns collected in the forest.

Between 1984 and 1985 nine additional behavioral patterns were recognized. Six are obvious variations of the five behaviors described earlier; pick up and drop (repeated over and over), rub on surface (of roof, cement, tree etc.), flinting (strike two stones together), pick up small stones, rub with hands, and grasp with hands. The remaining three behaviors can be considered to reflect a growing familiarity with the objects and their integration with locomotor activities: toss and walk, move and push, and grasp and walk.

The younger, rather than older more "experienced," individuals were responsible for the increase in behavioral patterns observed. In 1991, the only year for which this was tested, a statistically significant difference was recognized between 1 and 4 year olds and adults in the number of behavioral patterns displayed during a stone-handling session (Mann-Whitney, 1- to 4-year-olds $N1 = 32$, 5 years or older $N2 = 71$, $U = 661$, $p = 0.0006$). There was also a statistically significant negative correlation between age and the number of behavioral patterns exhibited ($N = 103$, Rho $= -.38$, $Z = -3.838$, $p = 0.0001$).

These trends were also statistically significant for the Takasakiyama data collected in 1989 (Mann-Whitney, 1- to 4-year-olds $N1 = 31$, 5 years and older $N2 = 23$, $U = 159$, $p = 0.0005$; $N = 34$, Rho $= -0.464$, $p = 0.0003$).

Not only did the number of behaviors exhibited per stone-handling session decrease with age, but also the types of behavioral patterns utilized changed with age. While no single individual was observed to display all of these behaviors, looking at the number of behavioral patterns exhibited by age, in both 1985 and 1991, 1- to 4-year-olds exhibited twice as many of these 17 behavioral patterns than did adults (1- to 4-year-olds in 1985: mean 14.5 per age, SD 2.38 n = 4; in 1991: mean

12.75, SD 1.5 n = 4; 5 years and older 1985: mean 5.60, SD 1.52, n = 5; 1991: mean 8.2, SD 3.29, n = 10). Again, the Takasakiyama data confirms this trend for the 16 behavioral patterns recognized there (1- to 4-year-olds in 1989: mean 12.5, SD 2.08, n = 4; 5 years and older in 1989: mean 6, SD 3.03, n = 6).

Thus, as they grew older individuals tended to become more conservative in their stone-handling behavior, and narrow their handling activities to a few fixed behaviors such as gather, pick up, grasp, rub with hands, and scatter. The more active behaviors, such as clacking, move and push, flinting, toss and walk, and rub on surface, were limited to 1- to 4-year-olds.

Most individuals had their own idiosyncratic repertoire of behaviors that remained stable over an extended period of days, weeks, and in the case of a few of the oldest stone-handling adults, years.

Firm conclusions about the ontogeny of these behaviors await the testing of future data (preferably video) collected from individuals throughout their lifetimes.

Patterns of Transmission

In June of 1984, 49% (115/236 individuals) of the troop had been seen to exhibit stone handling (Fig. 3a). As stone handling is a spontaneous behavior, there is no way to verify whether or not an individual has acquired it other than by persistent checking over time. The survey continued and by June of 1985, an additional 27 (60%, 142/236) individuals born before June 1984 were added to the list (indicated in gray on Fig. 3b). This increase in numbers between 1984 and 1985 is thought to be an effect of increased observation hours rather than an actual increase in the number of new stone handlers during this period. Eighty percent (92/115) of the individuals observed were born between 1980 and 1983, that is, after the first observed record of stone-handling behavior in this troop. The remaining 20% included 6 young adult males (between 4.5 and 8.5 years old), 11 adult females (5+ years), and 6 4-year-old young adult females.

The first female seen to exhibit stone handling, lower-middle ranking Glance-6476, was the only individual of her age group observed to acquire stone handling. Three older females, two her cousins of similar rank, Glance-6775 and Glance-6774, and lower ranking Blanche-596475 were the only monkeys older than Glance-6476 seen to exhibit the behavior. The likelihood that stone handling was first initiated in the group sometime around 1978 or 1979, if not by Glance-6476, then at least by one of the three females listed above, is strongly supported by these observations. Had stone handling begun earlier, we would expect to have found older stone handlers.

Unlike, sweet potato washing and wheat washing, stone handling was never

Arashiyama 1984

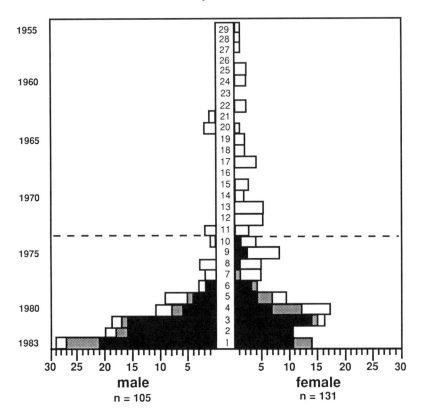

Fig. 3. Histograms of Arashiyama troops in 1984 (B troop) and 1991 (E troop). Black areas represent the number of verified stone handling individuals. Gray areas represent those additional individuals verified as stone handlers between June 1984 and June 1985 in B troop. White area represent those individuals not verified as stone handlers.

acquired by individuals after reaching the age of 5. In the July 1991 survey, no new individuals older than these three females born in 1974 and 1975 acquired the behavior. However, every individual under the age of 10 observed in E troop, was verified to be a stone handler (Fig. 3). In 1985, all offspring (n = 13) of the six mothers 10 years old or younger that exhibited stone handling (Glance-6774, Glance-6775, Blanche-596475, Glance-6476, Oppress-7078, Momo-5978), had also acquired it. It is likely that the infant is first exposed *in utero* to the "click-clacking" sounds of stones as its mother plays, and is then exposed visually to stone handling

Arashiyama 1991

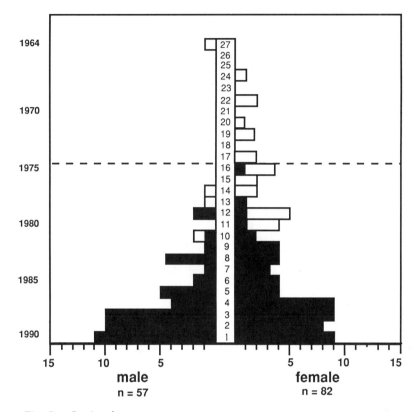

Fig. 3.—*Continued*

as one of the first activities it sees after birth, when its eyes begin to focus on objects around it (Fig. 4).

In 1985, 72% (28/39 pairs) of offspring 5 years old or younger, of nonstone-handling mothers 10 years old or older, had also been seen to handle stones. This observation clearly shows that stone handling can be acquired by a young individual even if its mother does not exhibit the behavior. For example, the infant of Glance-69 (who did not handle stones) was observed to attempt to pick up stones as early as 2 weeks after birth. The infant of Mino-636974 also began to pick up and scatter stones at about 10 weeks after birth even though its mother was not a stone handler. In both cases older siblings exhibited the behavior, suggesting that stone

Fig. 4. Momo-5978 and female infant (Momo-597885) during stone handling bout.

handling may have been acquired by watching the behavior of siblings or their sibling's playmates who spent much time with an infant after feeding and while engaged in stone handling. Observations made periodically during the fall and winter months between 1985 and 1989 strongly suggest that all infants acquired stone-handling behavior within the first 6 months after birth. The exact mechanism of transmission is not known, but social facilitation is likely to play an important role.

TRANSMISSION OF STONE HANDLING AND OTHER CULTURAL BEHAVIORS COMPARED

Figure 5 shows differences and similarities in the patterns of diffusion of stone handling and previously reported cultural behaviors in Japanese macaques. Unlike potato washing or wheat washing, which initially spread from offspring to mother (Fig. 5a), from younger to older kin and among sibs (Kawai, 1965), stone handling first spread only laterally among individuals of the same age class (Fig. 5c). Fish

Transmission Phase

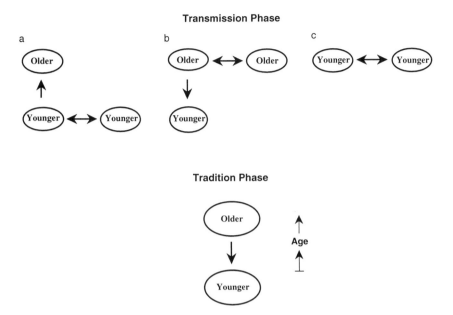

Fig. 5. Patterns of the transmission of Japanese macaque cultural behaviors. (a) Potato-washing, wheat-washing etc. (b) Fish eating and other newly acquired food sources. (c) Stone handling.

eating, on the other hand, initially spread from peripheral males to adult females within the Koshima troop (Fig. 5b). Once fish eating was acquired by adult females, it spread quickly from them to their offspring and among other members of the group (Watanabe, 1989). The pattern of the spread of fish eating is similar to other earlier reported cases of acquisition of new foods by Japanese macaques (Azuma, 1968; Kawamura, 1959, 1965; Yamada, 1958) in that the habit first spread from older to younger individuals. In all cases, this initial period of transmission is recognized as the transmission phase (Huffman, 1984). In common with all the behaviors is the fact that they were subsequently passed down from older to younger individuals in successive generations. This is called the tradition phase (Huffman, 1984, 1991b).

From the above discussion it can be seen that the route of transmission of a novel behavior or food item is in part determined by the nature of the behavior and the social networks in which it normally occurs. Those most likely to be together while engaging in behaviors of the same type should be likely to learn new variants of these behaviors from one another.

For example, while feeding, mothers, their offspring, and their close kin are more likely to feed together than nonkin, making the transmission of innovations concerning provisioned food more likely to occur within matrilines. The transmission patterns of both potato-washing or wheat-washing behavior and stone-handling behavior can be understood in these terms.

THE OCCURRENCE OF STONE HANDLING IN OTHER TROOPS

To the present, stone-handling behavior has been observed in five geographically isolated free-ranging troops of Japanese macaques besides the Arashiyama troop; Koshima (M. Kawai and S. Mito, personal communications), Takagoyama (Hiraiwa, 1975), Takasakiyama (Huffman, 1984; Matsui, 1984), Gagyuzan (F. Fukuda, personal communications) and Funekoshiyama (Itani, Huffman personal observations). See Fig. 1.

According to Kawai and Mito, during the early days of provisioning at Koshima, one individual was seen to clack stones together. However, no other monkeys were observed to replicate this behavior.

Stone handling was first observed at Takagoyama (troop 1) in 1974 by Hiraiwa (1975). However, the frequency of occurrence was much lower and the behavioral patterns less diverse; equivalent to gathering, cuddling, pick up, rubbing stones, and roll in hands as described for Arashiyama (Huffman, 1984). Diffusion of stone-handling behavior had not yet reached the tradition phase, as only individuals 3 years old or younger were observed engaged in the activity. After provisioning at Takagoyama was stopped in 1984, individuals were only occasionally observed handling stones (T. Fujita, personal communication cited in Huffman, 1984).

According to Matsui (1984), stone handling was first noticed at Takasakiyama in 1979 around the same time as it was first seen at Arashiyama. However, the behavior was not widespread enough for me to notice it while briefly observing the Takasakiyama troops (A, B, C) in December of 1979. By 1984, however when I next visited the Takasakiyama site, as at Arashiyama, stone handling had become a widespread, daily occurrence. In the survey conducted at Takasakiyama in 1989, the behaviors observed at Takasakiyama could be classified into the same patterns as those seen at Arashiyama (Huffman, 1991b). Now, at Takasakiyama, as at Arashiyama, stone handling is firmly established as a tradition, i.e., it is consistently passed from one generation to the next. All the behavioral patterns recorded at

Arashiyama (see above), except for pick up and drop, have been observed at Takasakiyama (Huffman, 1991c).

While written records do not exist, I. Narahara, a caretaker at the Funeko-shiyama Monkey Park, recalls seeing stone handling at this site perhaps as early as 1966 (I. Narahara, cited in K. Kaneko, unpublished report). Stone handling does not appear to have ever spread widely among the approximately 300-member troop. The frequency of occurrence of stone handling is noticeably lower at Fune-koshiyama than in either of the similar-sized Arashiyama troops, or the much larger Takasakiyama troops (A,B,C; approximately 2000 individuals total). How-ever, the behavioral patterns displayed are similar (K. Kaneko, unpublished report, Huffman, personal observations).

FACTORS POSSIBLY RESPONSIBLE FOR THE OCCURRENCE OF INNOVATIVE BEHAVIORS AT DIFFERENT SITES

Whether a given behavior spreads and is maintained as tradition within a troop depends on any number of factors, that have yet to be defined. One reason for the variation between sites in the expression of behaviors may be environmental differ-ences between study sties which act to enhance or suppress the transmission of certain new behaviors.

Like stone handling, potato-washing behavior has been observed in a number of troops at locations other than Koshima (Kawai, 1965). Since 1979 I have observed only two adult females, one at Arashiyama and another at Takasakiyama, wash off dirt from potatoes in a small watering pond. While the behavioral elements of brushing off dirt or sand from food items with the hand are typical of Japanese macaques, and nearly all provisioned Japanese macaques at one time or another have been given sweet potatoes, potato washing as a behavioral tradition has never been reported in any free-ranging troop in Japan other than Koshima.

Since the feeding stations of both Arashiyama and Takasakiyama are located on the steep slopes of mountains where there is little space around the limited water sources, opportunities for the behavior to be observed and practiced by many individuals are considerably less than they are at Koshima. This may partially explain why the behavior has never widely spread among monkeys in other provi-sioned troops in Japan.

When conditions are right, however, the behavior can and has been observed to spread in captive situations. Scheurer and Thierry (1985) report the spontaneous transmission of potato washing among Japanese macaques kept in a sandy enclosure with a pool and water spigot at the Burgers' Zoo in Arnhem, Netherlands. Also, Visalberghi and Fragaszy (1990) have artificially induced food-washing behavior in both hand-reared capuchins and crabeating monkeys by placing sandy foods and water basins in their cages. In doing so, Visalberghi and Fragaszy demonstrated that both species have an innate propensity to perform this behavior. However, while free-ranging crabeating monkeys frequently wash and rub food items in water (Wheatly, 1988), food washing in wild capuchins does not appear to be a naturally occurring habit (Visalberghi & Fragaszy, 1990). In the same light, gorillas and orangutans have not been reported to habitually use tools in the wild, but when either is raised in the presence of humans or in captivity they are quite skillful (Russon & Galdikas, 1993; see McGrew, 1992 for discussion).

The fact that we can observe a behavior in some populations, but not others, tells us that the behavior even if within the behavioral capacity of a species, is not always habitual. Environmental factors appear to be important in influencing the expression of such behaviors.

On the other hand, in places where the environmental conditions are quite similar, but behavioral patterns are different, these differences may be attributable to sociopsychological differences such as troop history or in some instances even individual personality traits. Itani and Nishimura (1973) attributed the rapid acquisition of wheat eating in the Minoo B troop to their long history of contact with humans and novel food and material objects.

LIFE HISTORY VARIABLES OF STONE HANDLING

Age and Gender

Two video surveys were conducted in the winter of 1989 at Takasakiyama and the summer of 1991 at Arashiyama. Only data from these video recordings of focal animals were used in the analyses described below. A highly significant negative correlation was found between subject age and total handling time/stone-handling session at both Takasakiyama and Arashiyama (Takasakiyama: n = 53, Rho = −.488, p = 0.0004; Arashiyama: n = 167, Rho = −.435, p = 0.0001; Fig. 6a, b). During any one stone-handling session, older monkeys spent less time stone handling than younger monkeys.

Because the Takasakiyama data set was not large enough to represent equally all age-sex classes, only the Arashiyama data set was used in the following analysis.

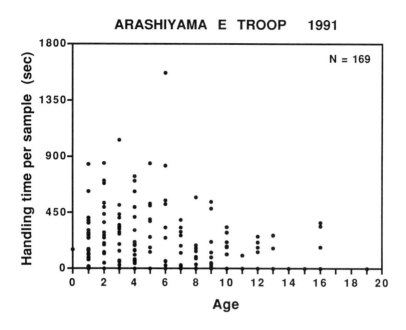

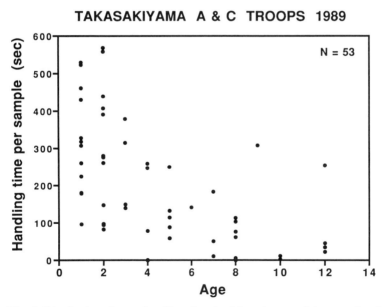

Fig. 6. Distribution of stone handling time (sec.)/bout by age of the stone handler at Arashiyama E troop in 1991 and Takasakiyama A and C troops in 1989.

Among immature males and females (1–4 years old), there was no significant difference in the amount of time spent stone handling during a given session. However, among adults (5 years and older), females spent significantly less time/stone-handling session than did males (Mann-Whitney, Males N1 = 45, Females N2 = 85, U = 1334.5, p = 0.0029).

Parity

Between 1985 and the birth season in the spring of 1986, particular attention was paid to possible changes in stone-handling behavior of females before and after giving birth. Some females were observed to engage in stone handling on the day they gave birth. For example, primiparous Oppress-7079 did not at first, appear to know what to do with her infant; she tried to shrug away from it at times and eventually forced the infant onto the third-ranking male, Deko-64, and moved away to handle stones. Multiparous females will often stone handle with their newborn infants clinging to their chests and nursing. Clearly, birth of an infant does not prevent adult females from stone handling. However, it was found that females with infants 6 months old or younger spent significantly less time/stone-handling session than did adult females without infants (Mann-Whitney with infant n1 = 54, without infant N2 = 31, U = 637, p = 0.04).

Changing Social Roles

In Japanese macaques, females become sexually mature at 3.5 years and males at 4.5 years. At about the time of sexual maturity both males and females stop exhibiting typical play behavior and start to devote more of their energies to the formation and maintenance of different social relationships both inside and outside of their immediate kin groups.

While females' lives may change drastically between the ages of 4.5 and 5.5 years with the birth of their first infants, males of the same age still spend most, if not all, their time in the periphery of the group with a few other males of similar age. This is an extension of the life males have lead since the age of 2 or 3.

Nonetheless, as demonstrated above, males too spend less and less time in stone handling as they mature. For males, the beginning of the resocialization process with adult males and females in the center of the troop, attainment of the highest rank in the periphery, or a move back into the central part of the troop appears to effect a decline in stone-handling time. Older, higher-ranking males both in the central part of the troop and in the periphery spend more time feeding or "policing" disturbances during feeding time than do younger, lower ranking counterparts.

At both Arashiyama and Takasakiyama, adult males between 8 and 13 years old engage in stone handling much less frequently than do females of the same age. Some older males have been observed to only grip briefly or scatter a few stones in passing (Huffman, 1990, unpublished observations). At Arashiyama, the high-ranking adult son of the alpha female, Mino-63697481 has not been seen stone handling since 1985. He has remained within the central part of the troop his entire life and has attained a high rank in E troop.

Unlike potato or wheat washing, in which monkeys stop exhibiting these behaviors only when they are not given potatoes or wheat to wash (Watanabe, 1994), the decline in stone handling by maturing individuals is closely correlated with social and biological life history variables.

THE ADAPTIVE VALUE OF STONE HANDLING CONSIDERED

To date, all newly acquired cultural behaviors reported in Japanese macaques have in common the fact that they are subsistence oriented and thus provide direct benefits to the user in daily feeding activities. Monkeys who wash sweet potatoes, dip them in salt, or separate sand from wheat, are benefiting from their efforts. It is possible that observations of direct benefits acquired by individuals from practice of these behaviors actually encourages their wide diffusion.

However, as the study of stone handling suggests, it is not necessary for a behavior to provide tangible benefits in order for it to be passed from one generation to the next. Stone handling can not be explained in terms of contribution to reproductive fitness or facilitation of subsistence activities. In this respect, stone handling differs from all other cultural behaviors previously reported in Japanese macaques. To the best of my knowledge, stone handling is the first documented case of object play (sic. Candland, French, & Johnson, 1978) as cultural behavior in Japanese macaques.

While stone handling, in its current state, provides no tangible benefits to the performer, it is possible that there are less direct benefits of the behavior. Reminiscent of worry beads or pocket-sized rubbing stones used by humans in some cultures, stone handling itself may be relaxing or stimulating and thus its benefits could be psychological or physiological in nature. Such benefits would go undetected by the human observer without the aid of sophisticated devices to monitor physiological variables such as pulse rate.

It has also been argued that if stone handling persists in any one of these troops material benefits may be acquired in the future if the behaviors undergo modification or are adapted to more practical applications (Huffman & Quiatt, 1986). For example, the initial experience of stone manipulation is an important precursor to the use of stones for adaptive purposes: as tools or as elements of display patterns (e.g., Eaton, 1972; Candland, 1978, 1981; Huffman & Quiatt, 1986).

While not every socially learned behavior is or has to be adaptive, the propensity to learn and adopt new behaviors surely is adaptive in most circumstances. Perhaps someday as a result of the experience gained from stone handling, a new behavior of adaptive value to the troop will arise. However, in its present state stone handling, rather than being the means to an end, appears to be rewarding in itself.

ACKNOWLEDGMENTS

I wish to give my sincere gratitude to Professor J. Itani for his guidance and never ending encouragement throughout the years and to F. Naito for sharing with me her friendship and a unique and sensitive perception of nature. I give my whole-hearted thanks to B. G. Galef and C. Heyes for their invitation to attend the Human Frontier Science Program Conference on Social Learning and Tradition in Animals at Madingly, Cambridge, in August of 1994, where this paper was presented. I am indebted to the Asabas, H. Suzuki, K. Kanbara, N. Iwahashi, and T. Arima; the management and staff of the Iwatayama Nature Park and to T. Matsui of the Takasakiyama Monkey Park for their assistance in the surveys and still more for their friendship and hospitality. My sincerest thanks go to C. Cagampang for her invaluable assistance and lively discussions during the 1991 survey at Arashiyama. Also, I thank the many other colleagues, far and wide, who have provided their critical advise, guidance, and enthusiasm. The research was supported in part by a Grant-in-aid for Special Project Research on Biological Aspects of Optimal Strategy and Social Structure, a full scholarship for graduate studies at Kyoto University by the Japanese Government-Monbusho, a postdoctoral fellowship from the Japan Society for the Promotion of Science, the Cooperative Research Fund of the Primate Research Institute, Kyoto University, and a grant from the Hayao Nakayama Foundation for Science, Technology, and Culture. Without the assistance and aid of these people and institutions this work would not have been possible.

REFERENCES

Azuma, S. (1968). Acquisition and propagation of food habit in a troop of Japanese monkeys. In C. R. Carpenter (Ed.), *Social regulators of behaviors in primates* (pp. 284–292). Lewisburg: Bucknell University.

Baldwin, L. A., Koyama N., & Teleki, G. (1980). Field research on Japanese monkeys: An historical, geographical, and bibliographical listing. *Primates*, **21**, 268–301.

Brueggeman, J. A. (1978). The function of adult play in free-ranging Macaca mulatta. In E. O. Smith (Ed.), *Social play in primates* (pp. 169–191). New York: Academic Press.

Caine, N., & Mitchell, G. (1979). A review of play in the genus *Macaca*, social correlates. *Primates*, **21**, 535–546.

Candland, D. G. (1981). Stone-grooming in *Macaca fuscata*. *American Journal of Primatology*, **1**, 464–468.

Candland, D. G., French, D. K., & Johnson, C. N. (1978). Object-play: Test of a categorized model by the genesis of object-play in *Macaca Fuscata*. In E. O. Smith (Ed.), *Social play in primates* (pp. 259–296). New York: Academic Press.

Dobzhansky, T. (1972). On the evolutionary uniqueness of man. *Evolutionary Biology*, **6**, 415–430.

Eaton, G. (1972). Snowball construction by a feral troop of Japanese macaques (*Macaca fuscata*) living under seminatural conditions. *Primates*, **13**, 411–414.

Fragaszy, D. M., & Visalberghi, E. (1989). Social influences on the acquisition of tool-using behaviors in tufted capuchin monkeys (*Cebus apella*). *Journal of Comparative Psychology*, **103**, 159–170.

Fragaszy, D. M., & Visalberghi, E. (1990). Social processes affecting the appearance of innovative behaviours in capuchin monkeys. *Folia primatologica*, **54**, 155–165.

Galef, B. G. (1990). Tradition in animals: Field observations and laboratory analyses. In M. Bekof & D. Jamieson (Eds.), *Interpretation and explanation in the study of animal behavior* (pp. 74–95). Boulder, CO: Westview Press.

Galef, B. G. (1992). The question of animal culture. *Human Nature*, **3**, 157–178.

Halloway, R. L. (1969). Culture: a *human* domain. *Current Anthropology*, **10**, 395–412.

Harris, M. (1979). *Cultural materialism*. New York: Vintage.

Higuchi, Y. (1992). *The cultural behavior of Japanese macaques*. Tokyo: Kawashima Shobo. (Japanese)

Hiraiwa, M. (1975). Pebble-collecting behavior by juvenile Japanese monkeys. *Monkey*, **19**, 24–25. (Japanese)

Huffman, M. A. (1984). Stone-play of *Macaca fuscata* in Arashiyama B troop: Transmission of a non-adaptive behavior. *Journal of Human Evolution*, **13**, 725–735.

Huffman, M. A., & Quiatt, D. (1986). Stone handling by Japanese macaques (*Macaca fuscata*): Implications for tool use of stone. *Primates*, **27**, 427–437.

Huffman, M. A. (1991a). History of Arashiyama Japanese Macaques in Kyoto, Japan. In L. M. Fedigan & P. J. Asquith (Eds.), *The monkeys of Arashiyama. Thirty-five years of research in Japan and the West* (pp. 21–53). Albany: SUNY Press.

Huffman, M. A. (1991b). Stone-play: A cultural play behavior invented by young Japanese macaques. In T. Nishida, K. Izawa, & T. Kano (Eds.), *Saruno Bunkashi* (pp. 495–504). Tokyo: Heibon Sha. (Japanese)

Huffman, M. A. (1991c). Long-term trends of stone handling, a culturally transmitted

behavior in two isolated populations of free ranging Japanese macaques. International
Ethological Conference, Kyoto Japan (conference paper).

Imanishi, K. (1952). Evolution of humanity. In K. Imanishi (Ed.), *Man* (pp. 36–94). Tokyo:
Mainichi-Shinbunsha. (Japanese)

Itani, J. (1958). On the acquisition and propagation of a new food habit in the troop of
Japanese monkeys at Takasakiyama. In K. Imanishi & S. Altmann (Eds.), *Japanese
Monkeys, a collection of translations* (pp. 52–65). Edmonton: University of Alberta Press.

Itani, J. (1959). Paternal care in wild Japanese monkeys. *Macaca fuscata fuscata. Primates, 2,*
61–93.

Itani, J., & Nishimura, A. (1973). The study of infrahuman culture in Japan. In E. W. Menzel
Jr. (Ed.), *Symposia of the Fourth International Congress of Primatology* (Vol. 1, pp. 26–60).
Basel: Karger.

Itani, J., & Tokuda, K. (1958). Monkeys on Koshima Islet. In K. Imanishi (Ed.), *Nihon
Dobutsuki* Vol. 3. Tokyo: Kobunsha. (Japanese)

Kaneko, K. (1994). Stone play behavior in the Funekoshiyama Japanese monkeys. Research
report submitted as a requirement for the Bachelor of Arts Degree in Humanities,
Kobe Gakuin University, Japan, March 1994. (Japanese)

Kawai, M. (1965). Newly acquired pre-cultural behavior of a natural troop of Japanese
monkeys on Koshima Island. *Primates, 6,* 1–30.

Kawamura, S. (1954). On a new type of feeding habit which developed in a group of wild
Japanese macaques. *Seibustu Shinka, 2,* 11–13. (Japanese)

Kawamura, S. (1959). The process of sub-human culture propagation among Japanese
macaques. *Primates, 2,* 43–60.

Kawamura, S. (1965). Sub-culture among Japanese macaques. In S. Kawamura & J. Itani
(Eds.), *Monkeys and apes—Sociological studies* (pp. 237–289). Tokyo, Chuokoronsha.
(Japanese)

Kroeber, A. L., & Kluckhohn, C. (1952). Culture: a critical review of concepts and definitions. *Papers of the Peabody Museum of American Archeology and Ethnology, 47,* 41–72.

Kummer, H. (1971). *Primate societies.* Chicago: Aldine-Atherton.

Kummer, H., & Goodall, J. (1985). Conditions of innovative behaviour in primates. *Philosophical Transactions of the Royal Society of London, Series B, 308,* 203–214.

Machida, S., (1990). Standing and climbing a pole by a members of a captive group of
Japanese monkeys. *Primates, 31,* 291–298.

Mann, A. (1972). Hominid and cultural origins. *Man, 7,* 379–386.

Matsui, T. (1984). *Takasakiyama, the land of the monkey.* Fukuoka: Nishinihon-Shinbunsha.
(Japanese)

McGrew, W. C. (1992). Chimpanzee material culture: Implications for human evolution.
Cambridge: Cambridge University Press.

Moore, J. H. (1974). The culture concept as ideology. *American Ethnologist, 1,* 537–549.

Scheurer, J., & Thierry, B. (1985). A further food-washing tradition in Japanese macaques
(*Macaca fuscata*). *Primates, 26,* 491–494.

Russon, A. & Galdikas, B. M. F. (1993). Imitation in free-ranging rehabilitant Orangutans (*Pongo pygmaeus*). *Journal of Comparative Psychology,* **107,** 147–161.

Visalberghi, E. (1993). Capuchin monkeys. A window into tool use activities by apes and humans. In T. Gibson & T. Ingold (Eds.), *Tools, language and cognition in human evolution* (pp. 138–150). Cambridge: Cambridge University Press.

Visalberghi, E., & Fragaszy, D. M. (1990). Food-washing behaviour in tufted capuchin monkeys, *Cebus apella,* and crabeating macaques, *Macaca fascicularis. Animal Behaviour,* **40,** 829–836.

Watanabe, K. (1989). Fish: A new addition to the diet of Japanese macaques on Koshima Island. *Folia Primatologica,* **52,** 124–131.

Watanabe, K. (1994). Precultural behavior of Japanese macaques: Longitudinal studies of the Koshima troops. In R. A. Gardner, A. B. Chiarelli, B. T. Gardner, & F. X. Plooji (Eds.), *The ethological roots of culture* (pp. 182–192). Dordrect: Kluwer Academic Publishers.

Weiss, G. (1973). A scientific concept of culture. *American Anthropologist,* **75,** 1376–1413.

Wheatly, B. P. (1988). Cultural behavior and extractive foraging in *Macaca fascicularis. Current Anthropology,* **29,** 516–519.

Yamada, M. (1958). A case of acculturation in a subhuman society of Japanese monkeys. *Primates,* **1,** 30–46. (Japanese with English summary)

14

Studies of Imitation in Chimpanzees and Children

ANDREW WHITEN

DEBORAH CUSTANCE

Scottish Primate Research Group
School of Psychology
University of St. Andrews
Fife KY16 9JU
Scotland

One might think that after a century of research, social learning would be well understood. Yet, we are still wrestling with the most fundamental issues one can think of. Recent papers show there are no concensual answers to the questions "Do monkeys ape?" (Visalberghi & Fragaszy, 1990) "Do rats ape?" (Byrne & Tomasello, 1995) or even "Do apes ape?" (Tomasello, this volume). This chapter describes experiments specifically designed to answer such questions in a comparative fashion across different primate species.

Monkeys and apes have long been thought to be the star performers with respect to imitation in the animal kingdom: Romanes (1882), for example, wrote that "monkeys . . . carry this principle to ludicrous lengths." To "ape" someone is to imitate them. Not satisfied to acquiesce in such everyday ideas, comparative psychologists and ethologists have conducted a profusion of experimental and observational studies in the century since Romanes' pronouncement (Whiten & Ham, 1992). Until recently, these studies were generally seen as supporting the significance of imitation, and the cultural transmission it could produce, in the primates. However, recent years have seen critiques which question the supposed important role of these phenomena, and in some cases even challenge their exis-

Social Learning in Animals: The Roots of Culture

tence. After a century's effort this is quite a scientific drama, conjuring up images of a house of cards collapsing.

The reasons for the reappraisal are various and several reviews document them in detail (Galef, 1990, 1992; Visalberghi & Fragaszy, 1990; Tomasello, 1990, 1994; Whiten & Ham, 1992; Heyes, 1993; Tomasello, Kruger, & Ratner, 1993; Byrne, 1994; and see Tomasello, and Visalberghi & Fragaszy, this volume). Scepticism has multiplied that what has passed for imitation may instead be the result of more simple social learning processes (see Zentall, this volume), or even individual learning, or some combination of the two. Perhaps the most widespread problem has been the failure to rule out the operation of *stimulus* (and/or local) *enhancement* (Spence, 1937), in which an observer B, watching a performer A, has its attention drawn to the most relevant aspects of the objects or environment for the task in hand, without ever copying anything of A's actions (or "learning from A some part of the form of a behavior" as Whiten & Ham formally defined imitation). This is a general problem for field observations of naturally occurring behaviour, illustrated perhaps most graphically in the famous case of sweet potato washing by Japanese macaques—the spread of which was more plausibly through stimulus enhancement than imitation, according to the scrutiny of Galef (1990). But laboratory experiments, typically resting on the "savings method" in which an individual B succeeds on a task more quickly having watched A than not, suffer the same ambiguity (Moore, 1992). This difficulty was made graphic when stimulus enhancement was shown so readily to produce the same outcomes as would imitation in experiments with birds (Sherry & Galef, 1984), and it plagues the earlier extensive range of monkey experiments reviewed by Whiten and Ham. None of this shows that monkeys do *not* imitate: maybe potato washing was learned by imitation, maybe not. What is now clear is that methods with the power to discriminate these processes are required, and we describe these below.

In apes the historical picture is characterized more by lack of experimentation per se. The first experiment in which chimpanzees were offered the benefit of watching another chimpanzee solve a task (raking in food) produced evidence not of imitation, but of what was subsequently called *emulation,* in which an observer copied the result rather than the form of the actions they had witnessed (Tomasello, Davis-Dasilva, Camak, & Bard, 1987; see also Tomasello, this volume). A similar lack of imitation was found in further, similar experiments with chimpanzees and orangutans (Nagell, Olguin, & Tomasello, 1993; Call & Tomasello, 1995a, b).

The scepticism about ape imitation which was thus engendered, and which is thoroughly reviewed by Tomasello in this volume, is a prime reason for the experiments described below. Two others deserve mention, however. One is that there is a wealth of nonexperimental evidence suggesting that apes *do* imitate, often in ways superior in their complexity to anything claimed for other nonhuman animals. The sources of this evidence all have their methodological difficulties (e.g., Hayes & Hayes, 1952; Russon & Galdikas, 1993; and numerous reports of spontaneous imitation collated by Whiten & Ham 1992, and Custance, Whiten, & Russon, in prep.), but together they suggest that a capacity for imitation exists in apes, awaiting experimental confirmation. The other reason for our studies was well illustrated at the Cambridge conference which gave rise to the present volume: there is now little consensus about the nature and existence of imitation in animals. Heyes (1993) argues that imitation has been demonstrated in rats and budgerigars, but not in primates. By contrast, Moore (1992) and Byrne and Tomasello (1995) dismiss the rat and budgerigar studies, but accept several primate studies. We have therefore aimed to refine methods which we believe should unambiguously identify imitation, and we have first applied these to chimpanzees and children, traditionally the most promising candidates.

THE DO-AS-I-DO EXPERIMENT

The question "Can chimpanzees imitate," the principal subject of this experiment, is worth distinguishing from the question "Do they imitate"—the focus of the experiment described much more fully below. The goal of the do-as-I-do experiment was to encourage subjects to show imitation if they were able, and to document this over a range of arbitrary, relatively novel gestures. Subjects were first exposed to a series of training actions performed by a familiar human and their responses were shaped to match what they had seen, on the call "do this!"; then they were presented with an extensive battery of new actions, each with the call to "do this!" Stimulus enhancement is ruled out, because the acts are arbitrary gestures, not object-directed problem solving.

Hayes and Hayes (1952) used a similar approach with a young home-reared chimpanzee, Viki, but unfortunately they did not list the demonstrated actions; they described only *two* of the critical novel responses; no statistical analysis was attempted, and no confirmation by independent observers was offered. Our replication with two, 4-1/2-year-old, nursery-reared chimpanzees therefore had to incorporate a number of refinements: training and test actions have been fully described; subjects' responses were coded by two independent observers; and the statistical probability of the observed number of matches being achieved by chance was computed (Custance, Whiten, & Bard, 1994, 1995).

One chimpanzee attempted 32 out of the 48 demonstrated actions, and 18 of these were correctly identified on the first choice by at least one observer. The second chimpanzee was judged to have attempted 35 actions, and 17 of these were correctly identified on at least one observer's first choice. The polynomial probability that this recorded "hit rate" would have occurred by chance was computed for each of eight tests (two observers × two series of test trials × two chimpanzees): it was less than 0.0001 in all cases. Correct matches were identified in each of eight classes of response (Table 1), including actions on parts of the body which could not be seen (see also Custance & Bard, 1994). We concluded that these chimpanzees, which were not home reared, had at last demonstrated imitation under controlled and fully documented conditions.

Artificial Fruit Experiments

The goal of the do-as-I-do approach is to discover whether a species *can* imitate, when actively encouraged to do so. Given a positive answer, we can still ask a different and important question about what the species typically *does* imitate. We thus become more concerned with what the natural function of imitation is, and what role it plays in the context of social learning in general. In the following experiment we permitted observers to watch the skilled performance of others, and we then asked just what use the observer spontaneously made of the information available.

We designed these experiments around the manipulation of an "artificial fruit." We thus avoided the approach which is common in primate social learning experiments, of incorporating a tool—such as a hammer or a rake—into the task. Although chimpanzees are recognized to be keen tool users, most monkeys are not, and we wished to develop a test which could be used more fairly to compare imitation across such taxa. Therefore, we focused on food-processing, which is a common feature of primate foraging. Once harvested, food items are further manipulated so that components with high nutritive value are separated from others by actions like peeling, tearing, poking, and pulling (e.g., Whiten, Byrne, Barton, Waterman, & Henzi, 1991; Barton & Whiten, 1994). Such skilled actions may take different forms for many different foods and if imitation does subserve cultural transmission in the wild, such actions are obviously likely candidates (Whiten, 1989).

We designed artificial fruits which would be novel to subjects, and which could be presented repeatedly in the same form, whether tackled by demonstrators or observers. A key feature of the fruits is that each part of the "shell" which must

TABLE 1

Matching Acts Identified in the Do-As-I-Do Experiment

Touch in sight	Shoulder	**	Single hand	Open hand	**
	Elbow	**		Wiggle fingers	**
	Stomach	***		Wave stiffly	**
	Thigh	*		Arch fingers	**
	Foot	**			
			Facial	Protrude lips	*
Touch out of sight	Back of head	***		Lip smack	***
	Top of head	**		Teeth chatter	**
	Nose	***		Puff cheeks	*
	Ear	*			
			Face-head	Mouth pop	**
Symm. hand	Clap	**		Lip wobble	**
	All digit touch	***		Pull mouth sides	*
	Interlink fingers	***		Look up	**
	Roll fists	**		Look right	*
	Peekaboo	**			
			Whole body	Jump	**
Asymm. hand	Clap back of hand	***		Flap arms	**
	Clap two digits	*		Hug self	***
	Grab thumb	***		Foot to foot	**

Note. Descriptions given here are intended to convey the range of action presented. Comprehensive descriptions are offered in Custance et al. (1995).

Table includes all actions identified at least once:
* Identified for at least one subject, one coder's second guess.
** Identified for one subject by at least one coder.
*** Identified for both subjects by at least one coder.

be removed can be manipulated in at least two main ways, only one of which is demonstrated to each subject. We can then ask whether subjects who have seen one method tend more often to use this, rather than the other method. Stimulus enhancement is ruled out, because the two different techniques are directed at the same location in space. This method was first applied in experiments with budgerigars (Dawson & Foss, 1965; Galef, Manzig, & Field, 1986), and later with mice and rats (Collins, 1988; Heyes & Dawson, 1990). Yet despite its apparent power to discriminate imitation from other learning processes, primatologists have been slow to adopt it (but now see also Call & Tomasello, 1995a, b). Methods and results we now describe are given in more detail in Whiten, Custance, Gomez, Teixidor, and Bard (1996).

Methods

The fruit used in the present experiment is shown in Figures 1A and B. It does not look much like a fruit, but should rather be thought of as a functional analogue of natural foods, in which parts have to be removed to gain access to an edible core. The latter included popular treats such as strawberries or grapes for the nonhuman primates, and favorite sweets for the children. The "shell" to be removed consisted of various catches restraining a hinged lid. In one series of experiments the catches were a pair of bolts, running through two sets of rings. Subjects saw one of two methods demonstrated to remove these: *poke*, in which the tip of an index finger was used to poke the bolts out; or *twist*, in which the bolt-end was grasped and pulled out with a repeated twisting motion. The twisting motion was not actually necessary, but was incorporated to investigate if such a nonfunctional element would be copied.

In a second series of experiments, the bolts were not in view, and instead the lid was restrained by a flange on a handle, which was located in a barrel fixed to the side of the fruit. Additionally, a pin held the handle in the barrel. Two methods were used to remove both pin and handle. The pin was removed either by the method *spin*, in which the protruding T-piece was spun by the fingers without gripping it, or by the method *turn*, in which the T-piece was gripped by fingers and thumb, then turned by rotation of the wrist and arm. In both cases it was then pulled out. As in the case of *twist* described above for the bolts, this rotation method was not actually necessary for removal, and was therefore unlikely to be performed later by observers unless they were influenced by what they had witnessed. The handle was dealt with either by a *pull*, in which the T-piece on its end was grasped and pulled up vertically, or by a *turn*, in which it was turned so the flange turned away from the lid.

This fruit has been opened using the same techniques in front of individual subjects belonging to several sample groups. The apes were eight chimpanzees of mean age 5.5 years (range approximately 3–8 years), six housed in Madrid Zoo and two at the Yerkes Center, Emory University. Three groups of children were used for comparison, with mean ages of 2.5, 3.5, and 4.5 years. In addition, two samples of capuchins and other groups of children with learning difficulties have been tested, the detailed results for which will be reported elsewhere. Subjects were

Fig. 1. The artificial fruit showing (A) turning the pin, (B) twisting a bolt.

necessarily tested in physically different conditions, but we do not believe this played any major part in the differences observed (see Whiten et al., 1996, for details).

In each of the first four trials the bolts were not present, and subjects witnessed a familiar human either twist the pin and pull out the handle, or turn the pin and then the handle. In a second set of four trials, the handle was not present and instead subjects witnessed the bolts being either poked out using the finger ends, or twisted and pulled out. Before these demonstrations, the subjects had not seen the fruit. Whenever the fruit was opened, the food inside was shown to the subject, and the nonhuman primates were allowed to sniff it. After each demonstration the fruit was reconstituted and approximately 30 s later the subject was allowed 2 min to manipulate the fruit itself. No help or feedback of any sort was offered at this time. After consumption of any food obtained, the fruit was again reconstituted and the next demonstration began. All tests took place in familiar surroundings and were videotaped for later analysis.

Results

Several approaches to analysis of subjects' responses were applied. Together these revealed a limited imitative tendency in the chimpanzees, and a more thorough imitative tendency in the child. In one analysis, two observers blind to which acts subjects had seen, but who had themselves seen demonstrations of the alternative techniques were asked to view videotapes of subjects' manipulation of the fruit and estimate which of the two actions the subject had watched, on a seven-point scale (1 = very confident subjects had watched technique A, 7 = very confident they had watched technique B). The two observers agreed on which act had been witnessed by subjects on 84% of the chimpanzees' bolt trials and 84% of 2-year-olds' pin and handle trials; in all other cases percentage agreements were higher, rising to 100% for 4-year-olds' bolt trials.

Mann-Whitney tests on these scales showed significant differences in the predicted direction between groups who had actually watched one technique rather than the other for the bolts (all samples, $p < 0.04$; Fig. 2) and for the handle (child samples, $p < 0.04$; chimpanzee sample n.s.; Fig. 3), but no sample showed significant differences with respect to the technique used on the pin.

This approach was complemented by a microanalysis of subjects' actions by Custance, in which the occurrence of each target action (twist, poke, etc.) was logged. Comparisons of the frequencies with which each of the specific target actions were performed showed a significant bias toward the observed action only for the 3-year-olds in the case of *poke* ($p = 0.01$), but for all child groups in the case

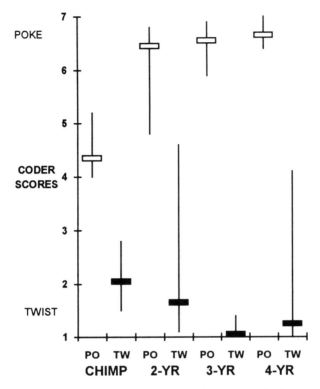

Fig. 2. Independent coders' scores for subjects' action on the bolts. A score of 7 represented confidence the subject had witnessed poke; a score of 1 that they had witnessed twist. Medians and interquartile ranges are shown for subjects who had actually witnessed poke (PO, median = light bar) or twist (TW, median = dark bar).

of *twist* (p = 0.01 in all cases). For both actions, the chimpanzee comparisons approached significance (p = 0.06; Figs. 4 and 5). However, the more powerful way to test for imitation in this design, using the same rationale as for the raters' scale described above, is to test for a predicted bias toward one act rather than the other. Comparing the proportion of target actions which were twists showed significant differences in favor of subjects who had watched twists, for the chimpanzees (p = 0.02)[1] as well as for all child samples (Fig. 6).

1. Because of small sample sizes we have taken the conservative approach of applying non-parametric tests. Such tests, of course, do not take account of the magnitude of the differences. To avoid a misleadingly cautious stance we should also note that application of a t test to this particular, critical result shows a significant difference with t = 12.2 (p < 0.0003).

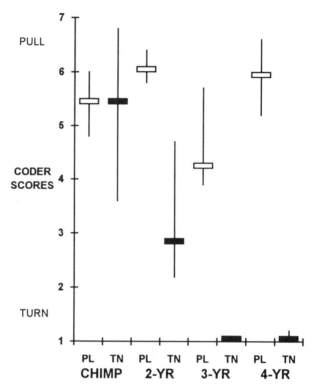

Fig. 3. Independent coders' scores for subjects' action on the handle. A score of 7 represented confidence the subject had witnessed pull; a score of 1 that they had witnessed turn. Medians and interquartile ranges are shown for subjects who had actually witnessed pull (PL, median = light bar) or turn (TN, median = dark bar).

Because subjects also performed acts which resembled those modeled in less precise ways, similar analyses were performed on a category of poke-like actions (which combined *poke* and *push,* the latter involving pushing on the bolts with the hand, but without the finger extended in the way required to force the bolts through the rings), and on a category of twist-like actions (which combined *twist* and *pull,* the latter involving pulling on the bolts without the twisting action). Twist-like actions showed significant bias in the predicted direction for all samples including the chimpanzees ($p < 0.03$), poke-like actions only for the 2- and 3-year-olds ($p < 0.03$).

If subjects were imitating, one might expect this to affect their first actions on the fruit. In fact, all samples showed a significant tendency to perform the act that had been observed before the alternative action (binomial tests, chimpanzees $p =$

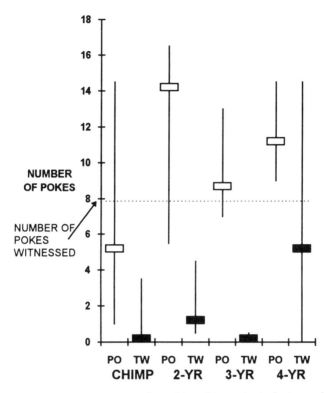

Fig. 4. Number of poke actions performed by subjects who had witnessed poke (PO, median = light bar) or twist (TW, median = dark bar) on the bolts. Interquartile ranges are shown.

0.008; 2-year-olds $p = 0.04$; 3-year-olds $p = 0.004$; 4-year-olds $p = 0.008$). Subjects were accordingly quite efficient in their efforts, the chimpanzees more so than the 2-year-olds. Median latencies to remove the bolts and open the lid for the first time were: chimpanzees 38 s; 2-year-olds 77 s; 3-year-olds 17 s; and 4-year-olds 23 s.

The results for the bolts, summarized in Table 2, thus showed quite extensive imitative matching for the children. Chimpanzees were shown to imitate also, but at a less fine level of accuracy. By contrast, parallel analyses to those described above for the bolts failed to discover any evidence of imitation with respect to the *spin* and *turn* actions on the pin. In the case of the handle, there was again no significant evidence for the chimpanzees, and limited evidence for the child samples, as summarized in Table 3. In this case, however, it was the chimpanzees that opened the fruit most quickly—median times taken to remove the pin, deal with the barrel

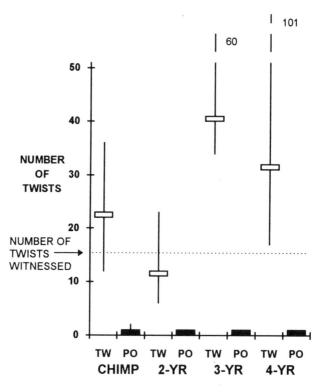

Fig. 5. Number of twist actions performed by subjects who had witnessed twist (TW, median = light bar) or poke (PO, median = dark bar) on the bolts. Interquartile ranges are shown.

and open the lid were: chimpanzees 47 s; 2-year-olds 70 s; 3-year-olds 124 s; and 4-year-olds 91 s. These long periods for the older children were often the result of extensive turning of the pin, as discussed below.

DISCUSSION

The do-as-I-do and artificial fruit experiments are a complementary pair which together show that chimpanzees which do not appear closely to fit the "enculturated" case described by Tomasello (this volume) can and do imitate. The artificial fruit experiment just described offers the first experimental evidence for ape imitation in a functional context designed as an analogue of foraging tasks, thought to be

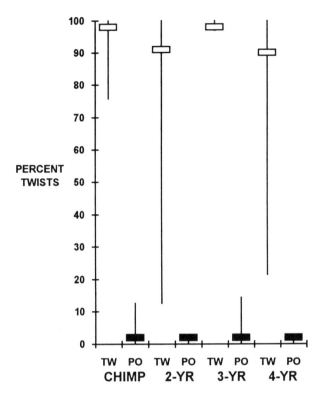

Fig. 6. Percentage of twist actions (relative to total twist + poke) performed by subjects who had witnessed twist (TW, median = light bar) or poke (PO, median = dark bar) on the bolts. Interquartile ranges are shown.

the principal focus of imitative, cultural transmission in the wild (McGrew, 1992; Wrangham, McGrew, de Waal, & Heltne, 1994). Tests using a scaled-down version of the same artificial fruit with capuchin monkeys, to be described in detail elsewhere, did not find comparable evidence of imitation. However, demonstration of imitation is itself no longer a straightforward enterprise. Several different kinds or levels of imitation (or imitation-like processes, according to one's predilection to split or lump categories) have been discriminated (see, in particular, Galef, 1988; Mitchell, 1989; Tomasello, 1990 and this volume; Whiten & Ham, 1992; Tomasello et al. 1993; Byrne, 1994, 1995; Heyes, 1994; Zentall, this volume). The behavior we have observed in chimpanzees and children appears to illustrate some of the most interesting distinctions, which we now discuss in some depth.

TABLE 2

Summary of Significant Differences in the Direction Predicted for Imitation, for Actions on the Bolts

	Chimps	2-yr-olds	3-yr-olds	4-yr-olds
No. of Pokes	+		**	
No. of Twists	+	**	**	**
Percent Twists	**	**	**	**
No. of Poke-like		*	**	
No. of Twist-like	*	*	**	**
Binomial - first act	**	**	**	**
Coders' scores	*	*	**	**

Note. Coders scores are those of the independent observers (see Fig. 2).
* $p < 0.05$
** $p < 0.02$
\+ $p < 0.06$

IMITATION, EMULATION, AND COPYING FIDELITY

Wood (1989) and Tomasello (1990) have drawn attention to a form of social learning which they call *emulation,* in which an observer copies the result of the model's actions, rather than their form. (The latter, which Whiten and Ham called imitation, was distinguished as *impersonation* by Wood. We are not inclined to adopt this because of its common meaning of mimicking a particular person, as in "she gave a good impersonation of the Prime Minister.") In the case of the artificial fruit, emulation would presumably involve learning from the results of the demonstrator's efforts that, to take one example, the bolts had to be slid out of the rings: but the particular method (poking versus twisting) would not be acquired. Emulation goes beyond stimulus enhancement. In the latter, attention is merely drawn to a locus, such as the bolts, whereas in emulation there is additional learning about what objects can *do.*[2] However, Whiten and Ham were careful to

2. On first sight, the distinction between learning about another's behavior versus learning about the environment seems a clear dichotomy (Heyes, 1993; Tomasello, this volume). However, as Whiten and Ham (1992) noted, the distinction can become arbitrary, perhaps most obviously in the case of tool use. In the case of hammering with a hammer, for example, one might replicate just the particular shape of the hammer movements. Could this be emulation and not imitation, because the hammer is classed as

TABLE 3

Summary of Significant Differences in the Direction Predicted for Imitation, for Actions on the Barrel Latch

	Chimps	2-yr-olds	3-yr-olds	4-yr-olds
No. of Turns		*	*	
No. of Pulls				*
Percent Pulls		**	NOT TESTABLE	**
Binomial - first act				
Coders' scores		*	**	**

Note. Coders' scores are those of the independent observers (see Fig. 3).

* $p < 0.05$

** $p < 0.02$

define imitation as involving the learning of "some part of the form" of a behavior, because no imitation will be exact in all its details—after all, even consecutive performances of the "same" action by a single individual will never be identical. An act may be recognized as imitative even if the replication is only "in outline," or if it involves just one or two of several features which could potentially be copied. Thus,

---

part of the environmental result of what one is doing with one's arm and hand? But why should the concept of "behavior" have to stop at the boundary provided by the skin in this way? In hammering, one causes one's limbs to move in a certain way, and the hammer to move in a certain way. Copying could occur at the level of either limb movement or hammer movement, so why should the former be imitation and the latter emulation? We might instead regard the hammer as just an extension of the limb (or conversely, the limb as a rather intimate tool), and copying of the form of the hammering could occur at this level and thus be imitation just as is copying of the limb movements per se (in which case the limb is the tool). However, if this argument is accepted, where does it stop? Dawkins (1982) advocated we extend the concept of the phenotypic effects of genotypes beyond the skin, and once one does this, all parts of "the environment" affected, like beaver dams, become part of the phenotype. By analogy, if beyond-the-skin effects like what a hammer can be made to do can easily be thought of as behavior and can thus be imitated rather than emulated, what of nontool effects, like the movement of the joy-stick in the experiments by Heyes et al, for example, or pin-rotating in our artificial fruit experiments? Perhaps the interesting issue here is not where we as academic investigators draw our semantic boundaries, but whether an animal copying, say, the hammering movements of another makes similar distinctions between what the demonstrator "does" with its limbs, tools, or other parts of the environment.

we can envisage a continuum of fidelity in imitation, from the close to the inexact. The inexactitude could take different forms as suggested above: The imitator might perform just a subset of the elements of the task, or it might perform them all but in a manner only vaguely like the model. Where the imitative match is minimal but the same outcome is nevertheless achieved, it becomes questionable whether a qualitative distinction (i.e., a dichotomy) between imitation (impersonation, in Wood's and Tomasello's terms) and emulation should be sustained. For example, if an observer in the artificial fruit experiment fails to copy the poke/twist variant they saw, but unlike a naive individual immediately gets the bolts out, we might think we should call this emulation. But insofar as they follow the model in using their *hands* to do this, we may be witnessing just a cruder level of imitation by the Whiten and Ham definition. After all, in another experiment we might have had a condition in which the model used their teeth to pull the bolts out, and found that those who had watched a manual poke or twist tended to use a manual action of some sort more than the observers of the oral technique.

Thus, it is perhaps most sensible to consider imitation as involving a continuum of copying fidelity, with the most exact copying at one extreme and emulation as one of the less exact possibilities at the other (Whiten & Ham, 1992; Byrne, 1994). One reason that this continuum is of interest is that, although there are reasons to think imitation can be a significant cognitive achievement for an animal (e.g., Whiten & Ham, 1992; Tomasello et al., 1993; Heyes, 1993), it is easy to appreciate that in the appropriate circumstances, emulation may represent the cognitively advanced option: An ability to select intelligently just those pieces of information which are useful, neglecting details of behavioral form judged to be redundant, perhaps because the requisite skills needed to bring about the observed result are already in place.

Much of the children's behavior we recorded exemplified the relatively high-fidelity end of the imitative continuum (as is obvious from Tables 2 and 3), even to the extent of detracting from ultimate efficiency in problem solution. A clear example was the extensive performance of twisting the bolts (up to 161 times), and rotating the pin (up to 209 times); both of which were actions they had witnessed the demonstrator perform no more than 16 times and which were not necessary for task completion. It is important to note that the children, like the chimpanzees, were not told or encouraged to imitate. Nagell, Olguin, and Tomasello (1993), who tested children and chimpanzees in a rake-using task, found the same contrast we did, with higher level of imitative copying in the children compared to the chimpanzees, sometimes producing relatively less efficient task completion. One may speculate that quite high-fidelity imitation of adults is such a powerful general-purpose strategy for children of these ages that occasional counterproductive over-shooting of its application is a small price to pay.

To the extent that human imitation represents a benchmark for the study of imitation in animals (Meltzoff, 1988; and see Zentall, this volume), the behavior of these children validates the artificial fruit as a potential naturalistic elicitor of imitation. Imitation was indeed demonstrated in the chimpanzees, but most obviously when the level of fidelity required was reduced from that at which it could be demonstrated for the children (i.e., "twist-like" rather than "twist"; Table 2). However, some of the chimpanzees, like the children, performed extensive numbers of twists (Fig. 5). The total number of twists performed by the demonstrator was 16, and all subjects who exceeded this in their own twisting actions were from groups who had witnessed twisting. This included three of the four chimpanzees, who performed 22, 23 and 48 twists respectively, comparable with the older children (3-year-olds: 28, 39, 42, 161 twists; 4-year-olds: 30, 34, 86 twists); just one 2-year-old exceeded the threshold, with 31 twists. Such results are unlikely to be attributable to emulation insofar as the results of the two different actions both involved the bolts moving out of the rings. Moreover, the bolts themselves could not be seen to rotate because they were homogenous in pattern and obscured by the hand performing the twisting actions; all that was visible was the twisting *action* of the hand.

Were cases of low fidelity associated with emulation? To clearly identify emulation we would need further controls, including a condition in which no demonstrator was available, but subject availability did not permit this. Certainly, chimpanzee subjects were quick to make relevant contacts (for example, all but one of the chimpanzees had touched or removed the pin within 6 s of presentation of the handle-and-pin version of the fruit), and without imitating *pull* versus *turn*, they solved the handle task faster than all the child samples (median 47 s for chimpanzees; 70, 124 and 91 s for children of ascending ages). These results could be interpreted as the effects of (rather efficient) stimulus enhancement. More suggestive of emulation were the actions on the pin. Even though rotation of the pin was not functionally necessary, all subjects (excepting one chimpanzee) replicated the pin movement they had witnessed (median rotations of 6 for chimpanzees and 14, 37, and 44 for children of ascending ages), which would not be expected by mere stimulus enhancement. Yet no sample showed an inclination to imitate the technique (either *turn* or *spin*) the demonstrator had used. This is what we would expect in the case of emulation, although without nonobserving controls (all our available subjects were devoted to the two-action design, specifically to test for imitation), we cannot prove that subjects would not have performed such actions spontaneously.

However, we should recall our earlier point, that imitation could still be occurring here but at a cruder level, such as using the hand rather than the mouth or (in the case of the chimpanzees), a foot. We can illustrate the same point about the relative frame within which a sensible contrast of emulation versus imitation must be made, but starting from the end of the continuum where we have documented imitation—the differential use of *twist* versus *poke* on the bolts. The point to be emphasized here is that an action which matches the model in having the form of a manual *twist* can nevertheless vary in several other ways. Among those observed were: (1) holding the bolt in an index-finger-and-thumb precision grip, (2) a power grip, (3) a chuck grip using the thumb and several fingers, or (4) twisting both bolts simultaneously with two hands. Thus, imitating "manual twisting action on the bolt" can be realized through actions which at finer grades of analysis are different in form. This means that *within the imitative continuum* we have exactly the same relationship commonly used to contrast "emulation" of a result, and "imitation" of different forms of behavior used to achieve it. The relative nature of the imitation/emulation distinction is thus emphasized again.

Three significant points emerge from the discussion so far:

1. In the right context, both chimpanzees and children will show behaviour illustrative of the concepts of imitation and emulation
2. Imitation and emulation offer us more insights if we treat them not as absolutes but as phenomena which need to be defined relative to each other within any particular frame of reference
3. Within such a frame of reference, provided by a standardised two-action test with the artificial fruit, the chimpanzee subjects in this study showed less imitative fidelity, and perhaps more emulation, than the children.

PROGRAM-LEVEL IMITATION?

As with other experimental tests of bird, rodent, and ape imitation (for summaries see Heyes, 1993; and Tomasello, this volume), we have focused in the foregoing on the replication of simple "unitary" actions, like poke and twist. We now discuss the possibility of imitation of *sequences* of such actions, for several reasons. First, the apparent imitation of such sequences is perhaps the major way in which some of the evidence of spontaneous imitation in apes, dolphins (Whiten and Ham, 1992), and a parrot (Moore, 1992) could be claimed to be more complex than that shown so far for other taxa (Heyes, 1993). Second, the artificial fruit was originally designed to assess imitation on this level. Third, Byrne (1994, 1995) has suggested that, in a case where imitation operated only at the level of a sequence of subgoals, rather than details of how these are achieved, we have yet to even recognize and label such a process, and he suggests "program-level imitation."

Sequence-Imitation and Program-Level Imitation

Byrne is right to attempt to focus attention on issues of complexity neglected by some recent debates about the existence of imitation in different animal taxa, but what the "program level" can mean in practice requires further analysis. Byrne (1994) attempts to define program-level imitation as "copying only the outline logical structure" of a task, in contrast to "slavish copying." However, we have already noted Whiten and Ham's recognition that all imitation is likely to be inexact to some degree. All imitative copies must have a *schematic* or program-like character, because not every muscle twitch is going to be copied. That is made graphic in interspecies imitation, such as that of a parrot using its foot or wing to wave goodbye, which a human had done with their arm (Moore, 1992 and this volume). Yet Byrne (1995) classifies this as impersonation, not emulation or program-level imitation.

Given that this "outline structure" may be apparent in all imitation, it is important to note that Byrne was careful to insert "logical" into the definition quoted above, remarking that the program-level idea "only becomes interesting when the behaviour involves some logical complexity, such as requiring several subgoals to be achieved in the process of achieving the main goal" (1994). Yet a problem similar to that discussed in the last paragraph remains: Any imitated act can be broken down into a number of sub-goals, the details of how each is achieved varying as noted above. Thus, opening the pin-and-handle version of our fruit involves steps dealing with the pin, handle, and lid; but in turn, dealing with the pin alone involves a "program" incorporating grasping the pin, rotating it in a "loop" of turn-release-grasp-turn, and pulling it out; and at an even lower level, just grasping it involves a program of reaching, orchestrating the digits to (say) a chuck grip adapted to the topography of the pin, and closing these on it. Naturally there is variation in the way each of these steps is achieved. Whiten and Ham thus talked of imitation *in general* when they said that that one individual imitating another, A, has in effect "to get the program for the behaviour out of A's head."

One way to resolve these ambiguities in the idea of program-level imitation is to focus just on differential imitation of the *sequencing* of actions, irrespective of the fidelity with which the form of each action is reproduced. This may or may not address the full scope of program-level imitation as envisaged by Byrne, but it has the merit of being clearly discriminated in practice. The minimum requirement is an act made of two elements, X and Y. If subjects who observe a sequence X-Y are more likely than otherwise to try to perform the acts in the sequence X-Y than Y-X, we can say they are doing "sequence imitation." Given it is just this sequencing we propose to be testable (see below for details) and in view of the ambiguities discussed above with respect to the concept of program-level imitation, we will here cautiously restrict ourselves to the expression *sequence imitation*. However, we *are* proposing that this approach is one way (probably the best, and possibly the only way) to operationalize identification of such programmatic levels of imitation as Byrne's approach focuses on.

A final matter which requires discussion at this point is the nature of X and Y with respect to the concepts of imitation and emulation. If X and Y are each imitated elements, then there appears to be no problem with the label of sequence imitation. However, what if X and Y are emulations? Byrne (1995) acknowledges that they could be, in the case of program-level imitation. But then, should this not be distinguished as program-level emulation (or sequence emulation, in our terms)? This would seem to depend on whether the sequence is learned from the other's *behavior*. If the observer is really only learning about the order of environmental changes X and Y (that the pin comes out, then the handle comes out, in the case of the artificial fruit, for example), then sequence emulation might be an appropriate term. If, however, it is necessary that the observer sees the model act to cause X and Y, then the sequence X → Y becomes "part of the form of a behaviour" in Whiten and Ham's terms, and it seems appropriate to talk of sequence imitation. Here, the observer *does* XY because the model *did* XY. Presumably the same distinctions would apply to program-level imitation/emulation.

Sequence Is One Component of Imitatable Form

Given that the definition of imitation offered by Whiten and Ham allows that only a certain part of the form of an act is replicated, it is important to recognize that sequence is just one of several parts or aspects which might be selectively copied in this way. Such aspects might include the following:

1. *Shape.* This would refer to movements of body parts relative to each other. A wave and a punch differ in shape, as do a smile and a sneer.
2. *Extent.* The copy of any one shape could incorporate the same extent of movement as the model, or less, or more (exaggeration).

3. *Speed.* The performance of any shape could replicate the original speed or vary it. (In addition, if elements are repeated, rate of performance could be subject to selective copying.)

4. *Laterality.* Left-right laterality could be preserved, or ignored.

5. *Egocentric orientation.* The copy of certain shapes might or might not retain their orientation with respect to self—a gesture like waving might vary in orientation with respect to the rest of the body, for example. (Laterality could be regarded as a sub-category of this.)

6. *Allocentric orientation.* Orientation of a behavioral shape might vary with respect to some locus or locii in the environment (as Whiten & Ham noted, it is not clear that this feature in isolation—without any imitation of shape—can be discriminated from the effects of stimulus enhancement).

7. *Sequence.* The sequencing of a series of actions might be maintained, or might not.

Testing for Sequence Imitation Using the Dawson and Foss Two-Action Technique

We suggest that the same approach already described for actions like *poke* versus *twist* with the artificial fruit could be used to test for sequence imitation. In fact, the fruit was originally designed for this with the test involving demonstration of operating first on the bolts, then the pin-handle complex for some subjects, and showing the reverse sequence to others. We can then ask if observers follow the sequential order they witnessed. At first sight, this seems to have the advantage of the proven power of the original experimental design used at the nonsequential level. However, there is a problem. When this design is applied to sequences, it depends on both sequences being feasible for task completion, and both elements being demonstrated. With sufficient mechanical understanding, a subject who might be quite capable of sequence imitation might nevertheless decide that performing the two acts in any order will work (which is true in a case like the artificial fruit), perhaps just doing the nearest one first.

In fact recent results suggest this is not a problem, at least using a human benchmark. In our first experiments with chimpanzees described above we did not incorporate the sequences test, because recent critiques of the imitative abilities of primates led us to favor testing using a relatively simple task in the first instance: We tested bolts and pin-handle separately. However, our colleague Julie Brown (1995) has recently presented the "whole fruit"—with bolts, pin and handle all in place—to several different samples of normal children and children with learning difficulties she found that 67% of 3- to 4 year-olds and 92% of 5- to 6 year-olds followed the sequence witnessed. Thus the older children, who might be most expected to follow the "aloof" attitude sketched above, were the ones who most clearly copied the sequence they had seen. We conclude that this does indeed provide us with a meaningful experimental test of sequence learning for further exploration with nonhuman species.

Testing for Sequence Imitation Using Logical Sequences

By referring to program-level imitation as copying "logical structure," Byrne (1994) appears to circumvent the potential difficulties discussed in the previous section, of applying tests involving XY versus YX sequences which subjects might perceive as arbitrary. The tasks he focuses on instead are those which necessarily require achievement of the observed series of subgoals in logical order. According to Byrne (1995) the best evidence for this comes from the foraging of mountain gorillas (Byrne & Byrne, 1993), and it takes two forms. The first is that for each of four major food types, virtually all mature individuals in the population used the same method when analyzed in terms of a sequence of several subgoals (e.g., for stinging nettles, "accumulation of a bundle of leaves, detaching the particularly virulent petioles from the more innocuous leaf blades, and then folding the blades so that the least stinging surface is presented to the lips"), yet the ways different individuals achieve these varies. But of course this combination (universal scheme, done with minor variations) also characterizes the behavioral product of trial-and-error learning under conditions where there is an optimal series of steps to take, and although Byrne believe that this is implausible in the case of mountain gorilla foraging, there is as yet no direct evidence to discount it, or other forms of social learning (see also Tomasello, this volume). The second line of evidence from the gorillas is that an immigrant female and her son, who came from an area with few nettles, do not fold the stinging surface under. This is circumstantial evidence of a role for imitation in acquisition of this technique, but at a single-element level (equivalent perhaps, to the level of twist versus poke, in the artificial fruit), rather than bearing on the question of program-level or sequence imitation.

Sequence imitation in children can take only minutes, so to isolate such a process within the years of opportunities for trial-and-error learning of optimal sequences in primates is clearly a considerable challenge. We suggest that experiments can complement the field research on this issue, by virtue of their power to discriminate cause and effect. However, the experimenter faces a problem: If the task has a logical structure such that it can only be done in a particular sequence (for example, removing pin before handle, in the artificial fruit), the scope for controls appears limited. For a start, one cannot apply the two-action approach described earlier, because only one sequence provides the logical solution to the task. And the alternative of a no-demonstrator control suffers a different difficulty, namely that it could not itself rule out a role of stimulus enhancement if animals who watch a demonstrator are then able to perform the sequence of actions more quickly than nonobservers. What seems to be required is a task in which subjects could be recorded as at least *trying* to perform elements in any order (without success, obviously, if the order is counter to the logical one required: trying to operate the handle of the artificial fruit, for example, before tackling the pin). If in the demonstrator condition they attempt the elements in the demonstrated order more than in the no-demonstrator control condition, we would have direct evidence of sequence imitation (and presumably, program imitation in Byrne's terms).

In our artificial fruit experiments with the pin and handle alone, there were three distinguishable subcomponents: (1) removing the pin, (2) pulling or turning the handle, and (3) raising the lid by grasping and pulling on the bolts rings. For six chimpanzee subjects *attempts* at each of these were recognizable, even when they could not be carried through because a necessary preceding step had not been completed (trying to pull open the lid before dealing with the handle, for example). Although we did not have sufficient subjects to perform no-demonstrator controls, we have made an exploratory study on the order of emergence of these three actions on the first trial to answer the initial question of whether the observed sequence is followed. In fact only two of the six chimpanzees followed the sequence of pin-handle-lid which they had seen demonstrated. This is obviously consistent with the outcome we might expect by chance. Three attempted the handle first, of whom two then tackled the pin next, the third switching instead to pull up on the bolt rings in a failed effort to open the lid. The sixth subject started by pulling up on the bolt rings, and then tried to manipulate the handle.

There was thus no sign of imitation at the sequence (or program) level in this preliminary posthoc analysis. However, it must be remembered that subjects had at this stage seen one demonstration only, and because we wished to avoid prior discovery of opening techniques by trial-and-error, this was also their first opportunity to explore the fruit. Observations of chimpanzee subjects attempting to open the fruit used in this experiment suggest that experiments explicitly directed at the sequence issue should involve a greater number of prior demonstrations, in a task which is sufficiently difficult that subjects would be unlikely to solve the first step themselves if allowed a period of exploration before any demonstration of opening.

IMITATIVE NOVELTY AND COMPLEXITY

Probably most workers in the area of social learning would agree that the most impressive evidence of imitation comes from the replication of acts which share the (often related) characteristics of novelty and complexity. Thorpe (1963) defined imitation in terms of reproducing "novel or improbable" acts. Byrne and Tomasello (1995) insisted that imitation must involve "the animal's learning a new response," and for this reason rejected the results of Galef et al. (1986) because the budgerigars' actions which were replicated were "part of the budgerigar's foraging repertoire."

But what is to count as novel? Any species can only do a limited number of actions with its effectors, and by the time it is mature it will likely have explored most of the basic motor patterns of which it is capable. We may think of human manual actions as particularly flexible: But what does it take for you to perform an "entirely novel" hand action (leaving aside environmental effects) that you have never performed before? Novelty is surely always relative, and what is accepted as an interestingly novel imitation is going to be constructed from actions already in the repertoire (in the do-as-I-do experiment, it was possible to observe this process taking place in some cases; Custance et al., 1995). Accepting this, there are perhaps two obvious bases for what is to be judged objectively as novel: (1) novel combinations (and/or adjustments) of acts already in the repertoire with each other, and (2) novel combinations (and/or adjustments) of acts already in the repertoire with some environmental interface. Possible examples of the first of these include the sequential combinations of gorillas' foraging actions described by Byrne and Byrne (1994), which may be built up from elements already in the repertoire (such as "pick out debris," an element incorporated in several food processing routines); sequences of foot and wing movements by Moore's parrot (Moore, 1992); and responses to our "do-as-I-do" items like peekaboo, which involved a symmetrical movement of both hands to temporarily obscure the eyes, like those which the animal must frequently perform unilaterally. Possible examples of the other basis of novelty, directing actions already in the repertoire to new aspects of the environment, include those performed on objects by the chimpanzees in the experiments by Tomasello et al. (1993), food processing skills like the "hold loosely in one hand and pick out debris with the other" applied by gorillas to several foods (Byrne & Byrne, 1992), and the actions of budgerigars and rats in the experiments by Galef et al. (1986) and Heyes, Dawson, and Nokes (1992). The actions of poking and twisting bolts studied in the present experiment are, we submit, novel in this sense.

Are any researchers claiming that the imitative acquisition of skills in the wild involve more than these "making new from old" steps?

Extended analysis of the subject of complexity of imitation is beyond the scope of this chapter. However, we suggest that it can be related to the preceding discussion, because one basis for comparison of complexity (across taxa, or ages, for example) will lie in the number of novel act-act or act-environment combinations acquired in any particular performance. This is surely what is impressive in orangutans' performance of action sequences they have seen done by humans, like lighting fires, as documented by Russon and Galdikas (1993 and in press).

IMITATIVE ABILITY AND ONTOGENETIC HISTORY

Tomasello (1990 and this volume) has pointed out that evidence of imitation in apes has been restricted to those reared in ways very like those experienced by children, experiences he describes as "enculturation." Experiments with nonenculturated apes have produced negative results. This leads to doubt about whether wild apes, who lack such support, imitate. An alternative outlined by Whiten (1993) is that it is the nonenculturated apes who are the odd ones out, often living in socially and physically impoverished conditions contrasting with the apparent opportunities of wild apes to acquire a variety of complex behaviors such as tool use and foraging skills. Whiten speculated that, as is true for many aspects of ontogeny, the first steps an infant takes toward imitation may require certain rewards to be forthcoming to advance the ability, and without these, it may atrophy.

The present experiments cannot resolve such questions, which for the moment must remain open, as Tomasello (this volume) agrees. But so far as we can see, our chimpanzees do not obviously fit the specification of enculturation that Tomasello outlined: although reared in captivity with some early care by humans, this has not involved any attempt to "humanize" them and there has been no encouragement to imitate. Yet we have provided evidence of imitation, in both the do-as-I-do and artificial fruit experiments, which supports the existence of a basic imitative capacity in the absence of intensive humanization. The relatively unimpressive nature of the imitation shown in these studies when compared with that shown by the children or claimed elsewhere for apes (e.g., Hayes & Hayes, 1952; Whiten & Ham, 1992; Russon & Galdikas, 1993 and in press) might turn out to be due to lack of human enculturation, but it remains to be determined whether it may instead be the result of a lack of an ontogenetic history of natural "chimpanzee enculturation."

REFERENCES

Barton, R. A., & Whiten, A. (1994). Reducing complex diets to simple rules: food selection by olive baboons. *Behavioural Ecology and Sociobiology, 35,* 285–385.

Brown, J. D. (1995). *Imitation, play and theory of mind in autism—an observational and experimental study.* PhD Thesis, University of St Andrews.

Byrne, R. W. (1994). The evolution of intelligence. In P. J. B. Slater & T. R. Halliday (Eds.), *Behaviour and evolution.* Cambridge: Cambridge University Press.

Byrne, R. W., & Byrne, J. M. E. (1994). The complex leaf-gathering skills of mountain gorillas (*Gorilla g. beringei*): Variability and standardization. *American Journal of Primatology, 31,* 241–261.

Byrne, R. W., & Tomasello, M. (1995). Do rats ape? *Animal Behaviour, 50,* 1417–1420.

Call, J., & Tomasello, M. (1995a). The social learning of tool use by orangutans (*Pongo pymaeus*). *Human Evolution, 9,* 297–313.

Call, J., & Tomasello, M. (1995b). The use of social information in the problem-solving of orangutans (*Pongo pygmaeus*) and human children (*Homo sapiens*). *Journal of Comparative Psychology, 109,* 308–320.

Collins, R. L. (1988). Observational learning of a left-right behavioural asymmetry in mice (*Mus musculus*). *Journal of Comparative Psychology, 102,* 222–224.

Custance, D. M., & Bard, K. A. (1994). The comparative and developmental study of self-recognition and imitation: the importance of social factors. In S. T. Parker, R. W. Mitchell, & M. L. Boccia (Eds.), *Self-awareness in humans and animals: developmental perspectives* (pp. 207–226). Cambridge: Cambridge University Press.

Custance, D. M., Whiten, A., & Bard, K. A. (1994). The development of gestural imitation and self-recognition in chimpanzees (*Pan troglodytes*) and children. In J. J. Roeder, B. Thierry, J. R. Anderson, & N. Herrenschmidt (Eds.), *Current Primatology: Selected Proceedings of the XIVth Congress of the International Primatological Society, Strasbourg, Vol. 2: Social Development, Learning and Behaviour.* (pp. 381–387). Strasbourg: Université Louis Pasteur.

Custance, D. M., Whiten, A., & Bard, K. A. (1995). Can young chimpanzees imitate arbitrary actions? Hayes and Hayes (1952) revisited. *Behaviour, 132,* 839–858.

Custance, D. M., Whiten, A., & Russon, A. E. (in prep.) A catalogue of observational reports of imitation in primates.

Dawkins, R. (1982). *The extended phenotype.* Oxford: Freeman.

Dawson, B. V., & Foss, B. M. (1965). Observational learning in budgerigars. *Animal Behaviour, 13,* 470–474.

Galef, B. G. (1988). Imitation in animals: History, definitions, and interpretation of data from the psychological laboratory. In T. Zentall & B. Galef (Eds.), *Social Learning: Psychological and biological perspectives* (pp. 3–28). Hillsdale, NJ: Erlbaum.

Galef, B. G. (1990). Tradition in animals: Field observations and laboratory analyses. In

M. Bekoff & D. Jamieson (Eds.), *Interpretations and explanations in the study of behaviour: Comparative perspectives* (pp. 74–95). Boulder, CO: Westview Press.

Galef, B. G. (1992). The question of animal culture. *Human Nature,* **3,** 157–178.

Galef, B. G., Manzig, L. A., & Field, R. M. (1986). Imitation learning in budgerigars: Dawson and Foss 1965 revisited. *Behavioral Processes,* **13,** 191–202.

Hayes, K. J., & Hayes, C. (1952). Imitation in a home-raised chimpanzee. *Journal of Comparative and Physiological Psychology,* **45,** 450–459.

Heyes, C. M. (1993). Imitation, culture, cognition. *Animal Behaviour,* **46,** 999–1010.

Heyes, C. M. (1994). Social learning in animals: Categories and mechanisms. *Biological Review,* **69,** 207–231.

Heyes, C. M., & Dawson, G. R. (1990). A demonstration of observational learning in rats using a bidirectional control. *Quarterly Journal of Experimental Psychology,* **42B,** 59–71.

Hayes, C. M., Dawson, G. R., & Nokes, T. (1992). Imitation in rats: initial responding and transfer evidence. *Quarterly Journal of Experimental Psychology,* **45b,** 59–71.

McGrew, W. C. (1992). *Chimpanzee material culture: Implications for human evolution.* Cambridge: Cambridge University Press.

Meltzoff, A. N. (1988). The human infant as *Homo Imitans.* In T. Zentall & B. Galef (Eds.), *Social Learning: Psychological and biological perspectives* (pp. 319–341). Hillsdale, NJ: Erlbaum.

Mitchell, R. W. (1987). A comparative developmental approach to understanding imitation. *Perspectives in Ethology,* **7,** 183–215.

Moore, B. R. (1992). Avian movement imitation and a new form of mimicry: Tracing the evolution of a complex form of learning. *Behaviour,* **122,** 231–263.

Nagell, K., Olguin, R. S., & Tomasello, M. (1993). Processes of social learning in the tool use of chimpanzees (*Pan troglodytes*) and human children (*Homo sapiens*). *Journal of Comparative Psychology,* **107,** 174–186.

Passingham, R. (1982). *The human primate.* New York: Freeman.

Romanes, G. J. (1882). *Animal intelligence.* London: Kegan Paul Trench & Co.

Russon, A. E., & Galdikas, B. M. F. (1993). Imitation in ex-captive orang-*utans* (*Pongo pygmaeus*). *Journal of Comparative Psychology,* **107,** 147–161.

Russon, A. E., & Galdikas, B. M. F. (1995). Constraints on great ape imitation: model and action selectivity in rehabilitant orangutan imitation (*Pongo pygmaeus*). *Journal of Comparative Psychology,* **109,** 5–17.

Sherry, D. F., & Galef, B. G. (1984). Cultural transmission without imitation: Milk bottle opening by birds. *Animal Behaviour,* **32,** 937–938.

Spence, K. W. (1937). Experimental studies of learning and higher mental processes in infra-human primates. *Psychological Bulletin,* **34,** 806–850.

Thorpe, W. H. (1963). *Learning and instinct in animals.* London: Methuen.

Tomasello, M. (1990). Cultural transmission in the tool use and communicatory signalling of chimpanzees? In S. Parker & K. Gibson (Eds.), *Language and intelligence in Monkeys and*

Apes: Comparative Developmental Perspectives (pp. 274–311). Cambridge: Cambridge University Press.

Tomasello, M. (1994). The question of chimpanzee culture. In R. W. Wrangham, W. C. McGrew, F. B. M. d. Waal, & P. G. Heltne (Eds.), *Chimpanzee cultures* (pp. 301–317). Cambridge, MA: Harvard University Press.

Tomasello, M., Davis-Dasilva, M., Camak, L., & Bard, K. (1987). Observational learning of tool-use by young chimpanzees. *Human Evolution, 2*, 175–183.

Tomasello, M., Kruger, A. C., & Ratner, H. H. (1993). Cultural learning. *Behavioral and Brain Sciences, 16*, 495–552.

Tomasello, M., Savage-Rumbaugh, S., & Kruger, A. C. (1993). Imitative learning of actions on objects by children, chimpanzees, and enculturated chimpanzees. *Child Development, 64*, 1688–1705.

Visalberghi, E., & Fragaszy, D. (1990). Do monkeys ape? In S. Parker & K. Gibson (Eds.), *Language and intelligence in monkeys and apes: Comparative developmental perspectives* (pp. 247–273). Cambridge: Cambridge University Press.

Whiten, A. (1989). Transmission mechanisms in primate cultural evolution. *Trends in Ecology and Evolution, 4*, 61–62.

Whiten, A. (1993). Human enculturation, chimpanzee enculturation and the nature of imitation. *Behavioral and Brain Sciences, 16*, 538–539.

Whiten, A., Byrne, R. W., Barton, R. A., Waterman, P. and Henzi, S. P. (1991). Dietary and foraging strategies of baboons. *Philosophical Transactions of the Royal Society of London. Series B, 334*, 27–35.

Whiten, A., Custance, D. M., Gomez, J.-C., Teixidor, P., & Bard, K. A. (1996). Imitative learning of artificial fruit processing in children (*Homo sapiens*) and chimpanzees (*Pan troglodytes*). *Journal of Comparative Psychology, 110*, 3–14.

Whiten, A. & Ham, R. (1992). On the nature of imitation in the animal kingdom: reappraisal of a century of research. *Advances in the Study of Behaviour, 21*, 239–283.

Wood, D. (1989). Social interaction as tutoring. In M. H. Bornstein & J. S. Bruner (Eds.), *Interaction in human development* (pp. 59–80). Hillsdale, NJ: Erlbaum.

Wrangham, R. W., McGrew, W. C., deWaal, F. B. M., & Heltne, P. G. (Eds.). (1994). *Chimpanzee cultures:* Cambridge, MA: Harvard University Press.

Do Apes Ape?

MICHAEL TOMASELLO

Department of Psychology and Yerkes Primate Center
Emory University
Atlanta, Georgia 30322

I n 1990, Visalberghi and Fragaszy asked the question: "Do monkeys ape"? After a review of the pertinent research literature, they answered in the negative. In this chapter I would like to ask the similar question: "Do apes ape"? Although many researchers assume that the answer to this question is more positive, there are two complicating factors. First, there are many different ways to ape. Beginning with Thorpe (1956), a number of theorists have demonstrated that the same behavioral outcome may result from very different processes of learning and social learning (e.g., Galef, 1988, 1992; Whiten & Ham, 1992). This potential ambiguity in what it means to "ape" thus makes the current question less than totally straightforward, and it means that a certain amount of theoretical work must be done before the relevant research literature may be usefully examined.

The second complicating factor is that research on the social learning of apes comes from three very different sources: (1) observational studies of apes in their natural habitats in which it has been discovered that there are population differences of behavior within the same species. These studies have the virtue of ecological validity, but their lack of experimental control means that they are of limited usefulness in determining with any precision the learning processes that are responsible for the observed population differences. (2) experimental studies of apes in captivity. These studies have the required experimental controls, but they investigate animals who, because they have been raised and tested in artificial circumstances, may or may not be representative of their conspecifics in the wild. (3) observational and experimental studies of apes that have been raised and/or trained by humans in various ways. These studies reveal the "potential" of apes given a certain kind of ontogenetic niche, but again they investigate animals who may or may not be representative of their conspecifics in the wild. The upshot is that because each of these types of studies has its own strengths and weaknesses, any attempt to provide an overall answer to the current question must take explicit account of the very different nature of the types of apes involved.

In my attempt to answer the question of whether apes ape, I proceed as follows. First, I make some theoretical distinctions between several different types of social learning that may be at work in various circumstances in which apes have been observed. Second, I provide an assessment of the research literature concerning the social learning skills of apes, being careful to keep separate the three types of ape specified above. Finally, I conclude by attempting an overall answer to the question of whether apes ape and what this might mean for our understanding of their behavioral traditions or "culture" in the wild.

EMULATION, MIMICKING, AND IMITATIVE LEARNING

As perhaps no other area in the study of animal behavior and cognition, the study of social learning is plagued by terminological issues. For current purposes I would simply like to make one broad distinction. An animal observing the instrumental behavior of another animal may learn something about either: (1) the environment and the changes in the environment that result from the observed behavior, or (2) the observed behavior itself. Heyes (1993a,b) makes a similar distinction, referring to the former as social learning and the latter as imitation. But because she is working within a theoretical framework of associative learning, Heyes must talk about what the observing animal learns in the awkward terminology of stimuli and responses and their derivatives such as "stimulus-response learning by observation," "action-outcome contingency learning," and "response-reinforcer learning." I would like to make this same very important distinction, but I would like to do so from the point of view of cognitive theorists such as Kohler (1927), Piaget (1962), and Vygotsky (1978).

Learning about the Environment in Social Situations

The classic term for learning about the environment as a result of the behavior of others is "local enhancement" (Thorpe, 1956) or "stimulus enhancement" (Spence, 1937). Local enhancement is a phenomenon widespread in the animal kingdom, and refers to situations in which animals are attracted to the locations at which conspecifics are behaving (or perhaps to stimuli conspecifics are interacting with). This then places them in a position to learn something that they would not otherwise have learned, and what they learn is often identical to what the conspecifics are learning. For instance, one animal may be attracted by conspecifics to a location where there is food, and then learn on its own to extract the food from a substrate (e.g., dig it out of the ground). The conspecifics may be digging also, but in local enhancement the observing animal is not influenced by their behavior at all.

In local or stimulus enhancement nothing is actually learned from the behavior of others; the learner is simply attracted to a location or object. In the cognitive view, however, animals may sometimes actually learn things about the environment by observing the manipulations of others. A simple example might be the classic experiments on observational learning by monkeys (summarized by Hall, 1963) in which one individual learns where food is located by watching another search under various objects (assuming that indeed they are learning a spatial relationship and not just an attraction to the occluding object). More complex examples would be cases in which individuals learn by observing others in such things as: a nut can be opened and food found inside, a log can be rolled over and food found under it, sand comes off food when it is in water, or a stick hitting a piece of fruit can cause it fall to the ground. In the terminology of Gibson (1979), by observing the manipulations of other animals individuals may learn all kinds of "affordances" of the environment that they would be unlikely to discover on their own. The individual observing and learning some affordances or changes of state of the inanimate world as a result of the behavior of another animal, and then using what it has learned in devising its own behavioral strategies, is what I have called emulation learning (Tomasello, 1990). Emulation learning might thus be thought of as the cognitivist's answer to local enhancement—an individual is not just attracted to the location of another but actually learns something about the environment as a result of its behavior.[1]

It must be emphasized that in the cognitive view just looking in the direction of environmental changes is not enough to learn about and reproduce them; the

1. Perhaps because in my original formulations of emulation learning I focused on one animal's reproduction of the "outcome" of another's behavior, Whiten and Ham (1992) recast the process of emulation as "goal emulation" in which the learner understands the goal of another and reproduces that. This was not and is not my idea of emulation learning. To reiterate, in emulation learning the learner observes and understands a change of state in the world produced by the manipulations of another— which may be its only way of learning that such a change of state is possible. On some occasions the individual will then want to produce that same change of state for itself.

cognitive capacities of the particular organism doing the observing must always be considered. Thus, Visalberghi (1993) has argued and presented evidence that even when capuchin monkeys "observe" the tool use of others they do not understand it because they do not understand the causal structure of tool use. Capuchins' observations of others may serve to attract them to the tool and goal (local or stimulus enhancement), but they learn to use the tool exclusively through their own trial-and-error efforts. They thus do not learn this task by means of emulation learning, although they might learn other tasks more suited to their cognitive abilities in this way. On the other hand, Tomasello, Davis-Dasilva, Camak, and Bard (1987) argued and presented evidence that chimpanzees who observe a conspecific using a tool *do* learn something about the task. By watching the tool touch and move the food in particular ways they learn about the cause–effect relations between tool and goal (the affordances of the tool vis-à-vis the goal), and they then incorporate this knowledge into their own attempts at tool use. They learned to use the tool via emulation learning because as a result of observation they understand something of the causal structure of the task.

The overall point is thus that animals may learn a variety of facts about entities and events in the environment, and the spatial-temporal-causal relations among entities and events in the environment, on the basis of observation—with what they are capable of learning depending on their cognitive capacities. The behavior of other animals often causes changes or transformations in the environment that expose individuals to learning experiences that they would very likely not encounter on their own. Emulation learning is thus a potentially very powerful means of social learning in which the individual discoveries of one individual facilitate the learning of others in the group. It should also be noted that if animals of the same species have similar cognitive abilities with which to understand their observations, and similar motoric abilities with which to produce behavior, then in similar circumstances their behaviors and behavioral strategies learned via emulation learning will appear highly similar.

Learning about Behavior

Despite the power of local enhancement, stimulus enhancement, observational learning, and emulation learning to help individuals benefit from the skills of others, all of these processes of social learning operate without the individual organism paying any attention whatsoever to the actual behavior of other organisms. But some animals in some circumstances may sometimes seek to reproduce the behavior of others. Again, how this is done varies as a function of their cognitive skills. In this case we must consider two aspects of the process: (1) the organism's ability to perceive or conceive a correspondence between the behavior of others and its own behavior, which allows some kind of reproduction, and (2) the organism's manner of understanding the behavior of others, which determines the manner in which the other's behavior will be understood an reproduced.

For example, the mimicking of human speech by some species of psittacine birds represents a case of behavioral reproduction that would seem to take place on the perceptual or sensory-motor level. These birds have some way of "mapping" the sounds they hear to their own skills of vocal production. Although we do not know precisely how they do this, it presumably does not involve cognitive processes of the type that help them to understand for what purpose the human is making those sounds, since their reproductions often occur in inappropriate circumstances from the point of view of the speaker's communicative intentions. In contrast, human children learn to use specific pieces of human speech for specific communicative purposes in a way that does seem to depend on such an understanding (as well as on their sensory-motor skills for mimicking the sound itself). The difference is that the child, but not the bird, understands something of the intentions of the person who produced the speech and thus attempts to reproduce "the same" speech (via mimicking) when it has "the same" communicative intentions. Since the bird does not understand communicative intentions in this same way, it cannot do this (although perhaps some species can be taught some relevant skills of this type; Pepperberg, 1990). More generally, Tomasello, Kruger, and Ratner (1993) have argued that humans perceive and understand the behavior of others in intentional terms: they simply see another's behavior as "cleaning the window" or "drawing attention to the ball," not as "moving his hand in a circular motion on the surface of the window while holding a cloth" or "making a noise while looking in the direction of the ball." When they then seek to reproduce such a behavior, their understanding of what the behavior was attempting to accomplish determines which aspects of the behavior are relevant for reproduction; for example, they may not reproduce a cough that takes place during the cleaning and they may not think it relevant which hand is used or how many times the surface is wiped.[2]

---

2. In lectures on imitation I sometimes wave my hand and ask the audience to imitate me. They invariably wave back. I then say "No, I was waving at the people passing by on the walkway outside the window" (or "No, I was waving my hand at 100 waves per second."). The reason that the audience did not turn in their seats to wave at the people outside the window is that they did not understand that this was what I was doing. If I were to look intently out the window as I waved, it is possible that this would lead them to such an understanding and so make the direction of the waving a relevant aspect of the behavior to be reproduced.

In the cognitive view, then, organisms reproduce the behavior of others at different levels and in different ways depending on their perception or understanding of the behavior to be reproduced. In my terminological scheme, the reproduction of behavior on the sensory-motor level (basically, without cognitive processes) may be called mimicking, as, for example, parrots mimicking human speech or a human child mimicking an adult behavior whose significance it does not grasp. This is clearly an important process in the way many songbirds learn the vocalizations characteristic of their species, among other behavioral phenomena, and it obviously requires some type of matching of perceptual and motoric processes (and some birds may even be able to do this in nonvocal behavior as well; Moore, this volume). But cases in which an individual understands another's behavior intentionally and attempts to reproduce that intentional behavior (entailing an understanding of both the behavior's goal and its strategy for achieving that goal) require more than sensory-motor matching. Imitative learning, my version of the ever-elusive "true" imitation, requires that the learner perceive and understand not just the bodily movements that another individual has performed (mimicking), and not just the changes in the environment that that individual's behavior has resulted in (emulation learning). The learner must also understand something of the "intentional" relations between these, that is, how the behavior is designed to bring about the goal. This then determines precisely what of the other's behavior it seeks to reproduce. (I should add that there are other levels of social cognition that may be involved in some forms of human cultural learning, for example, involving the thoughts and beliefs of others; Tomasello, Kruger, & Ratner, 1993.)

The Cognitive Perspective

The cognitive perspective thus changes in important ways how we look at processes of social learning. Rather than investigating these processes in terms of stimuli and responses, conceived similarly across species, we may investigate which aspects of a conspecific's instrumental behavior different species attend to and understand in particular contexts. The specific proposal here is that we distinguish between cases in which the learner is focused on the environment and its affordances from cases in which the learner is focused on the behavior of others. Within each of these, there are important further distinctions to be made in terms of how the learner understands what it is observing. In the environment the learner may understand the changes that the behavior of others brings about in terms of various properties and affordances of objects and events, or even in terms of various spatial-temporal-causal relations among objects and events. In the behavior of others the learner may understand simply in terms of motoric movements, or it may understand the intentional relations between the behavior produced and the outcome in the environment that behavior is designed to achieve. I should add that it has been my hypothesis for some time (Tomasello, 1990; Tomasello, Kruger, & Ratner, 1993) that how all of this works in the case of any particular species will be an important determiner of the nature of its behavioral traditions—how they originate, how they are maintained, how they respond to changes in the environment, and, in general, how they serve as repositories of information that may affect the evolution of the species in important ways.

APE BEHAVIOR IN THE WILD

In their natural habitats, different groups of chimpanzees behave differently and these differences often persist across generations. For a variety of reasons, genetic explanations are unlikely and so it is probably the case that some form of learning, perhaps social learning, is at work (Tomasello, 1990).

Chimpanzee Tool Use

The best known example is chimpanzee tool use. For example, members of the Kasakela community at Gombe (as well as some other groups elsewhere) fish for termites by probing termite mounds with small, thin sticks, whereas in other parts of Africa there are chimpanzees who simply destroy termite mounds with large sticks and scoop up the insects by the handful. Field researchers such as Goodall (1986), Nishida (1987), Boesch (1993a), and McGrew (1992) have claimed that specific tool use practices such as these are "culturally transmitted" among the individuals of the various communities, and they often use the term "imitation" as a key part of their explanation (although, to be fair, it should be noted that for the most part these fieldworkers have used the term imitation very broadly to include virtually all types of social learning).

The problem is that it is possible that chimpanzees in some localities destroy termite mounds with large sticks because the mounds are soft from much rain, whereas in other localities other chimpanzees cannot use this strategy because the mounds they encounter are too hard—with no type of social learning involved at all. This same type of explanation is even possible in the case of nut cracking, the case for which the most intense investigation of ecological factors has been conducted. Boesch, Marchesi, Marchesi, Fruth, and Joulian (1994) report that within western Africa (the only region of the continent in which the behavior is found), nut cracking is confined to a particular geographic region on one side of one particular river. In a very detailed investigation, they found no significant differences in the local ecologies of two populations—one nut cracking and one not—living less than 50 km apart but on opposite sides of the river. Both populations lived in the same kinds of forest and had available the same types and abundance of nuts and tools. Even in this case, however, there is no way to know for sure that all of the relevant ecological differences have been investigated (Tomasello, 1990). For example, for the population that does not crack nuts: Are the nuts in their region harder? Are their trees located near predators? Are there other more easily obtainable sources of protein? Such possibilities may be multiplied basically *ad infinitum*.

Let us assume, however, that more is involved in patterns of chimpanzee tool use in the wild than simple individual learning shaped by the local ecologies of different groups (this possibility is supported by experimental research; see next section). The problem now is that even if nut cracking and other forms of tool use are in some sense socially transmitted, the social-learning processes involved cannot be determined through naturalistic observations alone. For example, it is possible that one chimpanzee in western Africa invented nut cracking. Her behavior would have then left a stone hammer, some unopened nuts, and some opened nuts all in one place near a suitable substrate—very propitious learning conditions that might facilitate the individual learning of others. Moreover, her behavior would also be quite noisy and attract the attention of groupmates. Thus, the combination of propitious learning conditions and processes of local enhancement (or even emulation learning) might result in the acquisition of nut cracking by the inventor's groupmates. This interpretation is in fact supported by Sumita, Kitahara-Frisch, and Norikoshi (1985), who looked very closely at the acquisition of nut cracking by individual chimpanzees in a captive group setting. What they found was that learning for all individuals was gradual and that there were many idiosyncrasies in the way that individuals performed the nut-cracking behavior. They concluded that individual trial-and-error learning (along with local enhancement) was the learning process responsible for the spread of the behavior in the group: one creative individual made a discovery which left propitious learning conditions for others, made salient through processes of local enhancement and emulation learning. (And this same explanation may also apply to the other reported cases of the rapid spread of a tool-use behavior in relatively natural captive settings, e.g., as reported by Menzel, 1973; Hannah & McGrew, 1987.) It is also possible, of course, that the nut-cracking and other tool-use behaviors of chimpanzees in the wild are acquired via imitative learning (although this interpretation would not seem to be supported by the experimental literature; see next section).

Chimpanzee Gestural Communication

The other well-known case of relevance to the current discussion is the gestural communication of chimpanzees, for which there also seems to be some population-specific behaviors (Goodall, 1986; Tomasello, 1990). There are two reasonably well-documented examples from the wild. First, Nishida (1980) reported "leaf clipping" in the Mahale K group of chimpanzees—thought to be unique to that group but later observed by Suyigama (1981) in another group across the continent. The reporting of data for individuals in these studies showed that there were marked individual differences within the groups in how (toward what end) the signal was used, for example, for sexual solicitation, aggression toward groupmates, or aggression toward humans. One hypothesis is that after one individual used leaf clipping to make noise (the tearing of the rigid dead leaves makes a very loud noise) others learned via emulation to make the same noise (i.e., they learned the affordances of the leaf).

The second example is the "grooming hand clasp" reported by McGrew and Tutin (1978), also thought at one time to be unique to the Mahale K group, but later also found in one wild and one captive community (Ghiglieri, 1984; de Waal, 1994). But this behavior is not really an instrumental gesture at all; one individual simply raises the arm of another to access a new place to groom. Others could easily learn this behavior by having others approach with their arm raised in hopes of further underarm grooming (because on previous occasions they have enjoyed being groomed under the arms) or by local enhancement as they watch others groom under the arms of groupmates. The fact that both leaf clipping and grooming hand clasp have been observed in more than one group, who have not had the opportunity to observe one another, speaks to the possibility that in all groups the behavior is spontaneously invented by some kind of ritualization process (see next section).

Other Apes

It is interesting, and perhaps telling, that there are no well-documented cases of population-specific behavioral traditions in ape species other than chimpanzees (although other ape species have not been observed nearly as extensively in the wild). There is in fact only one relevant example that does not involve chim-

panzees. Byrne and Byrne (1993) and Byrne (1994) observed the complex manipulations that gorillas perform in gathering plant foods and preparing them for eating.
Different plants require different techniques due to a number of factors, among
them the size, location, and texture of the plants, and how the edible parts are
attached to the inedible parts. These researchers observed a number of very complex sequences of behaviors that are reliably used with different plants by all
individuals—raising the possibility that they might be learning from one another
in some way. They also observed, however, that despite the overall similarity
among individuals, there were also differences in the details of how the sequences
were performed, leading them to posit the existence of what they call "program-
level imitation," that is, reproducing the general structure of the behavioral sequence with the details being discovered individually (it thus bears some resemblance to emulation learning). The problem with this case for inferring processes of
social learning is that there are no variations of technique associated with particular
families or groups of individuals, or any comparisons of gorillas who did and did
not observe conspecifics, to rule out the possibility that each animal is learning
individually from its interactions with the plants themselves.

STUDIES WITH CAPTIVE APES

To help interpret what is happening in the wild, we need experimental studies in
which individual apes observe different behavioral demonstrations and then are
given the opportunity to behave in the same situations themselves. To my knowledge, there are five and only five experimental studies of the social-learning skills of
apes in instrumental situations in which control conditions are systematically used
(other than studies with human-raised apes; see next section). Also relevant in this
context are some longitudinal studies of how the youngsters of a relatively natural
captive colony of chimpanzees learn their gestural signals, which will provide
converging evidence for the experimental studies of instrumental behavior.

Tool Use

Tomasello et al. (1987) trained an adult chimpanzee demonstrator to rake food
items into her cage with a metal T-bar. When a food item was in the center of the
serving platform she learned simply to sweep or rake the item to within reach.
However, when the food was located either along the side or against the back of the
platform's raised edges, she had to employ more complex two-step procedures to be

successful. We then exposed young chimpanzees to this adult demonstrator as she employed all three of her strategies (the experimental group). Several other chimpanzees in this same age range were exposed to the demonstrator in an unoccupied state throughout (the control group). Results showed that experimental subjects learned to use the tool (after only a few trials in most cases) while control subjects mostly did not. The finding supports the hypothesis that in the wild chimpanzees acquire their tool-use skills via some form of social learning. Two additional pieces of information help to specify the type of social learning involved. First, chimpanzees in the control group manipulated the tool as often as chimpanzees in the experimental group. This would seem to rule out an explanation in terms of simple stimulus enhancement. Second, experimental subjects employed a wide variety of different raking-in procedures, and none of them learned either of the demonstrator's more complex, two-step procedures (even though they were trying and failing on these trials most of the time). This would seem to rule out full-fledged imitative learning in which the observer attempted to reproduce the demonstrator's behavior. Our hypothesis was that what the experimental subjects learned from the demonstrator was the affordances of the tool vis-à-vis the food (some causal relations between movements of the tool and movements of the food), that is, they learned via emulation learning (see also Paquette, 1992).

In a second experimental study, Nagell, Olguin, and Tomasello (1993) attempted to investigate the emulation hypothesis directly. We presented juvenile and adult chimpanzees and 2-year-old human children with a rake-like tool and a desirable but out-of-reach object. The tool was such that it could be used in either of two ways leading to the same end result. For each species one group of subjects observed a human demonstrator employ one method of tool use (less efficient) and another group of subjects observed the other method of tool use (more efficient). The point of this design was that the stimulus enhancement of the tool was the same in both experimental conditions; only the precise methods of use were different. What we found was that whereas human children in general copied the method of the demonstrator in each of the two observation conditions (imitative learning), chimpanzees used the same method or methods no matter which demonstration they observed. This was despite generally equal levels of tool use and tool success for the two species (again ruling out stimulus enhancement). It should be noted that the majority of children insisted on this reproduction of adult behavior even in the case of the less efficient method—leading to less successful performance than the chimpanzees (who ignored the demonstration) in this condition. From this pattern of results we concluded, once again, that chimpanzees in this task were paying attention to the general causal relations between tool and food, but they were not attending to the actual methods of tool use demonstrated; they were engaged in emulation learning.

Call and Tomasello (1994) gave this same task to orangutans. Again some subjects observed a human demonstrator use the tool the more efficient way, while other subjects observed the demonstrator use the tool in the less efficient way. Subjects behaved identically in the two experimental conditions, showing no effect of the type of demonstration observed. Moreover, analysis of individual learning curves suggested that a large component of individual trial-and-error learning was at work, and this was also true of the data from the chimpanzees of Nagell et al. (1993) when they were reanalyzed in this way. One novelty of this study was that a follow-up was performed with additional subjects using orangutan demonstrators familiar to the subjects (mothers or other cagemates). The same pattern of results was found, leading again to the hypothesis that the orangutans acquired their tool-use skills in this task via emulation learning.

One possible criticism of these three studies is that the tool-use task may have been too difficult for the apes to learn via observation. Although there is no evidence for this in these studies, and in fact the majority of subjects in all experimental conditions gained some success with the tool during the course of the experiment, Call and Tomasello (1995) nevertheless presented juvenile and adult orangutans with task involving very simple behavioral strategies. The trick was that the causal relations of the apparatus, the way the tool contacted and made the food move, were hidden from the subject, thus making emulation learning impossible. What the subject saw was a human demonstrator manipulate a stick protruding from a box in one of four ways and then receive a reward. When subjects were then given their turn to manipulate the stick (in four different sets of randomized and blocked trials) there were no signs that they imitatively learned any of the actions demonstrated to them. They basically performed the actions randomly and enjoyed only a chance rate of success. Providing subjects with an orangutan demonstrator did not change this result. A majority of human 3- and 4-year-old children did learn to imitate the requisite actions and thus performed at above-chance levels. This study, then, provides further evidence for the emulation hypothesis since taking away the possibility of emulation learning seems to make it very difficult for apes to effectively learn from the behavior of others anything useful at all about a tool-use task.

Finally, Whiten et al. (in press) recently presented chimpanzees with a transparent box containing fruit ("artificial fruit"). On any given trial, the artificial fruit could be opened by one of two mechanisms, each of which could be operated in two ways. For each mechanism, a human experimenter demonstrated one way of opening the box to some subjects and the other way to other subjects (with the other mechanism being blocked). Subjects were then given the chance to open the box themselves. Results were that for one mechanism there was no effect of the demonstration observed. For the other mechanism there was some evidence that chimpanzees were more likely to use the manner of opening demonstrated. However, in my analysis, they could easily have learned this via emulation learning: one method was to push a bolt through a clasp and the other was to pull it through the clasp. One group of chimpanzees saw that the bolt could be pushed through the clasp, while the other saw that it could be pulled through (twisting of the bolt, which accompanied pulling but was not actually necessary to open the box, may have had some effect on the ape subjects as well—though the results are somewhat equivocal). It is also noteworthy that 3-year-old human children given this same task reproduced the demonstrator's technique much more often and much more faithfully than did the chimpanzees.

Overall, what these studies suggest, I believe, is that apes are very intelligent and creative in using tools and understanding changes in the environment brought about by the tool use of others, but they do not understand the instrumental behavior of conspecifics in the same way as do humans. For humans the goal or intention of the demonstrator is a central part of what they perceive, and indeed the goal is understood as something separate from the behavioral means. This understanding thus highlights the demonstrator's method or strategy of tool use—the way she is attempting to accomplish the goal, given the possibility of other means of accomplishing it. For apes, on the other hand, what is salient in the demonstration is the tool, the food, and their spatial-temporal-causal relations, with the intentional states of the demonstrator, and thus her methods as distinct behavioral entities, being either not perceived or irrelevant. Apes may then use their own behavioral strategies to attempt to reproduce these environmental relations, and thus engage in what we have called emulation learning, not imitative learning.

Gestural Communication

Despite our best efforts to provide apes with tool-use tasks that are within their capabilities, using conspecific demonstrators, the criticism may still be made that the findings are artificial in two ways: the animals have been raised in artificial conditions and they have been presented with artificial tasks. It is therefore important to look at ape behavior in more natural social conditions, in a domain of interaction not involving human-posed tasks, to seek further evidence of ape proclivities for different types of social learning.

For the past 12 years we have been observing the gestural signaling of a captive colony of chimpanzees at the Yerkes Regional Primate Research Center Field Station. These apes have led lives that resemble the lives of their wild conspecifics to a much greater degree than the subjects of the tool-use experiments. The group has been relatively stable socially over many years and the physical setting is relatively diverse so that the environment of developing youngsters has been similar in many ways to the environment of developing wild chimpanzees. This is especially true with regard to peer interaction and play, which seem to occur very much as they do in the wild, and which were the contexts of use for the majority of the signals we observed and analyzed. In our first study, we observed the infants and juveniles of the group (1–4 years old), with special emphasis on how they used their signals (Tomasello, George, Kruger, Farrar, & Evans, 1985). Looking only for intentional signals accompanied by gaze alternation or response waiting (both indicating the expectation of a response), we found a number of striking developmental patterns. For example, we found that juveniles used many gestures not used by adults, that adults used some gestures not used by juveniles, and that some juvenile gestures for particular functions were replaced by more adult-like forms at later developmental periods. None of these patterns was consistent with the idea that infant and juvenile chimpanzees acquire their gestural signals by imitating adults.

In a longitudinal follow-up to these observations, Tomasello, Gust, and Frost (1989) observed the same juvenile chimpanzees in the same group setting 4 years later (at 5- to 9-years-old), this time with a more direct focus on learning processes. We sought various forms of evidence that the youngsters acquired their gestures by means of imitation or ontogenetic ritualization (what we once called "conventionalization"). In ontogenetic ritualization a communicatory signal is created by two organisms shaping each others' behavior in repeated instances of a social interaction. For example, an infant may initiate nursing by going directly for the mother's nipple, perhaps grabbing and moving her arm in the process. In some future encounter the mother might anticipate the infant's desire at the first touch of her arm, and so become receptive at that point—leading the infant to abbreviate its behavior to a touch on the arm with response waiting (cf. Tinbergen 1951, on "intention movements"). Note that there is no hint here that one individual is seeking to reproduce the behavior of another; there is only social interaction that results eventually in a communicative signal.

Two observations suggest that indeed ontogenetic ritualization, and not imitation, was responsible for at least some of the gestural signals observed. First, there were a number of idiosyncratic signals that were used by only one individual

(Goodall, 1986, also reports this for her wild group). These signals could not have been learned by imitative processes and so must have been individually invented and ritualized. Second, many youngsters also produced signals that they had never had directed to them, for example, others never begged food, or solicited tickling or nursing from youngsters. The youngsters' signals for these functions thus could not have been a product of their imitation of signals directed to them (second-person imitation), and in many cases it was also extremely unlikely that they were imitated from other infants gesturing to conspecifics (third-person imitation or "eavesdropping") because many were only produced in close quarters between mother and child with little opportunity for others to observe (e.g., for nursing). I should also note that at least some of the gestures that were widespread in the group may have been learned by response facilitation (Byrne, 1994), for example, an individual slapping the ground to make a noise and attract the attention of others could easily be reproduced by another individual who hears the sound and already knows from past experience how to make it.

We have recently completed a third set of longitudinal observations of the gestures of the youngsters of this group (Tomasello, Call, Nagell, Olguin, & Carpenter, 1994). In this study, a completely new generation of youngsters was the object of study, that is, the 1- to 4-year-olds who were not even born (with one exception) at the time of the last observations. In order to investigate the question of potential learning processes, we made systematic comparisons of the concordance rates of all individuals with all other individuals in terms of the gestures produced across the three longitudinal time points. The analysis revealed quite clearly, by both qualitative and quantitative comparisons, that there was much individuality in the use of gestures, especially for nonplay gestures, with much individual variability both within and across generations. As before, there were also a number of idiosyncratic gestures used by single individuals at each time point. Also notable is the fact that there was a fairly large gap between the two generations of this study (4 years), and many of the gestures learned by the youngsters of the younger generation were ones that youngsters of the older generation would have been using infrequently during the crucial learning period. It is also important that the gestures that were shared by many youngsters are gestures that are also used quite frequently by captive youngsters raised in peer groups with no opportunity to observe older conspecifics (e.g., Berdecio & Nash, 1981).

This overall pattern of results suggests that chimpanzee youngsters acquire the majority, if not the totality, of their gestures by individually ritualizing them with one another. The explanation for this is analogous to the explanation for emulation learning in the case of tool use. Like emulation learning, ontogenetic ritualization does not require individuals to understand the intentions of others in the same way as does imitative learning. Ritualizing an "arm raise" requires only an anticipation of the future behavior of a conspecific and the ability to translate this anticipation into an instrumental action of one's own; to imitatively learn an arm raise as a play solicit would require that an individual understand the intentions of a conspecific when it raises its arm (as does the imitative learning of linguistic conventions demonstrated by human children, which in many cases require in a very specific way the understanding of the intentions of others; Tomasello, 1992). My conclusion, therefore, is that in this domain as well chimpanzees display much individual inventiveness and creative intelligence, but they do not employ skills of imitative learning. They do not imitatively learn gestures from one another because they do not understand the behavior of others intentionally.

I will mention, very tentatively, that we have also recently completed a small pilot study in which we removed an individual from the Yerkes group and taught her two different arbitrary signals by means of which she obtained desired food from a human (Tomasello, Ashley, & Nagell, unpublished data). When she was then returned to the group and used these same gestures to obtain food from a human in full view of other group members (usually several groupmates were in close proximity and watching) there was not one instance of another individual reproducing either of the new gestures. The study has several limitations (e.g., this is a situation in which most chimpanzees had already used other gestures successfully in the past, only one demonstrator was used and she had no offspring, and other limitations), and so I do not want to rely on it heavily until further studies of this type are conducted. However, the fact that its findings are in general agreement with those of the naturalistic observations provides at least some further validity to the findings.

Summary

These two sets of studies of captive apes provide us with two very different sources of evidence about ape social learning. In the case of tool use, it is very likely that captive apes acquire the tool-use skills they are exposed to by a process of emulation learning. In the case of gestural signals, it is very likely that the chimpanzees of a socially stable captive colony acquire their communicative gestures through a process of ontogenetic ritualization. Both emulation learning and ontogenetic ritualization require skills of social learning, each in its own way, but neither requires skills of imitative learning. In both of these domains apes rely mainly on their considerable skills of individual cognition and learning.

STUDIES WITH HUMAN-RAISED APES

It is an interesting and important finding that apes raised and/or trained by human beings seem to acquire more human-like skills of social learning than their conspecifics raised in more typical captive conditions (and perhaps than those raised in the wild). Thus, almost all of the anecdotes of ape imitation and mimicking reported by Whiten and Ham (1992) were of apes raised in various ways by humans. There are some more systematic studies of human-raised apes as well, and these have been conducted with: (1) arbitrary body movements, (2) object manipulations, and (3) communicative signals and symbols.

Arbitrary Body Movements

Hayes and Hayes (1952) trained their human-raised chimpanzee Viki to imitate (mimic) various body movements and gestures, for example, blinking the eyes or clapping the hands. They trained her throughout her daily life in their home for a period of more than 17 months before systematic testing began. The training consisted of a human performing a behavior and then using various shaping and molding techniques, with rewards, to get Viki to "Do this." After she had become skillful at this, some novel behaviors were systematically introduced to Viki. In general, she reproduced them faithfully and quickly; she had clearly "gotten the idea" of the mimicking game. Recently, Custance and Bard (1994; Whiten, this volume) have demonstrated in a more rigorous fashion similar abilities in two nursery-reared chimpanzees after they were trained for a period of several months in a manner similar to Viki.

With regard to other ape species, Miles also found that the human-raised orangutan Chantek could be taught to imitatively reproduce a number of arbitrary movements and gestures (Miles, Mitchell, & Harper, 1992). On a "Do this" command in American Sign Language, Chantek learn to produce many different bodily configurations that humans produced for him, and we have since observed the same skill repeatedly in this same individual in his new home at Yerkes (Call & Tomasello, 1995). The nature and extent of the training involved in this case is not known.

Actions on Objects

After she had been trained to mimic actions, Hayes and Hayes (1952) exposed Viki to six problem-solving tasks that she would be unlikely to solve for herself. She was given a 2-min baseline period with each task on her own, and then up to three human demonstrations were given. Viki performed one of the target behaviors during baseline, and had trouble with two others, but for three problems she solved them only after demonstrations. The difficulty is that because there were no stringent control conditions, that is, there were no tasks given after the baseline period without demonstration to see what Viki would do on her own with repeated opportunities. Drawing firm conclusions about the precise social-learning processes involved is therefore very risky. Nevertheless, it is important to note that another chimpanzee tested by Hayes and Hayes, Franz, who had not been raised or trained in the special ways that Viki had, showed no signs of any type of social learning in this set of problem-solving tasks.

Tomasello, Savage-Rumbaugh, and Kruger (1993) conducted a more systematic study of the skills of chimpanzees and bonobos to reproduced modeled actions. In that study we compared the social learning of mother-reared captive apes, enculturated apes (raised like human children and exposed to a language-like system of communication), and 2-year-old human children. Each subject was shown 24 different and novel actions on objects and encouraged to reproduce them: children were told to "Do this", and the apes were pretrained to reproduce modeled actions (as well as having been informally encouraged to imitate earlier in their lives with some regularity). Each subject's behavior on each trial was scored as to whether it successfully reproduced (i) the end result of the demonstrated action, and/or (ii) the behavioral means used by the demonstrator. The major result was that the mother-reared apes reproduced both the end and means of the novel actions (i.e., imitatively learned them) hardly at all. In contrast, the enculturated apes and the human children imitatively learned the novel actions much more frequently, and they did not differ from one another in this learning. It is important to note, however, that many of the actions could have been learned by emulation learning (with the correspondence in behavioral means resulting from the similar motoric skills of the three species involved), and some may have been cases of simple mimicking.

A different type of evidence for imitative learning in human-raised apes is provided by Russon and Galdikas (1993). The subjects of their study were rehabilitant orangutans who had been raised in various ways by humans (mostly as pets), and then introduced to a human camp on the edge of a forest where they were free to come and go as they pleased. What Russon and Galdikas observed was a number of behaviors that were so unlike the typical behaviors of orangutans, and so unlikely to have been learned by individuals on their own, that imitative learning was inferred. These behaviors almost all involved the manipulation of human artifacts in human-like ways—such things as "brushing the teeth," "applying insect repellant," "using a knife," and so forth. The problems in interpreting this study stem from: (1) there was no documentation of precisely what the orangutans observed of these behaviors by humans; (2) there was no documentation of what orangutans could learn from exploring the objects by themselves; and (3) there was no information available on the nature of their earlier experiences with humans (and this is important since there were only a handful of orangutans who produced these intriguing behaviors with regularity). It is thus unclear precisely what these animals learned from humans, and how they learned it.

Finally, Call and Tomasello (1995) gave the orangutan Chantek a number of arbitrary body movements to mimic (along with the sign to "Do this"), which he did quite readily (thus replicating the findings of Miles et al., 1992). The interesting twist is that he was now given this same request ("Do this") in the context of the problem-solving apparatus used by Call and Tomasello (1995), which in the original study he had basically failed to cope with successfully even after a human demonstration. Unexpectedly, the command to "Do this" given before the demonstration of an action on the apparatus did not lead Chantek to above-chance performance. Apparently, Chantek had caught on to the mimicking game in the context of arbitrary body movements, but he did not apply that knowledge to a problem-solving context in which he was seriously trying to obtain a reward. One interpretation is that mimicking behaviors on the sensory-motor level is something that apes can be trained to do, but that understanding what another is doing in instrumental, problem-solving situations, in a way that is relevant for one's own problem-solving attempts, requires an understanding of the intentions of others— which apes may not be able to do without certain specific types of experience and training from humans (or at all).

Communication Skills

The "language" learning of home-raised apes is also of interest here. A number of early studies showed that apes learning gestural signs from American Sign Language learned them much easier through shaping and molding than through demonstration and imitation (e.g., Fouts, 1972). Recently, however, the bonobo Kanzi was raised in a more natural, human-like way and has shown some learning skills that seem to be based on some form of imitation (e.g., Savage-Rumbaugh et al., 1986). Kanzi spent his first 2-1/2 years holding onto and observing his mother while she was interacting with humans around a communicative keyboard, which contained lexigrams that were touched by humans in association with certain communicative intentions. The mother learned very little. After some time, however, Kanzi began spontaneously to demonstrate that he had learned the appropriate uses of some lexigrams. Because he had not been trained directly on the keyboard, though he had been encouraged to interact with humans and to reproduce their behavior in other contexts, the researchers concluded that he could only have learned his lexigrams through observational learning of some sort. Although Kanzi's ability to learn lexigrams observationally has never been systematically tested, if indeed that is how he is learning them, it goes beyond mere mimicking as he uses them in appropriate and novel communicative contexts (and there are no causal relations between lexigrams and the world to emulate).

Summary

What these findings with human-raised apes mean for our understanding of ape imitation is unclear at this point. One hypothesis is that human beings can, through various processes of enculturation and/or training over a relatively extended period of time in early ontogeny, lead apes down a developmental path that includes various human-like skills in the domain of social learning. This hypothesis would seem to be true in the case of the mimicking of arbitrary bodily movements. It is less clear, however, whether it is also true in the domain of instrumental behaviors. What is needed are tasks given to human-raised apes in which the possibility of learning through stimulus enhancements or emulation is controlled. In the one situation in which this was done, an ape who was an excellent mimicker did not transfer his skills to a problem-solving situation.

WHAT IS THE ANSWER?

The data reviewed suggest that if our ultimate interest is in apes in their natural habitats, the question of whether apes ape is in large measure dependent on which group of captive apes we take to be more representative of wild apes: captive or enculturated. Do enculturated apes display more species-typical imitative learning skills than captive apes because their more enriched rearing conditions more closely resemble those of wild apes than do the impoverished rearing conditions of other captive apes (Whiten, 1993)? Or might it be the case that the human-like socialization process experienced by enculturated apes differs significantly from the natural state and, in effect, helps to create a set of species-*atypical* abilities more similar to those of humans? There is no definitive answer to these questions at this point, but, for the moment at least, I cast my lot with the captive apes as better representatives of wild apes than those raised and trained by humans. There are two basic reasons for my choice.

The first reason concerns what the enculturated and captive apes are doing in the experiments. First, human-raised apes have shown some interesting skills of social learning in three experimental paradigms. In the first place apes can be trained to mimic the arbitrary body movements of humans. This is done only on human command, and teaching apes to understand and follow this command takes many weeks of training with external reinforcement. What results is a copying of behavior on the sensory-motor level, for no discernable goal other than to please the human trainer who will then give some reward. This behavior is interesting, and it is important to know the extent of ape abilities in this experimental paradigm, but it is hard to see what it might correspond to in the lives of apes in their natural habitats.

In the second place some enculturated apes have acquired human-like communicative symbols by means of some form of observational learning, presumably imitative learning. The ape for whom this is most systematically documented is Kanzi, and the actual behavior required of him is simply to touch a key that he has seen humans touch—a fairly simple motoric performance, although knowing when to touch which key is a far from simple affair. Again it seems that a good deal of human instruction (in this case of a more informal variety) is involved in bringing the ape to a point where it can learn in this way, as evidenced by the fact that Kanzi's mother, who was born and spent her early life in the wild, did not learn to acquire communicative symbols via observation and imitative learning. And again, it is difficult to see that this behavior corresponds exactly to anything in the life of wild apes, as their communicative gestures consist not of arbitrary human-like symbols but of natural instrumental behaviors ritualized into signals (mostly involving intention movements).

In the third place some apes can be trained by humans to reproduce on command certain actions on objects. As argued above, it is not entirely clear what they are learning to do in this situation—whether they are truly imitating the actions of humans or rather emulating the results. In addition, in this case as well it is important that the behavior must be specifically elicited by human training and commands: both Hayes and Hayes (1952) and Tomasello et al. (1993) had apes set up in situations where they had been trained to do as the humans had just done. The one possible exception is the orangutans of Russon and Galdikas (1993). In their observations some orangutans reproduced human actions on objects without any known human training or prompting from experimenters. But, as argued above: (1) it is unclear exactly what they are doing—emulating, mimicking, or imitatively learning; and (2) we do not know what other humans taught them prior to the scientists' observations.

I am thus not convinced that when human-raised, imitation-trained apes show signs of imitative learning that are simply being enabled to do what they do in the wild. Rather, it seems to me that they are being trained to do some things that they do not do in the wild, and there is at least some evidence that there is a critical period in their ontogenies when such training must take place if they are to acquire these human-like skills (Rumbaugh et al., 1991). The behavior shown by the captive apes, on the other hand, seems much more "natural." They have not been specifically trained and they are not behaving on human command. In the case of tool use they simply watch a human or ape conspecific do something to get a reward and are then given the opportunity to do the same themselves. It is possible that their tendency to ignore the demonstrator's behavior in these studies is due to the fact that in captivity some skills do not develop that would develop in a more natural environment. However, this argument applies much less forcefully to the case of gestural communication, in which the chimpanzees observed have lived much more natural lives and were not presented with any tasks or problems designed by human beings.[3]

The second reason for thinking that captive apes are more representative of wild apes than are their enculturated conspecifics has to do with the nature of the behavioral traditions that chimpanzees display in the wild (and among apes such traditions have been documented only for chimpanzees). First of all, as I have argued previously (Tomasello, 1994; Tomasello et al., 1993) the behavioral traditions of chimpanzees have different characteristics than those of humans, which might imply that the social-learning skills that support them are different as well. In particular I have focused on three characteristics that human cultural traditions have that chimpanzee behavioral traditions either do not possess or possess in a different way: (1) universality—in all human societies there are some traditions that are practiced by virtually everyone in the society; (2) uniformity—many human cultural traditions show uniformity of technique among all members of the culture, even when such uniformity is not driven by anything instrumental (e.g., social-conventional behaviors such as linguistic symbols or religious rituals); and (3) history—human cultural artifacts often show an accumulation of modifications

---

3. It is also relevant that apes in captivity, even without extensive contact with humans, show more complex skills of tool use than apes in the wild. This would seem to argue against their being cognitively impoverished by their captive lives. Their gestures in captivity are very much like those in the wild (Tomasello, 1990).

over generations (a.k.a., the ratchet effect) as novel exigencies arise and change in the history of how the artifact is used.

The evidence is that the behavioral traditions of chimpanzees do not really fit this pattern, especially not the third characteristic. With regard to universality, there are only a few tool-use behaviors (e.g., termite fishing at Gombe, ant fishing at Mahale, and nut cracking at Tai) that are practiced by virtually all noninfant members of their respective communities. Moreover, on the basis of their survey of the innovative behaviors of a number of primate species, including chimpanzees, Kummer and Goodall (1985) conclude that "only a few will be passed on to other individuals, and seldom will they spread through the whole troop." With regard to uniformity, it is clear that individual chimpanzees often use their own creative techniques in all kinds of instrumental tasks from termite fishing to nut cracking (e.g., Goodall, 1986; Hannah & McGrew, 1987; Sumita et al., 1985). Even more telling for current purposes is chimpanzees' use of idiosyncratic techniques in social conventional behaviors for which humans show such marked uniformity (Goodall, 1986; Tomasello et al., 1994). With regard to history, Boesch (1993b) reports two examples from chimpanzees in the Tai Forest that he believes show the ratchet effect, but, as I have argued in more detail previously (Tomasello, 1994) what changed in both of these cases was not the behavior in question (e.g., leaf clipping) but only the context of its use. With respect to all three of these characteristics, then, the evidence is that chimpanzee behavioral traditions are different in important ways from those of humans. This suggests that the social-learning processes underlying them may be different too. Chimpanzee behavioral traditions have a "looser" structure than human cultural traditions because when chimpanzees learn from one another they employ to a much greater extent than do humans processes of individually based social learning (e.g., emulation and ritualization) rather than processes of cultural learning such as imitative learning (Tomasello, 1994).

CONCLUSION

My conclusion is that captive chimpanzees raised by their mothers are a better model for wild chimpanzees than are chimpanzees raised in human-like ways. This means that my answer to the more general question of whether apes ape is: only when trained by humans, either formally or informally, to do so (and then perhaps in only some ways). The learning skills that chimpanzees develop in the wild in the absence of human interaction—skills involving individual learning as well as social-learning processes such as local enhancement, emulation learning, and ontogenetic ritualization—are perfectly sufficient to create and maintain their population-specific behavioral traditions. The imitative learning skills that humans train in apes might be sufficient to maintain more human-like cultural traditions (we simply do not know), but the point is moot since there is no evidence that such skills are a part of the repertoires of apes in their natural habitats. Researchers who maintain that any time we see a skill from any animal with any history in any environment we must say that the species has the "potential" to develop that skill are simply stating a truism. Much more interesting is systematic investigation of the ontogenetic conditions in which different types of skill develop. These conclusions are buttressed by two recent reviews of the literature in which it was concluded that: (1) the social cognition and social learning of apes is not fundamentally different from that of monkeys (Tomasello & Call, 1994), and (2) differences in the social cognition and social learning of monkeys and apes that have been reported in the past are due to the fact that some apes have been raised and trained in ways that monkeys have not (Call & Tomasello, in press).

On a more general theoretical level, I would argue that the cognitive view of social learning leads us down productive research pathways that are not possible in more associationistic theoretical paradigms. In the particular case of primates it allows us to investigate rather subtle differences of learning process that depend on the ability of different species to perceive and understand: (1) the causal structure of the physical environment, and (2) the intentional structure of the behavior of social partners. It then allows us to go on to investigate some of the factors in ontogeny that might contribute to the development of such cognitive skills. For example, we may investigate the possibility that to reproduce another's behavioral strategies via imitative learning requires some understanding of their intentions and that this understanding develops in, and only in, the context of certain kinds of social interactions with others in early ontogeny (Tomasello et al., 1993).

The question of whether apes ape is thus a complex one and I am sure that the last word on this question has yet to be spoken. I do know that we will make further progress only if we investigate systematically, and relatively dispassionately, both the similarities and differences of all of the various primate species, including humans, and only if we take into account the ontogenetic histories of the particular individuals involved. Moreover, to make the needed comparisons in sensitive and meaningful ways, we are going to need to develop further the theoretical tools for analyzing the cognitive capacities of the various primate species, particularly with respect to their social cognitive skills at understanding the intentional behavior of their conspecifics. It is possible that there are further important distinctions to be made in various types of social understanding and learning that we have yet to recognize.

ACKNOWLEDGMENTS

Thanks to Josep Call and Celia Heyes for helpful comments on an earlier version of the manuscript.

REFERENCES

Berdecio, S., & Nash, V. (1981). Chimpanzee visual communication. *Anthropological Research Papers, Vol. 26.* Arizona: Arizona State University.

Boesch, C. (1993a). Transmission aspects of tool use in wild chimpanzees. In T. Ingold & K. Gibson (Eds.), *Tool, language, and intelligence: An evolutionary perspective.* Cambridge: Cambridge University Press.

Boesch, C. (1993b). Toward a new image of culture in wild chimpanzees. *Behavioral and Brain Sciences,* **15,** 149–50.

Boesch, C., Marchesi, P., Marchesi, N., Fruth, B., & Joulian, F. (1994). Is nut cracking in wild chimpanzees a cultural behavior? *Journal of Human Evolution,* **26,** 325–338.

Byrne, R. (1994). The evolution of intelligence. In P. Slater & T. Halliday (Eds.), *Behavior and evolution.* Cambridge: Cambridge University Press.

Byrne, R., & Byrne, J. (1993). Complex leaf-gathering skills of mountain gorillas: Variability and standardization. *American Journal of Primatology,* **31,** 241–261.

Call, J. & Tomasello, M. (1994). The social learning of tool use by orangutans (*Pongo pygmaeus*). *Human Evolution,* **9,** 297–313.

Call, J. & Tomasello, M. (1995). The use of social information in the problem-solving of orangutans and human children. *Journal of Comparative Psychology,* **109,** 301–320.

Call, J., & Tomasello, M. (in press). The effect of humans on the cognitive and social-cognitive development of apes. In A. Russon, K. Bard, & S. Parker (Eds.), *Reaching into thought: The Minds of the great apes.* Cambridge: Cambridge University Press.

Custance, D. & Bard, K. (1994). The comparative and developmental study of self-recognition and imitation: The importance of social factors. In S. Parker, M. Boccia, & R. Mitchell (Eds.), *Self-awareness in animals and humans: Developmental perspectives.* Cambridge: Cambridge University Press.

de Waal, F. (1994). Chimpanzees' adaptive potential: A comparison of social life under wild and captive conditions. In R. Wrangham, W. McGrew, F. de Waal, & P. Heltne (Eds.), *Chimpanzee Cultures.* Cambridge, MA: Harvard University Press.

Fouts, R. (1972). Use of guidance in teaching sign language to a chimpanzee (*Pan troglodytes*). *Journal of Comparative Psychology,* **80,** 515–522.

Galef, B. (1988). Imitation in animals. In B. Galef & T. Zentall (Eds.), *Social learning: Psychological and biological perspectives.* Hillsdale, NJ: Erlbaum.

Galef, B. (1992). The question of animal culture. *Human Nature,* **3,** 157–178.

Ghiglieri, M. (1984). *The chimpanzees of the Kibale Forest.* New York: Columbia University Press.

Gibson, J. (1979). *The ecological approach to visual perception.* Boston: Houghton Mifflin.

Goodall, J. (1986). *The chimpanzees of Gombe.* Cambridge, MA: Harvard University Press.

Hall, K. (1963). Observational learning in monkeys and apes. *British Journal of Psychology,* **54,** 201–226.

Hannah, A., & McGrew, W. (1987). Chimpanzees using stones to crack open oil palm nuts in Liberia. *Primates* **28,** 31–46.

Hayes, K. & Hayes, C. (1952). Imitation in a home-raised chimpanzee. *Journal of Comparative and Physiological Psychology,* **45,** 450–459.

Heyes, C. (1993a). Imitation, culture, and cognition. *Animal Behavior,* **46,** 999–1010.

Hayes, C. (1993b). Anecdotes, training, trapping, and triangulating: Can animals attribute mental states? *Animal Behavior,* **46,** 177–188.

Kohler, W. (1927). *The mentality of apes.* New York: Harcourt Brace.

Kummer, H. & Goodall, J. (1985). Conditions of innovative behavior in primates. *Philosophical transactions of the Royal Society of London,* **308,** 203–214.

McGrew, W. (1992). *Chimpanzee material culture.* Cambridge: Cambridge University Press.

McGrew, W., & Tutin, C. (1978). Evidence for a social custom in wild chimpanzees? *Man,* **13,** 234–251.

Menzel, E. (1973). Further observations on the use of ladders in a group of young chimpanzees. *Folia Primatologica,* **19,** 450–457.

Miles, L., Mitchell, R., & Harper, S. (1992). Imitation and self-awareness in a signing orangutan. Paper presented at the XIV Congress of the International Primatological Society. Strasbourg, August.

Nagell, K., Olguin, R., & Tomasello, M. (1993). Processes of social learning in the imitative learning of chimpanzees and human children. *Journal of Comparative Psychology,* **107,** 174–186.

Nishida, T. (1980). The leaf-clipping display: A newly discovered expressive gesture in wild chimpanzees. *Journal of Human Evolution,* **9,** 117–128.

Nishida, T. (1987). Local traditions and cultural transmission. In B. Smuts, D. Cheney, R. Seyfarth, R. Wrangham, & T. Struhsaker (Eds.), Primate Societies. Chicago: University of Chicago Press.

Paquette, D. (1992). Discovering and learning tool-use for fishing honey by captive chimpanzees. *Human Evolution,* **7,** 17–30.

Pepperberg, I. (1990). Referential mapping. *Applied Psycholinguistics* **11,** 23–44.

Piaget, J. (1962). *Play, dreams, and imitation.* New York: Norton.

Rumbaugh, D., Hopkins, W., Washburn, D., & Savage-Rumbaugh, S. (1991). Comparative perspectives of brain, cognition, and language. In N. Krasnegor, D. Rumbaugh, R. Schiefelbusch, & M. Studdart-Kennedy (Eds.), *Biological and behavioral determinants of language development.* Hillsdale, NJ: Erlbaum.

Russon, A., & Galdikas, B. (1993). Imitation in ex-captive orangutans. *Journal of Comparative Psychology,* **107,** 147–161.

Savage-Rumbaugh, E., Mcdonald, K., Sevcik, R., Hopkins, W., & Rubert, E. (1986). Spontaneous symbol acquisition and communicative use by pygmy chimpanzees (*Pan paniscus*). *Journal of Experimental Psychology: General,* **115,** 211–235.

Spence, K. (1937). Experimental studies of learning and higher mental processes in infrahuman primates. *Psychological Bulletin,* **34,** 806–850.

Sumita, K., Kitahara-Frisch, J., & Norikoshi, K. (1985). The acquisition of stone tool use in captive chimpanzees. *Primates,* **26,** 168–181.

Suyigama, Y. (1981). Observations on the population dynamics and behavior of wild chimpanzees at Bossou, Guinea, 1979–1980. *Primates,* **22,** 432–444.

Thorpe, W. (1956). *Learning and instinct in animals.* London: Methuen.

Tinbergen, N. (1951). *The study of instinct.* Cambridge: Oxford University Press.

Tomasello, M. (1990). Cultural transmission in the tool use and communicatory signaling of chimpanzees? In S. Parker & K. Gibson (Eds.), *Language and intelligence in monkeys and apes: Comparative developmental perspectives.* Cambridge: Cambridge University Press.

Tomasello, M. (1992). The social bases of language acquisition. *Social Development,* **1,** 67–87.

Tomasello, M. (1994). The question of chimpanzee culture. In R. Wrangham, W. McGrew, F. de Waal, & P. Heltne (Eds.), *Chimpanzee cultures.* Cambridge, MA: Harvard University Press.

Tomasello, M. & Call, J. (1994). The social cognition of monkeys and apes. *Yearbook of Physical Anthropology,* **37,** 273–305.

Tomasello, M., Call, J., Nagell, K., Olguin, R., and Carpenter, M. (1994). The learning and use of gestural signals by young chimpanzees: A trans-generational study. *Primates,* **35,** 137–154.

Tomasello, M., Davis-Dasilva, M., Camak, L., & Bard, K. (1987). Observational learning of tool use by young chimpanzees. *Human Evolution* **2,** 175–183.

Tomasello, M., George, B., Kruger, A., Farrar, J., & Evans, E. (1985). The development of gestural communication in young chimpanzees. *Journal of Human Evolution,* **14,** 175–186.

Tomasello, M., Gust, D., & Frost, T. (1989). A longitudinal investigation of gestural communication in young chimpanzees. *Primates,* **30,** 35–50.

Tomasello, M., Kruger, A., & Ratner, H. (1993). Cultural learning. *Behavioral and Brain Sciences,* **16,** 495–592.

Tomasello, M., Savage-Rumbaugh, S., & Kruger, A. (1993). Imitative learning of actions on objects by children, chimpanzees, and enculturated chimpanzees. *Child Development,* **64,** 1688–1705.

Visalberghi, E. (1993). Tool use in a South American monkey species: An overview of characteristics and limits of tool use in *Cebus apella.* In A. Bethelet & J. Chavaillon (Eds.), *The use of tools by human and nonhuman primates.* Cambridge: Oxford University Press.

Visalberghi, E., & Fragaszy, D. (1990). Do monkeys ape? In S. Parker & K. Gibson (Eds.), *Language and intelligence in monkeys and apes: Comparative developmental perspectives.* Cambridge: Cambridge University Press.

Vygotsky, L. (1978). *Mind in society.* Cambridge, MA: Harvard University Press.

Whiten, A. (1993). Human enculturation, chimpanzee enculturation, and the nature of imitation. *Behavioral and Brain Sciences, 16,* 538–539.

Whiten, A., & Ham, R. (1992). On the nature and evolution of imitation in the animal kingdom: Reappraisal of a century of research. In P. Slater & J. Rosenblatt (Eds.), *Advances in the study of behavior.* New York: Academic Press.

Whiten, A., Custance, D., Gomez, J., Teixidor, P., and Bard, K. (in press). Imitative learning of artificial fruit processing in chimpanzees and children. *Journal of Comparative Psychology.*

16

The Human Infant as Imitative Generalist: A 20-Year Progress Report on Infant Imitation with Implications for Comparative Psychology

Department of Psychology
University of Washington
Seattle, Washington 98195-7920

INTRODUCTION

I propose that there are three key features of imitation in human infants. First, human infants are imitative generalists. The hallmark of normal infants is that they imitate a range of novel and arbitrary acts. Second, imitation is its own reward for the human young. Normal human infants are intrinsically motivated to "act like" other humans, a species-typical trait manifest in newborns. Human infants do not imitate solely, or most readily, as a means of obtaining food. Even if not obtaining food, praise, or other extrinsic rewards, normal human infants are driven to imitate the acts of conspecifics. Third, infant imitation is a bidirectional activity. Human parents are prolific imitators of their young infants. Experiments show that infants smile, increase their gaze at the imitator, and react distinctively to being matched. It will be suggested that reciprocal imitation games between parent and offspring provide ontogenetic roots for the growth of "theory of mind" in human children.

When I began work on imitation in human infants in the early 1970s, it was thought that human infants were poor imitators. Human infants were compared, perhaps unconsciously by observers, to adults or 5-year-olds. One cannot ask a human infant to "do as I do," and accordingly their imitative abilities were profoundly

Social Learning in Animals: The Roots of Culture

underestimated by traditional norms until the time they began to understand language. Infants may not imitate with the facility of a 5-year-old, but imitation in human infants is a useful touchstone or reference point for imitation in other animals, because language can be excluded. Infants provide a unique opportunity for investigating human imitation before language, before advanced symbolic development, and in the case of newborn imitation, even before "enculturation."

Learning more about human prelinguistic imitation should assist those interested in the biology, evolution, and comparative psychology of imitation. Reviews of animal imitation have sometimes not considered the new research on imitation in human infancy (e.g., Galef, 1988; Whiten & Ham, 1992, but see Russon & Galdikas, 1995; Visalberghi & Fragaszy, 1990). However, human infants are animals too. A comprehensive evolutionary picture needs to incorporate the human data. This chapter reviews two decades of progress made at both the empirical and theoretical level in understanding imitation in human infants.

At the empirical level, investigations have concerned:

a. methods used to distinguish nonverbal imitation from other forms of social learning
b. imitation of facial gestures
c. vocal imitation
d. imitation of novel acts
e. imitation of object-related acts
f. imitation after a memory delay
g. peer imitation
h. recognition of being imitated oneself

At the more theoretical level, there has been progress in:

a. relating imitation to other aspects of nonverbal cognition, especially cross-modal matching
b. connecting infant imitation to developing "theories of mind"

METHODOLOGY: DISTINGUISHING IMITATION FROM OTHER FORMS OF SOCIAL LEARNING

Twenty years ago, Meltzoff and Moore (1977) posed three questions about infant imitation: methods (what controls are necessary for distinguishing imitation from other forms of social learning?), existence (do newborns imitate?), and mechanism (what psychological processes underlie this behavior?). The same series of questions about methods, existence, and mechanism are now being worked through in studies on nonhuman animals (e.g., Byrne, in press; Byrne & Byrne, 1993; Custance, Whiten, & Bard, in press; Heyes & Dawson, 1990; Russon & Galdikas, 1995; Tomasello, this volume; Tomasello, Kruger, & Ratner, 1993; Tomasello, Savage-Rumbaugh, & Kruger, 1993; Visalberghi & Fragaszy, 1990; Whiten and Custance, this volume; Whiten, Custance, Gómez, Teixidor, & Bard, in press). The methodological progress made with human infants may be useful to those studying other animals, because similar problems arise whenever linguistic directions are excluded.

Cross-Target Procedure

Meltzoff and Moore (1977, 1983a,b) developed the "cross-target" design to distinguish imitation from other forms of social learning. In the cross-target design, infants' responses are compared across two (or more) different demonstrations by the same adult model. The power of the cross-target design is that it uses the same adult model, at the same distance, moving at the same rate, to perform different actions. For example, subjects are shown both a mouth-opening display and a tongue-protrusion display in a repeated-measures test using infants as their own controls. It is assessed whether the infants respond with more mouth openings to the mouth display than to the tongue display, and conversely respond with more tongue protrusions to the tongue display than to the mouth display. Imitation is demonstrated if the subject responds differentially and with high-fidelity matches to the same adult model demonstrating two different motor movements.

Using this design, the 1977 study reported imitation of four different body actions: lip protrusion, mouth opening, tongue protrusion, and sequential finger movement. The displays were selected to allow stringent cross-target comparisons. Note that there were two types of lip movements (mouth opening vs lip protrusion) and two types of protrusion actions (lip protrusion vs tongue protrusion). The results showed that when the body part was precisely controlled, when lips were used to perform two subtly different movements, infants still responded differentially. Moreover, when the same general movement pattern was demonstrated (protrusion) but with two different body parts (lip vs tongue protrusion), infants also responded differentially. This documented that infants were matching particular *acts,* not responding solely to the presence of a conspecific (the adult was present in all cases), activating a region of the body (oral region), or producing a general category of movements (e.g., protrusions).

Although we were the first to use the cross-target design with human infants, this design has also been used in the best studies trying to distinguish imitation from other forms of social learning in nonhuman animals (e.g., Dawson & Foss, 1965; Heyes & Dawson, 1990; Tomasello, Savage-Rumbaugh, & Kruger, 1993; Whiten et al., in press). Perhaps the cross-target design was independently invented by investigators in response to the same issue (how does an investigator separate imitation from other forms of social learning?).

Blind Scoring

A major difficulty with the work on humans before 1977 was that it used judges who were not free of observer bias; observers scored the infants' responses live while knowing what the subject had been shown. In all our studies infants responses were videotaped using two video cameras and tape decks. One system captured the infant, with no record whatever of the adult's behavior, and the other captured just the adult's display without record of the infant. The infants' responses were scored by observers who were uniformed about what display the infants had been shown.

FACIAL IMITATION: CROSS-MODAL MATCHING

It is useful to provide a brief review of the initial work on facial imitation, because it highlights several aspects of human imitation that became central to our theoretical position. In particular, the work on facial imitation led to the proposal that human infants were imitating on the basis of *cross-modal* matching (Meltzoff, 1990b; Meltzoff & Moore, 1977, 1983a, 1989, 1992, 1994).

To understand why facial imitation supports this inference, consider manual and vocal imitation. In imitating manual movements the infant can see the adult's hand and can also see their own hands. It is possible for infants to use visual guidance of their own motor movements to bring them into line with the movements they see. The same occurs in vocal imitation. The infant can use auditory guidance to achieve a perceived match to the auditorially specified target.

Of course, manual and vocal imitation both depend on generalizations by the infant. There are visual perspective differences between the infant's view of his own hands and the hands of others, and the hands are of different sizes and texture (Meltzoff & Moore, 1983b), but the possibility of a visual-visual match still exists. Infants will have seen their hands before (even if they do not look at them during the experiment), because they spend hours engaged in hand regard (Piaget, 1952).

The same argument applies to vocal imitation. Infants can hear themselves during the test and will no doubt have heard their own vocalizations before. Generalization is involved because the infants' own vocalizations will be of higher frequency than the adult's, will be perceived from a different place in space, and will partly be conveyed through bone conduction. Nonetheless, the comparison between demonstration and response can be made within a single modality, audition.

Both manual and vocal imitation are cross-modal in the sense that the subject must perform a motor act on the basis of what they saw (or heard). However, facial imitation is cross-modal in an additional, stronger sense. The additional sense is that within-modality comparisons between the external target and the infant's own response are completely excluded. In facial imitation the infant can see the adult's face, but their own face is invisible to them. If the subjects are young enough they will never have seen their own faces. Thus facial imitation demonstrates a double cross-modality: (a) it is cross-modal in the sense that all imitation involves cross-modal integration (motor responses based on perception), and (b) it is cross-modal in the stronger sense that no aspect of the model and the imitative response can be perceived via the same modality, only different modalities.

Based on this logical analysis, and certain empirical considerations concerning the organization of the response, Meltzoff and Moore (1977, 1983a, 1994) proposed that early facial imitation is mediated by "active intermodal mapping" (the AIM hypothesis). According to AIM, human infants are engaged in a matching-to-target process in which they actively compare the visual information about the seen body movements with the proprioceptive feedback from their own movements in space.

The AIM hypothesis supposes that young infants can detect and utilize cross-modal matches in form or structure (not merely timing) across different modalities. To pursue this hypothesis, we next investigated other instances of such cross-modal matching. One involved a test of cross-modal shape perception (Meltzoff & Borton, 1979). Infants were given a shape to feel in their mouths, but were prevented from seeing the shape. The shape was then withdrawn and the infants presented with a visual paired comparison between that shape and a different one. The results showed that human infants as young as 29-days-old could recognize shapes across modalities: they systematically looked longer at the shape they had previously explored orally. It was also shown that 18-week-old infants can perform auditory-visual matching between speech sounds and the particular lip movements that cause them (Kuhl & Meltzoff, 1982). Infants were visually presented a film loop of two faces articulating different speech sounds (/a/ as in "pop" and /i/ as in "peep") in synchrony with one another. The infants were auditorially presented with a tape loop of one of these phonetic units played over a loudspeaker located midway between the two faces. The results showed that infants preferentially looked at a face whose mouth movements matched these sounds: When listening to the /a/ sound, infants looked longer at the visual faces articulating /a/ versus /i/, and vice . versa when listening to the /i/ sound.

It is interesting that the same species that imitates facial movements, also matches felt shapes to seen shapes, and matches the auditory and visual aspects of speech sounds. Meltzoff (1990b) and Meltzoff and Moore (1983a,b, 1994, 1995b) provide an detailed analysis of how cross-modal matching mediates facial imitation in human infants.

Range of Acts Imitated

The existence of early facial imitation initially surprised developmental psychologists (it had not been predicted by the then-dominant Piagetian stage theory). The effect has now been replicated in 12 independent laboratories in more than 24 experiments. Tongue protrusion has been the most widely publicized example, but over the past 20 years imitation of a wide range of other gestures has been documented, including lip, head, and finger movements, emotional expressions, cheek movements, and brow movements (Abravanel & Sigafoos, 1984; Field, Woodson, Cohen, Greenberg, Garcia, & Collins, 1983; Field, Goldstein, Vaga-Lahr, & Porter, 1986; Field, Woodson, Greenberg, & Cohen, 1982; Fontaine, 1984; Heimann, 1989; Heimann, Nelson, & Schaller, 1989; Heimann & Schaller, 1985; Jacobson, 1979; Kaitz, Meschulach-Sarfaty, Auerbach, & Eidelman, 1988; Kugiumutzakis, 1985; Legerstee, 1991; Maratos, 1982; Meltzoff & Moore, 1977, 1983a, 1989, 1992, 1994; Reissland, 1988; Vinter, 1986). If imitation in early infancy had been limited to one or two privileged acts (e.g., if tongue protrusion alone had been demonstrated, as was mistakenly believed for some time), this would weigh against the AIM hypothesis, which predicts more generative and productive imitation. However, it is now established that a wide range of gestures can be imitated and that young infants imitate strangers as well as mothers (Meltzoff & Moore, 1992, 1994).

Novel Acts

Early imitation is not limited to one or two privileged acts, but the question of response novelty is a complex one that merits further investigation (the reader is referred to Meltzoff, 1988c for a fuller discussion of the problem: "what is novelty"?).

Meltzoff and Moore (1994) investigated whether young infants can imitate novel/unfamiliar facial acts. They showed 40 6-week-olds the following gestures: tongue protrusion to the side (TP$_{side}$, the novel act), tongue protrusion, mouth opening, and a baseline in which no facial gesture was shown. As expected, the spontaneous probability of producing TP$_{side}$ was very low, and yet infants produced this behavior in response to an adult who did so. A microanalysis of the response also revealed that infants made errors that were self-corrected, especially in imitating the novel TP$_{side}$ gesture. Adults make similar errors imitating novel acts—a first try that is close, combined with errors that are self-corrected. The discovery of error correction in the imitative response of infants is compatible with the AIM hypothesis, the notion that a true matching-to-target process is involved.

VOCAL IMITATION

Humans are the only mammals known to display "vocal learning"—the ability to acquire a species-specific vocal repertoire by hearing the vocalizations of adults and mimicking them (Konishi, 1989; Nottebohm, 1975). We share this ability with a few avian species (songbirds), and evolutionary biologists and ethologists have argued that it may represent a critical step in the evolution of speech and language (Hauser & Marler, 1992; Marler, 1974).

It has long been assumed that at some point during development young children become adept at mimicking the speech patterns they hear others produce. When do they display such an ability? There is circumstantial evidence that human infants can vocally imitate before the onset of meaningful speech, inasmuch as infant babbling sounds differ across cultures (see review by Kuhl & Meltzoff, 1996). Even to the casual ear, French babies sound distinctively French and different from American babies by about 24 months of age. It seems plausible that the linguistic input heard by infants in different cultures is the root cause of their different vocal productions.

We conducted an experimental study to investigate vocal imitation in human infants. Using the cross-target design, infants between 12 and 20 weeks of age were randomly assigned to three independent groups (Kuhl & Meltzoff, in press). Infants listened to a prerecorded signal of a female talker repeatedly producing one of three phonetic units, /a/, /i/, or /u/ (as in "hop," "heap," or "hoop"). The model was not live, but on tape, thus the possibility of unconscious shaping by the adult was excluded. Two types of scoring were conducted. The first was a computer-spectrographic analyses of the infants' vocalizations. This was a painstaking process that involved statistical analysis of more than 5500 acoustic measurements (for details, see Kuhl & Meltzoff, 1995). The second was perceptual scoring conducted by a trained phonetician.

The perceptual scoring was done by having a trained phonetician listen to each infant's productions and classify the productions as falling into the /a/, /i/, or /u/ category. The phonetician was kept uninformed as to the auditory stimulus presented to the infants being scored. The results showed imitation. Infants produced more /a/ vowels when listening to /a/ than when listening to /i/ or /u/; infants produced more /i/ utterances when listening to /i/ than when listening to /a/ or /u/; and finally infants produced more /u/ utterances when listening to /u/ than when listening to /a/ or /i/. It is informative for theories of imitation that Kuhl and Meltzoff's spectrographic analyses showed that infants were not duplicating the absolute frequencies of the adult, but rather the relationship among the frequencies, the internal structure or "pattern" of the vowel sounds.

Kuhl and Meltzoff (1996) have argued that vocal imitation of phonetic units plays a critical role in infants' acquisition of language. Different human languages use different phonetic units and prosodic structure. The endpoint of infancy is that infants "sound like" a speaker of their language, producing both the sound units and "accent" of their native language. Infants efficiently acquire the language-specific inventory of phonetic units, words, and prosodic features partly through imitation. Chomsky (1968, 1988) famously used syntax to argue his case for linguistic nativism. Our proposals about the role of imitation in language acquisition do not concern syntax, but the sound patterns of language that appear to be learned.

IMITATION OF OBJECT MANIPULATIONS: NOVEL ACTS AND DEFERRED IMITATION

For imitation to be of far-reaching significance in human development, infants will need to imitate not only body acts and speech sounds, but also tool use and other object-related behaviors. Moreover, they will need to imitate novel acts after significant memory delays. Human parents engage in purposeful pedagogy of the type "watch what I do," often demonstrating a new skill at a time and place far removed from when the infant has an opportunity to imitate. If the human young (or any species for that matter) could imitate acts immediately but *not* imitate after a lengthy memory delay, this would necessarily constrain theories about the role of imitation in the transmission of culture for that species (Galef, 1992; Heyes, 1993; Meltzoff, 1988a,e; Premack & Premack, 1994; Tomasello et al., 1993). Thus if we want to draw inferences to cultural transmission, we need to know about imitative generalization across time and space.

Meltzoff (1988b–d) conducted a series of studies relevant to these concerns.

One study with 14 month olds had three features: (a) imitation was tested after a 1-week delay, (b) infants were required to remember not just one demonstration but to keep in mind multiple different demonstrations, and (c) novel acts were used. One of the acts, bending forward from the waist and banging a panel with one's forehead, was not observed in 100 infants in free play, and certainly qualified as a novel display (baseline measures were also taken in the experiment, Meltzoff, 1988c).

Infants in the imitation group were shown six different acts on different objects on the first day of testing. Importantly, they were not allowed to touch or handle the objects. They were confined purely to watching the displays. Infants were then sent home for the 1-week delay. On returning to the laboratory, the infants in the imitation group were presented with the objects and their behavior videotaped for subsequent analysis. For infants in the adult manipulation control group, the adult manipulated the same objects during session 1, and produced the same "results" or "affordances" as he did in the imitation group; but he did so using different movement patterns (for further discussion about controlling for the affordances and results of the display, see Meltzoff, 1985, 1988c,d,e; Nagell, Olguin, & Tomasello, 1993; Tomasello, this volume; Tomasello, Savage-Rumbaugh, & Kruger, 1993; Whiten and Custance, this volume; Whiten et al., in press). We also used a second control group, a baseline control, in which the adult did not manipulate the test objects; he simply talked pleasantly to the mother and child on session 1 before the 1-week delay. This assessed the spontaneous likelihood of the infants producing the target acts when they returned to the laboratory for a second session.

The videotaped response periods were scored by coders who remained blind to the infants' experimental group. The results provided clear evidence for deferred imitation from memory after the delay. Fully 67% of the infants in the imitation condition leaned down and touched the panel with their heads, as compared to 0% in the other conditions ($p < .0001$). It was found that some infants could retain and imitate up to five different target acts after the delay, and 92% remembered and imitated at least three different acts.

Next investigated was deferred imitation after even longer delay intervals. Meltzoff (1995b) investigated imitation after retention intervals of 2 and 4 months. A total of 192 infants were tested and the effects of immediate and deferred imitation assessed. Importantly, in the deferred imitation condition infants were confined simply to observing the adult and were not allowed to engage in immediate imitation during the first session. The results again showed accurate imitation, including imitation of the novel head-touch display. Other work demonstrated deferred imitation when the features of the object were altered as compared to the adult's original, documenting an interesting kind of imitative generalization (Barnat, Klein, & Meltzoff, in press). That infants imitate *multiple* targets, including *novel* ones, after *lengthy delays* suggest that imitation is capable of playing a signifi-cant role in human development.

TUTOR INFANTS AND PEER IMITATION

The ecology of child rearing is changing in the United States. With the increase of women in the work force, infants are spending increasingly more time with peers in day-care settings. In all the previous experiments, adults were used as models. Do infants learn from and imitate their peers in day-care centers and other sites? In this series of studies we moved into the "field," examining peer imitation in day-care centers and homes (Hanna & Meltzoff, 1993).

The first study developed a controlled procedure for assessing peer imitation. Fourteen-month-old infants observed "tutor infants," 14-month-olds previously trained to play with the toys in novel ways. After observing the peer play with five objects, subjects left the test room. They returned 5 min later and were presented with the test objects in the absence of the peer. The results showed imitation. A second study used a day-care setting. The "tutor infant" was strapped into a car seat and driven to a variety of day-care sites. As the naive infants sat around a table, drinking juice, sucking their thumbs, and generally acting in a baby-like manner, the tutor picked up and acted on novel toys in particular ways. The naive infants were not allowed to approach or touch the toys. After a 2-day delay, a new experimenter (not the one who had accompanied the tutor) brought a bag of objects to the infants' homes and laid them out on a convenient table or floor. Neither the parent nor this new experimenter had been present in the day-care center 2 days earlier. The only person who knew what actions had been demonstrated was the subject. The results showed significant imitation, providing the first evidence for deferred imitation across a change in context (a shift from day care to home site).

Evidently even prelinguistic infants are influenced by their peer groups at school. The fact that human infants can transfer their imitative learning to a different environment from the one in which they observed the model, can do so after a long delay, and will imitate peers as well as adults, again supports the idea that imitation may play a role beyond the laboratory.

PERCEPTION VERSUS PRODUCTION: INFANTS RECOGNIZE WHEN THEY ARE IMITATED

Ethological studies have shown that human parents tend to imitate their young. Human parents shake objects when the infant does, slide them when they slide, and coo when they coo. Dyads often engage in long bouts of reciprocal imitation at the highchair or kitchen table—first the infant performs an act, then the parent, then the infant, and so on. Do infants notice they are being imitated? What function does reciprocal imitation serve in human ontogeny?

Experiments were set up in which adults purposely imitated infants, acting like biological mirrors. Three converging experiments were conducted using a total of 140 14-month-olds (Meltzoff, 1990a). The first study investigated whether infants showed any interest in their own behavior being reflected back to them. The infants sat at a table, across from two adults who sat side by side. All three participants were provided with replicas of the same toys. Everything the infant (I) did with his or her toy was mimicked by one of the adults (I'). It was as if I' was tethered to the infant, a marionette under the child's control. The second adult, the nonimitator (NI), was not so tethered. It was hypothesized that infants would prefer to look at I' and also smile at this adult more. We also thought that infants would test or probe the relationship with I' by manipulating their own behavior in special ways. For example, infants might perform sudden and unexpected movements to probe the connection between their own acts and those of I'. Adults sometimes do this when they see themselves in a store video camera. The experiment was videorecorded and was subsequently scored by observers who were kept appropriately blind to the side of the I' and NI. The results showed that infants looked significantly longer at I', smiled more at I', and directed more test behavior at I' (Meltzoff, 1990a).

In the next study, the NI actively manipulated the toys in a particular way. The adult did "baby-like" things with the toys so that no preference for the imitating experimenter could be based solely on a preference for infantile actions (Lorenz, 1943, postulated preference for baby-like visual forms). This was achieved by using a yoked-control procedure. Two television monitors were situated behind the infants and in view of the adults. One monitor displayed the actions of the current infant, live. The other displayed the video record of the immediately proceeding infant. Each adult mimicked one of the infants on television. Both adults acted like perfect babies, but only one matched the particular actions of the subject himself. The results again showed that infants looked longer at I', smiled more often at the adult, and directed more testing behavior toward that person.

How did the babies detect this relationship? Two kinds of information are available, what I will call *temporal contingency* versus *structural congruence*. According to the first alternative the infant need only detect that whenever he does x the adult does y. The infant need not detect that x and y are in fact equivalent, only that they are temporally linked. The second alternative is that the infant can do more than recognize temporal contingencies. In particular, the infant may be able to recognize that the actions of the self and other have the same form, that they are structurally congruent.

A third study was conducted in which the purely temporal aspects were controlled by having both experimenters act at the same time. Both experimenters sat with neutral expressions until the infant performed a target act from a predetermined list. If and only if the infant exhibited one of these target actions, both experimenters began to act in unison. I′ performed the infant's act; NI performed a different behavior from this list. What differentiates the two experimenters is not timing, but the structure of their actions vis-à-vis the subject. The results showed that the infants looked, smiled, and directed more testing behavior at I′ than at NI.

Evidently infants do not just recognize that another moves *when* they move (temporal synchrony), but recognize that another moves in the same *manner* as they do (structural congruence). Thus infants not only imitate others, but can recognize when the form of their own behavior is being matched. The fact that infants literally prefer (look longer at) adults who imitate them has social-developmental implications, and the discovery that infants can process the equivalence in the behavior, not merely the temporal contingencies, has potential developmental sequella which are elaborated in the final section of this chapter.

IMITATION AND THEORY OF MIND

From Piaget (1929) on, child psychologists have been interested in how children come to understand the mental states of their social partners. Recently, this has been dubbed "theory of mind" (ToM) research. The recent wave of ToM research was sparked by Premack and Woodruff's (1978) work with chimpanzees. It has been greatly expanded in both the comparative literature (e.g., Byrne & Whiten, 1988; Cheney & Seyfarth, 1990; Gómez, Sarriá, & Tarnarit, 1993; Povinelli & Eddy, in press a,b; Povinelli, Nelson, & Boysen, 1990, 1992; Povinelli, Parks, & Novak, 1991; Premack, 1988; Whiten, 1991) and child-development literature (e.g., Astington, Harris, & Olson, 1988; Gopnik, 1993; Wellman, 1990; Perner, 1991).

Human adults do not see others as mindless entities, but rather as beings, just like the self, who have beliefs, desires, and other mental states. Fodor (1987, 1992) argued that the adult belief-desire psychology is innately available to humans. Others suggest that in humans the attribution of mental states such as beliefs develops out of an ontogenetically earlier understanding of more primitive states such as attention and intention (e.g., Gopnik, Slaughter, & Meltzoff, 1994; Gopnik & Meltzoff, 1994; Meltzoff & Moore, 1995b; Tomasello, 1995; Tomasello, Kruger, & Ratner, 1993).

I have been particularly interested in the attribution of purposiveness and intention. Purposiveness is far "downstream" as it were, close to the action (behavior); it is not as far "upstream" as other psychological states such as beliefs. If there is ontogenetic change in understanding other minds, it seems likely that recognizing and understanding the purposiveness in other's behavior may be a foundational. (For related work in nonhuman primates the reader is referred to Menzel & Halperin, 1975 and Premack & Woodruff, 1978; for other work with human infants, see Tomasello & Barton, 1994.)

As adults, we believe that other people are not merely tracing physical movements in space but are aiming towards goals. We don't suppose that inanimate objects have such things as goals and intentions. Do prelinguistic infants interpret human behavior as intentional, purposive acts? If so, do they make the distinction between the movements of people and inanimate objects?

To address these questions, I traded on the infant's proclivity for imitation. However, this proclivity was used in a new, more abstract way. This time I was not interested in whether the infants imitated the literal surface behavior shown to them (established in many studies), but whether they read beyond the literal surface behavior to reenact something more abstract—the aim, intention, or goal of the act—even if it was not seen.

Infants were shown an unsuccessful act (Meltzoff, 1995a). For example, the adult tried to perform a behavior, but his hand slipped. Thus the object was not transformed in any way, and the goal state was not achieved. For other acts, the adult accidentally under- or overshot his target. To an adult, it was easy to read the actor's intentions. The experimental question was whether infants also read through the surface behavior. The infants, who were too young to provide verbal reports, informed us how they interpreted the event by what they imitated. The results showed that infants could infer the goal of the act, even though it was never seen or achieved. Most infants reenacted what the adult "meant to do," (the deep structure of the behavior), not what the adult actually did do (the surface structure).

Infants were randomly assigned to one of four independent groups. In the Demonstration$_{(target)}$ group, the adult demonstrated five target acts on different objects. The infants were presented with the objects, and it was measured whether they reproduced the target acts. This was straightforward imitation. The Demonstration$_{(failed\ attempt)}$ group provided the chief condition of interest. For this group the adult did not demonstrate the end state; the child merely saw the adult try but fail to achieve the target acts. Completed target acts were thus not observed. For example, the adult would try to pull apart an object, but his hand would accidentally slip off the end and the object would remain completely untransformed. In the

Control$_{\text{(adult maipulation)}}$ group, the adult showed neither the target acts nor the failed attempt to achieve them. Instead the adult manipulated the same object for the same length of time as in the Demonstration groups; moreover, the control acts were carefully designed to equate for spatial proximity, direction of movement, and other physical parameters with the Demonstration groups (see Meltzoff, 1995a for details). The Control$_{\text{(baseline)}}$ group assessed the likelihood that the target acts would occur spontaneously independent of the adult model.

The results showed that there were significantly more target acts produced in the two Demonstration groups than in the Control groups. Moreover, the Demonstration$_{\text{(target)}}$ and Demonstration$_{\text{(failed attempt)}}$ did not differ from one another ($M = 3.80$ and 4.00 of the 5 possible target acts respectively) and both significantly differed from the controls. Interestingly, the data showed that infants were as likely to perform the target after seeing the adult "trying but failing" as they were after seeing the full demonstration. It was as if the human infants saw directly through the surface behavior. Further analyses also showed that infants could readily imitate arbitrary surface behaviors with these objects. Infants produced more of the control actions in the adult-manipulation group than in the other groups. This confirmed that the Demonstrations were not simply prompting the infants to use an "affordance" of the objects, because infants used the *same* objects in two *different* ways, depending on the act demonstrated (cross-target design).

The foregoing experiment indicated that infants can pick up information from the failed attempts of human actors. What if infants see the same movements produced by an inanimate device? Do the spatial transformations in and of themselves "suggest" the target act? A device was built that did not look human but nonetheless could mimic the movements of the actor in the Demonstration$_{\text{(failed attempt)}}$ group. The device had a pincers that "grasped" the dumbbell on the two ends (just as the human hand did) and then pulled outward. These pincers then slipped off the cubes (just as the human hand did). The pattern of movements and the slipping motions were closely matched to the human hand movements.

A total of 60 infants were tested, and the videotapes scored by observers who were uninformed about group assignment. It was found that infants were visually riveted by both displays; visual attention to the displays exceeded 98% for both treatment groups. The infants were not more frightened by one display than the other. There was no social referencing (turning around toward the parent) during any of the displays. There was no fussing by any of the subjects during the test. Infants did not seem to behave differently when watching the human versus machine. However, the groups significantly differed in their tendency to produce the target act. Infants who saw the human's failed attempt were six times more likely to produce the target than infants in the other group, $p < .0005$. In fact the infants who saw the movements of the inanimate device behaved virtually identically to those in the first study who saw no demonstration at all.

Evidently, infants can infer an aim or goal from human action, but do not make this same ascription when similar movements in space are traced by an inanimate device. That human infants read human behavior but not machine movements in this way, has implications for developing a "theory of mind," as delineated below.

DEVELOPING THEORIES OF IMITATION

Three ideas can be abstracted from the empirical work with infants that might be useful reference points for work in evolutionary biology and comparative psychology. These ideas admittedly go beyond the data per se. They summarize at a more theoretical level what is known and generate predictions about future findings.

Human Infants Are Imitative Generalists

A species-typical, perhaps species-specific aspect of infant imitation is that it is ubiquitous: infants imitate a wide variety of acts in varied situations. Facial, manual, vocal, and object-related imitation has been documented; familiar and novel acts are imitated; both immediate and deferred imitation occurs; imitation can take place in the original setting or be transferred to novel contexts. Each adds a different piece to the puzzle.

Young infants imitate facial gestures such as mouth opening, lip protrusion, tongue protrusion, and head movements. Meltzoff and Moore proposed that facial imitation is mediated by active intermodal mapping (the AIM hypothesis). The crux of this view is that infants can, at some level of processing, apprehend the equivalences between body transformations they see and body transformations of their own whether they see them or not. On this account infant imitation, even early imitation, is a matching-to-target process. The goal or behavioral target is specified visually. Infants' self-produced movements provide proprioceptive information that is compared to the visually specified target.

Infants also imitate vocalizations. We suggested intramodal comparisons come into play in vocal imitation (Kuhl & Meltzoff, 1995, 1996). Infant cooing and babbling, which begins at about 4 weeks of age, allow extensive exploration and elaboration of a kind of auditory-articulatory map. During cooing, auditory events are related to the motor movements that caused them. Infants learn from cooing and babbling that articulatory movements of a particular type have specific auditory consequences. This experience then contributes to an ability to accurately achieve an auditorially specified target. We found developmental changes in vowel production between 12 and 20 weeks of age supporting the view that experience hearing others, hearing oneself, or both, makes a difference.

Infants also imitate actions on objects. The imitation of object-related acts raises a special methodological concern not presented by the imitation of pure body movements and vocal imitation. In object-related imitation one needs to be vigilant as to whether infants are simply striving to recreate the end state, result, or object transformation versus the specifics of the motor act itself (i.e., the form of body movement exhibited by the demonstrator). Tomasello and colleagues call the former "emulation" as opposed to imitation (Tomasello this volume; Tomasello et al., 1993; Tomasello, Savage-Rumbaugh, & Kruger, 1993; see also Wood, 1989). I concluded that human infants were not limited to emulation on the basis of experiments using controls aimed at distinguishing emulation from imitation, and by scoring the fidelity of the match (Meltzoff, 1985, 1988c,d). Perhaps the best existence of proof that infants were not limited in this way was provided by our novel object-related act. The adult leaned down and pushed a panel with his head. Infants imitated by bending forward and pushing the panel with their own heads. Further research in my laboratory has shown that 2-year-old infants will differentially push the panel with their heads, hands, or even their elbows depending on which was modeled. This demonstrates that they are able to imitate the body movements performed with an object and are not limited to recreating an affordance or result. Human infants can imitate the *means* used as well as the goal achieved.

Infants also demonstrate both immediate and deferred imitation. It has been found that human infants will imitate the behavior of a conspecific after delays of up to 4 months. Deferred imitation was demonstrated when the infant was barred from picking up or handling the test object during the initial exposure period. The infant simply watched the adult act during the first session. Memory and differential imitation of multiple acts after 4-month delays (with no immediate imitation) attest to the robust nature of imitative learning in the human infant. Other research found that infants will readily imitate peers as well as adults, strangers as well as mothers.

Human Infants Have an Inbuilt Drive to "Act Like" Their Conspecifics

For chimpanzees, it has been proposed that experience with human-like social interaction is essential for the emergence of imitation; Tomasello et al. (1993) reported that imitation was restricted to "enculturated" and not mother-reared chimpanzees. Whether enculturation is, in fact, a necessary condition for imitation by chimpanzees is a matter of current debate (Whiten and Custance, this volume; Whiten, et al., in press; see also Russon & Galdikas, 1993, 1995 for orangutans). Regardless of how this is resolved, it is clear that enculturation is *not* necessary in human infants: 42-min-old newborns imitate. Meltzoff and Moore (1983a) found imitation in a sample of 40 newborns with the mean age of 32-h-old and no infant was older than 72 h at the time of test. Human infants have an intrinsic motivation to imitate. Imitation may be a mechanism for enculturation, but it does not derive from enculturation in the human case.

The human infant's push to imitate is not restricted to the newborn. In the tests of imitation in older infants I was careful not to use food as a reward or goal to be obtained. This contrasts with the bulk of the studies on nonvocal imitation in animals. For example, in a recent experiment reporting imitation in a nonhuman primate, an artificial fruit (mechanical box) was constructed and presented to chimpanzees (Whiten et al., in press). The subjects could retrieve food from the center of the fruit box by duplicating the adult's retrieval strategy. In human infants, the recovery food is not necessary to motivate imitation. Imitation is an end in itself. Infants struggle to match the adult, self-correct if they do not get it right, and smile upon producing a matching behavior. Human infants derive joy in matching per se. Imitation is its own reward.[1]

Imitation and the Roots of Theory of Mind

A central topic in developmental cognitive science is to investigate how and when children develop a "theory of mind," the understanding of others as psychological

---

1. Ongoing research in my laboratory with G. Dawson suggests that children with autism deviate from this species-typical pattern. Very young children with autism have an imitative deficit (for reviews, see Meltzoff & Gopnik, 1993; Rogers & Pennington, 1991). Interestingly, prelinguistic children with Down syndrome are adequate imitators (Rast & Meltzoff, 1995) and later do not exhibit the profound theory-of-mind deficit shown in autism (Baron-Cohen, 1989; Baron-Cohen, Leslie, & Frith, 1985). I have presented a theory of how imitation may be linked to the emergence of a theory of mind in humans (Gopnik & Meltzoff, 1994; Meltzoff, 1990a; Meltzoff & Gopnik, 1993; Meltzoff & Moore, 1995a,b).

beings having mental states such as beliefs, desires, emotions, and intentions. Although it is sometimes supposed that infants are born with an adult-like understanding of other people (Fodor, 1992), developmental psychologists have sought roots for the adult's understanding of the mental states of others.

I suggest that the ontogenetic foundation from which a theory of mind grows is the perception that others are "like me." Infants' primordial like-me experiences are based on their understanding of spatiotemporal patterns of body movements. Infants monitor their own body movements through proprioception and can detect cross-modal equivalents between movements as felt and movements they see performed by others. This opens up an interesting path of development, especially when coupled with the species-typical adult behavior of reciprocal imitation.

In humans, imitation is a bidirectional activity. Human adults not only adopt an explicit "do what I do" pedagogical style (which requires infant imitation to be fulfilled), they also are rabid imitators of their young for the first several years of an infant's life—sliding objects when their infants slide, banging when their infants bang, and cooing when they coo (Bruner, 1983; Stern, 1985; Trevarthen, 1979).

I suggest that when parents "mark" certain infant behaviors by imitating them, when they selectively mirror certain acts back to them, this has special significance to the infant not only because of the temporal contingencies, but because infants can recognize the structural congruence between the adult's acts and their own. Reciprocal imitation games are not only a form of nonverbal communication, but also serve as private tutorials during which infants consolidate and elaborate knowledge about self and other, and the fundamental identity between the two (Meltzoff & Gopnik, 1993; Meltzoff & Moore, 1995a,b).

Such experiences may delimit the class of entities to which infants ascribe psychological properties. Infants may come to see other people, but not things, as purposive beings because people can be imitated, are perceived to be like themselves, and engage in reciprocal imitation. The experimental findings suggest that 18-month-olds see people, but not things, in purposive terms.

The raw fact that infants can make sense of a failed attempt indicates that they have begun to distinguish surface behavior (what people actually do) from another deeper level. This is only an embryonic structure, but it is of critical importance for human development. This differentiation is fundamental to our "theory of mind," underwriting some of our most cherished human traits. Such a distinction is necessary for fluid linguistic communication, which requires distinguishing what was said from what was intended (Grice, 1957, 1969). It is the basis for our judgments of morality, responsibility, and culpability, all of which require distinguishing intentions from actual outcomes. In civil human interaction it is not solely, or even primarily, the actual behavior of our social partners that carries weight, but their underlying intentions. Research indicates that 18-month-olds have begun to understand the acts of other humans in terms of a psychology involving goals, aims, or intentions, not solely in the physics of the literal movements in space. In this sense they already have a primitive, nonverbal building block for developing a theory of mind.

ACKNOWLEDGMENTS

Preparation of this chapter was supported by a grant from NIH (HD-22514). I thank M. Keith Moore for long-term collaboration on the infant work, Alison Gopnik for provocative discussions about theory of mind, and Pat Kuhl for helpful comments on an earlier draft of this chapter. I am indebted to Cecilia Heyes for setting me the goal of summarizing the human infancy work for those interested in animal behavior, and for her patience and encouragement along the way.

REFERENCES

Abravanel, E., & Sigafoos, A. D. (1984). Exploring the presence of imitation during early infancy. *Child Development,* **55,** 381–392.

Astington, J. W., Harris, P. L., & Olson, D. R. (1988). *Developing theories of mind.* New York: Cambridge University Press.

Barnat, S. B., Klein, P. J., & Meltzoff, A. N. (in press). Deferred imitation across changes in context and object: Memory and generalization in 14-month-old infants. *Infant Behavior and Development.*

Baron-Cohen, S. (1989). The autistic child's theory of mind: A case of specific developmental delay. *Journal of Child Psychology and Psychiatry,* **30,** 285–297.

Baron-Cohen, S., Leslie, A. M., & Frith, U. (1985). Does the autistic child have a "theory of mind"? *Cognition,* **21,** 37–46.

Bruner, J. S. (1983). *Child's talk: Learning to use language.* New York: Norton.

Byrne, R. W. (in press). Skill learning by imitation: Understanding hierarchical organization. In S. Bråten (Ed.), *Intersubjective communication and emotion in ontogeny.* New York: Cambridge University Press.

Byrne, R. W., & Byrne, J. M. E. (1993). Complex leaf-gathering skills of mountain gorillas (*Gorilla g. beringei*): Variability and standardization. *American Journal of Primatology,* **31,** 241–261.

Byrne, R. W. & Whiten, A. (Eds.). (1988). *Machiavellian intelligence: Social expertise and the evolution of intellect in monkeys, apes and humans.* Oxford: Clarendon Press.

Cheney, D. L., & Seyfarth, R. M. (1990). *How monkeys see the world: Inside the mind of another species.* Chicago: University of Chicago Press.

Chomsky, N. (1968). *Language and mind.* New York: Harcourt Brace Jovanovich.

Chomsky, N. (1988). *Language and the problem of knowledge: the Managua lectures.* Cambridge, MA: MIT Press.

Custance, D. M., Whiten, A., & Bard, K. A. (in press). Can young chimpanzees (*Pan troglodytes*) imitate arbitrary actions? Hayes & Hayes (1952) revisited. *Behavior.*

Dawson, B. V., & Foss, B. M. (1965). Observational learning in budgerigars. *Animal Behaviour,* **13,** 470–474.

Field, T., Goldstein, S., Vaga-Lahr, N., & Porter, K. (1986). Changes in imitative behavior during early infancy. *Infant Behavior and Development,* **9,** 415–421.

Field, T. M., Woodson, R., Cohen, D., Greenberg, R., Garcia, R., & Collins, E. (1983). Discrimination and imitation of facial expressions by term and preterm neonates. *Infant Behavior and Development,* **6,** 485–489.

Field, T. M., Woodson, R., Greenberg, R., & Cohen, D. (1982). Discrimination and imitation of facial expressions by neonates. *Science,* **218,** 179–181.

Fodor, J. A. (1987). *Psychosemantics: The problem of meaning in the philosophy of mind.* Cambridge, MA: MIT Press.

Fodor, J. A. (1992). A theory of the child's theory of mind. *Cognition,* **44,** 283–296.

Fontaine, R. (1984). Imitative skills between birth and six months. *Infant Behavior and Development,* **7,** 323–333.

Galef, B. G., Jr. (1988). Imitation in animals: History, definition, and interpretation of data from the psychological laboratory. In T. R. Zentall & B. G. Galef (Eds.), *Social learning: Psychological and biological perspectives* (pp. 3–28). Hillsdale, NJ: Erlbaum.

Galef, B. G. (1992). The question of animal culture. *Human Nature,* **3,** 157–178.

Gómez, J. C., Sarriá, E., & Tamarit, J. (1993). The comparative study of early communication and theories of mind: ontology, phylogeny, and pathology. In S. Baron-Cohen, H. Tager-Flusberg, & D. J. Cohen (Eds.), *Understanding other minds: Perspectives from autism* (pp. 397–426). New York: Oxford Medical Publications.

Gopnik, A. (1993). How we know our minds: The illusion of first-person knowledge of intentionality. *Behavioral and Brain Sciences,* **16,** 1–14.

Gopnik, A., & Meltzoff, A. N. (1994). Minds, bodies, and persons: Young children's understanding of the self and others as reflected in imitation and theory of mind research. In S. T. Parker, R. W. Mitchell, & M. L. Boccia (Eds.), *Self-awareness in animals and humans* (pp. 166–186). Cambridge: Cambridge University Press.

Gopnik, A., Slaughter, V., & Meltzoff, A. (1994). Changing your views: How understanding visual perception can lead to a new theory of the mind. In C. Lewis & P. Mitchell (Eds.), *Children's early understanding of mind: Origins and development* (pp. 157–181). Hillsdale, NJ: Erlbaum.

Grice, H. P. (1957). Meaning. *Philosophical Review,* **66,** 377–388.

Grice, H. P. (1969). Utterer's meaning and intentions. *Philosophical Review,* **78,** 147–177.

Hanna, E., & Meltzoff, A. N. (1993). Peer imitation by toddlers in laboratory, home, and day-care contexts: Implications for social learning and memory. *Developmental Psychology,* **29,** 701–710.

Hauser, M. D., & Marler, P. (1992). How do and should studies of animal communication affect interpretations of child phonological development? In C. A. Ferguson, L. Menn, & C. Stoel-Gammon (Eds.), *Phonological development: Models, research, implications* (pp. 663–680). Timonium, MD: York Press.

Heimann, M. (1989). Neonatal imitation, gaze aversion, and mother-infant interaction. *Infant Behavior and Development,* **12,** 495–505.

Heimann, M., Nelson, K. E., & Schaller, J. (1989). Neonatal imitation of tongue protrusion and mouth opening: Methodological aspects and evidence of early individual differences. *Scandinavian Journal of Psychology,* **30,** 90–101.

Heimann, M., & Schaller, J. (1985). Imitative reactions among 14–21 days old infants. *Infant Mental Health Journal,* **6,** 31–39.

Heyes, C. M. (1993). Imitation, culture and cognition. *Animal Behavior,* **46,** 999–1010.

Heyes, C. M., & Dawson, G. R. (1990). A demonstration of observational learning in rats using a bidirectional control. *The Quarterly Journal of Experimental Psychology,* **42B,** 59–71.

Jacobson, S. W. (1979). Matching behavior in the young infant. *Child Development,* **50,** 425–430.

Kaitz, M., Meschulach-Sarfaty, O., Auerbach, J., & Eidelman, A. (1988). A reexamination of newborn's ability to imitate facial expressions. *Developmental Psychology,* **24,** 3–7.

Konishi, M. (1989). Birdsong for neurobiologists. *Neuron,* **3,** 541–549.

Kugiumutzakis, J. (1985). *Development of imitation during the first six months of life* (Uppsala Psychological Reports No. 377): Uppsala, Sweden: Uppsala University.

Kuhl, P. K., & Meltzoff, A. N. (1982). The bimodal perception of speech in infancy. *Science,* **218,** 1138–1141.

Kuhl, P. K., & Meltzoff, A. N. (1995). Vocal learning in infants: Development of perceptual-motor links for speech. In K. Elenius & P. Branderud (Eds.), *Proceedings of the XIIIth International Congress of Phonetic Sciences* (pp. 146–149). Stockholm: Stockholm University.

Kuhl, P. K., & Meltzoff, A. N. (1996). Evolution, nativism, and learning in the development of language and speech. In M. Gopnik (Ed.), *The biological basis of language.* New York: Oxford University Press.

Kuhl, P. K., & Meltzoff, A. N. (in press). Infant vocalizations in response to speech: Vocal imitation and developmental change. *Journal of the Acoustical Society of America.*

Legerstee, M. (1991). The role of person and object in eliciting early imitation. *Journal of Experimental Child Psychology,* **51,** 423–433.

Lorenz, K. Z. (1943). Die angeborenen Formen möglicher Erfahrung. *Zeitschrift für Tierpsychologie,* **5,** 235–409.

Maratos, O. (1982). Trends in the development of imitation in early infancy. In T. G. Bever (Ed.), *Regressions in mental development: Basic phenomena and theories* (pp. 81–101). Hillsdale, NJ: Erlbaum.

Marler, P. (1974). Constraints on learning: Development of bird song. In N. F. White (Ed.), *Ethology and Psychiatry* (pp. 69–83). Toronto: University of Toronto Press.

Meltzoff, A. N. (1985). Immediate and deferred imitation in fourteen- and twenty-four-month-old infants. *Child Development,* **56,** 62–72.

Meltzoff, A. N. (1988a). Imitation, objects, tools, and the rudiments of language in human ontogeny. *Human Evolution,* **3,** 45–64.

Meltzoff, A. N. (1988b). Imitation of televised models by infants. *Child Development,* **59,** 1221–1229.

Meltzoff, A. N. (1988c). Infant imitation after a 1-week delay: Long-term memory for novel acts and multiple stimuli. *Developmental Psychology,* **24,** 470–476.

Meltzoff, A. N. (1988d). Infant imitation and memory: Nine-month-olds in immediate and deferred tests. *Child Development,* **59,** 217–225.

Meltzoff, A. N. (1988e). The human infant as *Homo imitans.* In T. R. Zentall & B. G. Galef (Eds.), *Social learning: Psychological and biological perspectives* (pp. 319–341). Hillsdale, NJ: Erlbaum.

Meltzoff, A. N. (1990a). Foundations for developing a concept of self: The role of imitation in relating self to other and the value of social mirroring, social modeling, and self practice in infancy. In D. Cicchetti & M. Beeghly (Eds.), *The self in transition: Infancy to childhood* (pp. 139–164). Chicago: University of Chicago Press.

Meltzoff, A. N. (1990b). Towards a developmental cognitive science: The implications of cross-modal matching and imitation for the development of representation and memory in infancy. In A. Diamond (Ed.), *Annals of the New York Academy of Sciences. Vol. 608: The development and neural bases of higher cognitive functions* (pp. 1–31). New York: New York Academy of Sciences.

Meltzoff, A. N. (1995a). Understanding the intentions of others: Re-enactment of intended acts by 18-month-old children. *Developmental Psychology,* **31,** 838–850.

Meltzoff, A. N. (1995b). What infant memory tells us about infantile amnesia: Long-term recall and deferred imitation. *Journal of Experimental Child Psychology,* **59,** 497–515.

Meltzoff, A. N., & Borton, R. W. (1979). Intermodal matching by human neonates. *Nature,* **282,** 403–404.

Meltzoff, A. N., & Gopnik, A. (1993). The role of imitation in understanding persons and developing a theory of mind. In S. Baron-Cohen, H. Tager-Flusberg, & D. Cohen (Eds.), *Understanding other minds: Perspectives from autism* (pp. 335–366). New York: Oxford University Press.

Meltzoff, A. N., & Moore, M. K. (1977). Imitation of facial and manual gestures by human neonates. *Science,* **198,** 75–78.

Meltzoff, A. N., & Moore, M K. (1983a). Newborn infants imitate adult facial gestures. *Child Development,* **54,** 702–709.

Meltzoff, A. N., & Moore, M. K. (1983b). The origins of imitation in infancy: Paradigm, phenomena, and theories. In L. P. Lipsitt (Ed.), *Advances in infancy research* (Vol. 2, pp. 265–301). Norwood, NJ: Ablex.

Meltzoff, A. N., & Moore, M. K. (1989). Imitation in newborn infants: Exploring the range of gestures imitated and the underlying mechanisms. *Developmental Psychology,* **25,** 954–962.

Meltzoff, A. N., & Moore, M. K. (1992). Early imitation within a functional framework: The importance of person identity, movement, and development. *Infant Behavior and Development,* **15,** 479–505.

Meltzoff, A. N., & Moore, M. K. (1994). Imitation, memory, and the representation of persons. *Infant Behavior and Development,* **17,** 83–99.

Meltzoff, A. N., & Moore, M. K. (1995a). A theory of the role of imitation in the emergence of self. In P. Rochat (Ed.), *The self in early infancy: Theory and research* (pp. 73–93). New York: North Holland-Elsevier Science Publishers.

Meltzoff, A. N., & Moore, M. K. (1995b). Infants' understanding of people and things: From body imitation to folk psychology. In J. Bermúdez, A. J. Marcel, & N. Eilan (Eds.), *The body and the self* (pp. 43–69). Cambridge, MA: MIT Press.

Menzel, E. W., & Halperin, S. (1975). Purposive behavior as a basis for objective communication between chimpanzees. *Science,* **189,** 652–654.

Nagell, K., Olguin, R. S., & Tomasello, M. (1993). Processes of social learning in the tool use of chimpanzees (*Pan troglodytes*) and human children (*Homo sapiens*). *Journal of Comparative Psychology,* **107,** 174–186.

Nottebohm, F. (1975). A zoologist's view of some language phenomena with particular emphasis on vocal learning. In E. H. Lenneberg & E. Lenneberg (Eds.), *Foundations of language development* (Vol. 1, pp. 61–103). New York: Academic Press.

Perner, J. (1991). *Understanding the representational mind.* Cambridge, MA: MIT Press.

Piaget, J. (1929). *The child's conception of the world.* New York: Harcourt Brace.

Piaget, J. (1952). *The origins of intelligence in children.* New York: International Universities Press.

Povinelli, D. J., & Eddy, T. J. (in press a). Chimpanzees: Joint visual attention. *Psychological Science.*

Povinelli, D. J., & Eddy, T. J. (in press b). What young chimpanzees know about seeing. *Monographs of the Society for Research in Child Development.*

Povinelli, D. J., Nelson, K. E., & Boysen, S. T. (1990). Inferences about guessing and knowing by chimpanzees (*Pan troglodytes*). *Journal of Comparative Psychology,* **104,** 203–210.

Povinelli, D. J., Nelson, K. E., & Boysen, S. T. (1992). Comprehension of role reversal in chimpanzees: Evidence of empathy? *Animal Behavior,* **43,** 633–640.

Povinelli, D. J., Parks, K. A., & Novak, M. A. (1991). Do rhesus monkeys (*Macaca mulatta*) attribute knowledge and ignorance to others? *Journal of Comparative Psychology,* **105,** 318–325.

Premack, D. (1988). "Does a chimpanzee have a theory of mind?" revisited. In R. W. Byrne & A. Whiten (Eds.), *Machiavellian intelligence: Social expertise and the evolution of intellect in monkeys, apes and humans.* Oxford: Oxford University Press.

Premack, D., & Premack, A. J. (1994). Why animals have neither culture nor history. In T. Ingold (Ed.), *Companion Encyclopedia of Anthropology* (pp. 350–365). New York: Routledge & Kegan Paul.

Premack, D., & Woodruff, G. (1978). Does the chimpanzee have a theory of mind? *Behavioral and Brain Sciences,* **4,** 515–526.

Rast, M., & Meltzoff, A. N. (1995). Memory and representation in young children with

Down syndrome: Exploring deferred imitation and object permanence. *Development and Psychopathology, 7,* 393–407.

Reissland, N. (1988). Neonatal imitation in the first hour of life: Observations in rural Nepal. *Developmental Psychology, 24,* 464–469.

Rogers, S. J., & Pennington, B. F. (1991). A theoretical approach to the deficits in infantile autism. *Development and Psychopathology, 3,* 137–162.

Russon, A. E., & Galdikas, M. F. (1993). Imitation in free-ranging rehabilitant orangutans (*Pongo pygmaeus*). *Journal of Comparative Psychology, 107,* 147–161.

Russon, A. E., & Galdikas, B. M. F. (1995). Constraints on great apes' imitation: Model and action selectivity in rehabilitant orangutan (*Pongo pygmaeus*) imitation. *Journal of Comparative Psychology, 109,* 5–17.

Stern, D. N. (1985). *The interpersonal world of the infant.* New York: Basic Books.

Tomasello, M. (1995). Joint attention as social cognition. In C. Moore & P. Dunham (Eds.), *Joint attention: Its origins and role in development.* Hillsdale, NJ: Erlbaum.

Tomasello, M., & Barton, M. E. (1994). Learning words in nonostensive contexts. *Developmental Psychology, 30,* 639–650.

Tomasello, M., Kruger, A. C., & Ratner, H. H. (1993). Cultural learning. *Behavioral and Brain Sciences, 16,* 495–552.

Tomasello, M., Savage-Rumbaugh, E. S., & Kruger, A. C. (1993). Imitative learning of actions on objects by children, chimpanzees, and enculturated chimpanzees. *Child Development, 64,* 1688–1705.

Trevarthen, C. (1979). Communication and cooperation in early infancy: A description of primary intersubjectivity. In M. Bullowa (Ed.), *Before speech* (pp. 321–347). New York: Cambridge University Press.

Vinter, A. (1986). The role of movement in eliciting early imitations. *Child Development, 57,* 66–71.

Visalberghi, E. & Fragaszy, D. (1990). Do monkeys ape? In S. Parker & K. Gibson (Eds.), *Language and intelligence in monkeys and apes: Comparative developmental perspectives* (pp. 247–273). New York: Cambridge University Press.

Wellman, H. M. (1990). *The child's theory of mind.* Cambridge, MA: MIT Press.

Whiten, A. (Ed.) (1991). *Natural theories of mind: Evolution, development and simulation of everyday mindreading.* Cambridge, MA: Basil Blackwell.

Whiten, A., Custance, D. M., Gómez, J.-C., Teixidor, P., & Bard, K. (in press). Imitative learning of artificial fruit processing in children (*Homo sapiens*) and chimpanzees (*Pan troglodytes*). *Journal of Comparative Psychology.*

Whiten, A., & Ham, R. (1992). On the nature and evolution of imitation in the animal kingdom: Reappraisal of a century of research. In P. Slater, J. Rosenblatt, C. Beer, & M. Milinski (Eds.), *Advances in the study of behavior* (Vol. 21, pp. 239–283). New York: Academic Press.

Wood, D. (1989). Social interaction as tutoring. In M. H. Bornstein & J. S. Bruner (Eds.), *Interaction in human development* (pp. 59–89). Hillsdale, NJ: Erlbaum.

17

Genuine Imitation?

CECILIA M. HEYES

Department of Psychology
University College London
London WC1E 6BT
United Kingdom

Grindley (1932), a student of Conwy Lloyd Morgan, used an apparatus of his own construction (see Fig. 1), four guinea pigs (Jim, Henry, Tom, and Roger), and a bidirectional control procedure to address a pressing question of his time: Can behavior be modified, not only through contact with relationships among stimuli (i.e., by Pavlovian conditioning) but also through exposure to contingencies between actions and their consequences (i.e., by instrumental learning)? He first trained each of his guinea pigs to turn its head either to the left or right when a buzzer sounded by rewarding the animals with a bite of carrot. The time between buzzer onset and head turning decreased over trials, suggesting that the animals had learned something, but it was not clear to Grindley how or what they had learned. The buzzer may have come to elicit head turning simply because the two events were contiguous, and thus head turning may have been strengthened through Pavlovian conditioning. Alternatively, the consequence of the act, access to the carrot, may have been important. The food reward may have "reinforced," retroactively, the association between the buzzer stimulus and head-turning response, or, as we would now think more likely in cases of instrumental learning, the animals may have discovered that head turning was followed by food.

To distinguish the Pavlovian and instrumental hypotheses, Grindley reversed the contingency. After the animals had acquired head turning in one direction, he withheld the reward for that response, and required them to turn their heads in the opposite direction to get the carrot. Under the new contingency, the guinea pigs

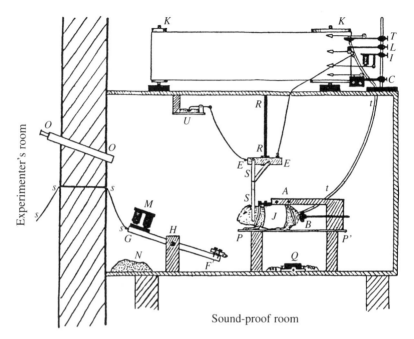

Fig. 1. The bidirectional control apparatus used by Grindley (1932) to demonstrate instrumental learning.

began to turn their heads in the opposite direction, suggesting that head turning was not merely a conditioned response, that it was affected by its consequences.

It occurred to my colleagues and me, working in the same laboratory as had Grindley, but more than 50 years later, that investigators of social learning have a problem similar to the one that Grindley tackled using a bidirectional control. In asking whether animals are capable of observational learning or imitation, in addition to stimulus or local enhancement, students of social learning are, like Grindley, trying to find out whether animals can learn about behavior per se, or whether behavioral change is merely a by-product of learning about events in the environment. Consequently, we developed a bidirectional control procedure to test for imitation in rats.

This chapter reviews briefly the results of our bidirectional control experiments, and then considers in some detail whether they provide evidence of imitation. The purpose of the discussion is not, however, to insist that the bidirectional control evidence of imitation is completely secure. In the course of the last century, most, if not all, putative demonstrations of imitation in animals have been challenged (e.g., Galef, 1988; Visalberghi & Fragaszy, 1992; Whiten & Ham, 1992; Tomasello, this volume; Zentall, this volume; Moore, this volume), usually with good reason. Therefore, no one with an eye to history could be confident that recent studies are decisive. Rather, the present discussion uses data from the bidirectional control experiments as a test case to examine how and why imitation is distinguished from other kinds of social learning, to examine our conception of "genuine imitation."

BIDIRECTIONAL CONTROL STUDIES OF IMITATION IN RATS

Apparatus and Basic Procedure

The bidirectional control procedure is carried out in an operant chamber divided into two parts by a wire-mesh partition (see Fig. 2), and it has two stages. In the first stage, each of a number of hungry observer rats watches a trained rat, a demonstrator, pushing a joystick to the left (toward the viewer of Fig. 2) or to the right (away from the viewer of Fig. 2) 50 times, and each time the demonstrator

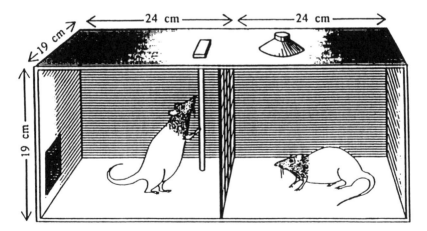

Fig. 2. The bidirectional control apparatus used by Heyes and colleagues to demonstrate observational learning. (Reprinted with permission of the Experimental Psychology Society.)

pushes the joystick, a tone sounds and he receives a food pellet. When the demonstrator has made 50 reinforced responses, he is removed and the observer is immediately placed in the chamber with the joystick. The observers have been magazine training in this compartment, but they have not previously encountered the joystick. During the second stage of the procedure, which begins when the observer is placed in the joystick compartment, and ends when it has pushed the joystick 50 times, the observers are rewarded with food each time they make a response, regardless of its direction.

The (Putative) Imitation Effect

In the basic procedure, observer rats show a reliable tendency to push the joystick in the same direction as their demonstrators; those who observe left pushing, push predominantly to the left, and those who observe right pushing, push predominantly to the right (Heyes & Dawson, 1990; Heyes, Dawson, & Nokes, 1992). The left-hand panel of Fig. 3 provides an example of this effect. It shows that, when all 50 test responses are taken into account, observers of left responding showed a stronger left bias than observers of right responding. In other experiments (e.g., Heyes et al., 1992), the effect has also been found early in the test session, when the observers have made just a few responses.

This result made us think that something interesting was happening, but it was far from conclusive evidence that the observer rats were being influenced by the directionality of the demonstrators' behavior. Clearly, they were not pushing the joystick so that it moved in the same direction within their visual field as it had during observation. If that were the case, then observers would have pushed in the opposite direction to their demonstrators, e.g., observers of left pushing would have pushed more to the right. However, rather than being influenced by the demonstrators' behavior, the observers' bias may have been due to what they saw of the *joysticks'* behavior. For example, they could have been acting such that the joystick moved in the same direction in space, or toward the same part of the chamber, as it had when they watched it before testing. This movement may have become attractive to them, a conditioned reinforcer, because it was immediately followed during the observation phase by food-related cues, the tone and sound of the magazine.

However, the results of two subsequent experiments suggested that, in the bidirectional control procedure, it is the demonstrators' action which influences the observers' behavior. In one of these experiments (Heyes, Jaldow, Nokes, & Dawson, 1994), we tested rats that had observed the joystick moving automatically to the left or to the right, driven by an invisible mechanism. A demonstrator was in the

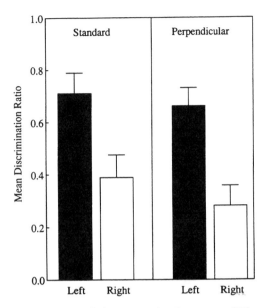

Fig. 3. Mean discrimination ratios [left responses/total responses (50)], and their standard errors, for observers of left and right pushing tested with the joystick in the same position as for the demonstrator (left panel) and when it was displaced to a perpendicular wall between observation and testing (right panel). (From Heyes et al. 1992. Reprinted with permission of the Experimental Psychology Society.)

joystick compartment when the observer witnessed these ghostly movements, but all the demonstrator did was collect the food pellets as they arrived after each joystick displacement. Under these conditions, the observers did not show a systematic directional preference.

In the second experiment, demonstrator rats pushed the joystick during the observation phase, but between observation and testing the joystick was moved to the front wall of the chamber (Heyes et al., 1992, see Fig. 4). Thus, its plane of movement on test was perpendicular to its plane of movement during observation. In spite of this transposition, there was still behavioral concordance between the demonstrators and their observers. For example, observers that had seen demonstrators pushing to the left (L1 in Fig. 4), pushed more to the left (L2) than observers of right pushing (R1), although when an observer pushed to the left, the joystick moved in the opposite direction, in absolute space and relative to cues within the chamber, to that in which it moved when a demonstrator pushed to the left.

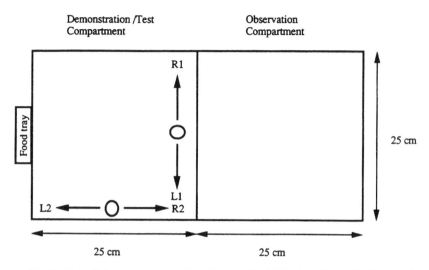

Fig. 4. Plan of the apparatus used by Heyes et al. (1992, Experiment 2), showing the position and plane of movement of the joystick during the observation phase and test phase for Standard groups (L1, R1), and during the test for Perpendicular groups (L2, R2). Responses effecting movement of the joystick toward locations marked L were scored as "left", and those effecting movement towards locations marked R were scored as "right". (Reprinted with permission of the Experimental Psychology Society.)

These data, which are shown in the right-hand panel of Fig. 3, therefore suggest that the rats are not merely learning by observation that food arrives when the joystick moves in a particular direction, or to a particular location. This would be stimulus learning by observation, or "observational autoshaping" (Hogan, 1988), a variety of Pavlovian conditioning. Instead, the subjects appear to be learning by observation to move the joystick in one of two directions relative to the actor's body. At minimum, they seem to be using the demonstrator's body, perhaps its vertical body axis, as a point of reference defining direction of joystick movement, and identifying their own body as being at that reference point on test. This looks to us like response learning by observation; a social-learning equivalent of instrumental learning.

Further Findings

In subsequent experiments we have begun to investigate the psychological processes mediating the bidirectional control effect by examining the conditions in which it occurs. We have found, for example, that it seems to require that the demonstrators are rewarded with food and a tone, but that these stimuli need not be contiguous with the demonstrator's responses; a delay of at least 5 sec is tolerated (Heyes, Jaldow, & Dawson, 1994).

Furthermore, it appears that the direction of a demonstrator's behavior can influence that of an observer's behavior even when the observer has learned to push the joystick prior to observation. For example, observers that have been rewarded consistently for pushing in one direction, say left, learn to push to the right faster if they have observed a right-pushing demonstrator than if they have seen a left-pushing demonstrator immediately before their first test session in which right responses are rewarded (Heyes & Dawson, 1990). In a procedurally similar experiment we found that observation of conspecific responding can facilitate not only acquisition, but also extinction, of a joystick response in rats. When rats are rewarded for pushing in one direction, say left, and then reward is withheld, the response extinguishes faster if they have observed a demonstrator responding to the left without reward than if they have observed a demonstrator responding to the right without reward (Heyes, Jaldow, & Dawson, 1993). We call this effect "observational extinction."

IS THE BIDIRECTIONAL CONTROL EFFECT "IMITATION"?

Definition by Exclusion

In the literature on social learning in animals, imitation has been defined largely by exclusion, by characterizing other forms of social learning and saying "Imitation is *not* that" (e.g., Thorndike, 1898; Thorpe, 1963; Galef, 1988; Whiten & Ham, 1992; Heyes, 1994a; Zentall, this volume). Therefore, in deciding whether the basic bidirectional control effect demonstrated imitation in rats, our first step was to check whether it belonged to any other, established category of social learning. (The history and membership of these categories is discussed in Galef, 1988; Heyes, 1994a; Zentall, this volume.)

Local or Stimulus Enhancement

Observation of the demonstrator pushing the joystick may have drawn the observer's attention to the joystick and thereby resulted in the observers approaching and pushing the joystick sooner than they would have done if they had not seen the demonstrators. However, this kind of local or stimulus enhancement process is not sufficient to explain why the observers pushed the joystick in the same direction as the demonstrators.

Byrne and Tomasello (1995) have suggested that the rats may have learned during observation that the joystick should be moved toward a particular part of the chamber, for example, toward location L1 in Fig. 4, and that in the perpendicular test condition (Heyes et al., 1992) the rats thought they were moving it to L1 when they were in fact pushing it the other way, toward L2. Although it was presented as a local enhancement explanation of the bidirectional control effect, this hypothesis suggests that observational conditioning (see below) is responsible for the effect. Regardless of the label we assign to it, the hypothesis is implausible. As Byrne and Tomasello (in press) pointed out themselves, rats generally have an "excellent sense of space," and, even if they were confused, the hypothesis does not explain why any error that occurred should have been so systematic (Heyes, 1995).

Social Facilitation

The bidirectional control effect does not seem to be an example of social facilitation as that term was used by either Zajonc (1969) or Thorpe (1963). According to Zajonc, social facilitation occurs when the presence of other animals "energizes all responses made salient by the stimulus situation confronting the individual at the moment" (Zajonc 1969). Even if one overlooks the fact that demonstrators are not present when observers are tested in the bidirectional control procedure, a social facilitation account would not explain why the most "salient" response for observers was the response they had seen the demonstrator making.

Thorpe's definition of social facilitation resembles Morgan's (1900) characterization of "instinctive imitation," and refers to phenomena that some authors call "mimesis" or "contagion." It specifies that social facilitation occurs when "the performance of a more or less instinctive pattern of behaviour by one [animal] will tend to act as a releaser for the same behaviour in others" (Thorpe 1963). The bidirectional control effect does not fit this definition because, not only is the demonstrator absent when the observer is tested, but it is highly unlikely that right and left joystick pushing are discrete, instinctive responses, *and* that rats are "prewired" such that the sight of each of these acts is a releaser of the same behavior in the observer. Rather, joystick pushing seems to be just the kind of arbitrary response that Thorndike (1898) urged us to use in studying learning.

Observational Conditioning

The bidirectional effect would be an example of observational conditioning (Cook, Mineka, Wolkenstein, & Laitsch, 1985) if it were due to the observers learning during observation an association between movement of the joystick in a particular direction in absolute space, or relative to features of the operant chamber, and reward. However, our transfer data showed this could not have been the case. When the two were put in opposition, the observers reproduced direction of movement relative to the actor's body, not direction of movement in absolute space or relative to cues inside the chamber.

Matched-Dependent Behavior (or Following)

If the observers in the bidirectional control experiments had been rewarded on test only when they pushed the joystick in the same direction as their demonstrators, the results would have provided an example of matched-dependent behavior (Miller & Dollard, 1941). In fact, the observers were rewarded with food for each joystick response, regardless of direction.

Vocal Mimicry/Copying

Imitation was distinguished from vocal imitation or mimicry by Morgan (1900), Thorndike (1898), and Thorpe (1963) among others. The bidirectional control effect obviously does not involve a vocal behavior, nor does it conform to Miller and Dollard's (1941) characterization of a related phenomenon, copying. According to Miller and Dollard, copying occurs when the demonstrator, or another agent, deliberately rewards responses that are the same as those of the demonstrator and punishes those that are different. In our experiments, the demonstrators were not present when the observers had access to the joysticks, and the experimenters rewarded both same and different responses.

Thus, the bidirectional control effect does not seem to belong to any established category of nonimitative social learning.

Positive Definitions

Although, in practice, imitation has been defined largely by exclusion, as a default category, there have been some influential positive definitions, attempts to say what imitation is, rather than what it is not. Therefore, a second natural step in deciding whether the bidirectional control effect, or any other set of observations, provides evidence of imitation, is to consider how it measures up to these positive definitions.

Thorndike (1898) characterized imitation as learning to do an act from seeing it done, and thus emphasized that, in imitation learning, it is experience of the demonstrator's specific act or action (not his mere presence, or presence in a particular place, or activity of some general kind) that is the critical input, or independent variable, and it is execution of the same act (not location or level of activity) that is the critical output, or dependent variable. The bidirectional control effect seems to meet Thorndike's criteria. Observers of left and right pushing are exposed to an equal extent to the presence of the demonstrator, in the same location, and engaging in the same degree of activity. The variable that makes a difference to the observer is the direction of the demonstrator's action, and the effect that it has on the observer is to bias him in favor of responding in the same direction.

Novelty and Lack of Instinctive Tendency

Thorpe's (1963) definition of imitation was more precautionary than Thorndike's: "By true imitation is meant the copying of a novel or otherwise improbable act or utterance, or some act for which there is clearly no instinctive tendency". Acutely sensitive to the risk of confounding "true imitation" with social facilitation, Thorpe thus tried to indicate in his definition how one would know when an animal had learned an act from seeing it done; the act learned would be novel and noninstinctive. As a consequence, however, there are practical and conceptual problems in deciding whether *any* example of social learning meets Thorpe's criteria for true imitation.

On the conceptual side, Thorpe's insistence that an imitated act have no instinctive tendency is compromised by subsequent developments in evolutionary theory which show that instinctiveness or innateness must be regarded as a degree property e.g., Mayr (1974); Plotkin, & Odling-Smee (1981). Similarly, Thorpe's emphasis on the novelty of imitated behavior apparently calls for a crisp conception of behavioral novelty which neither he, nor any user of his definition, has provided. What, for example, is the critical dimension of novelty: context, magnitude, force, topography? Can a behavior be entirely novel on any or all of these dimensions, or should it be assumed that all behavioral novelty derives from the recombination of existing elements? Even if these questions had been addressed, and answered in a principled way, it would be difficult to apply Thorpe's definition without two further, practical resources: some sort of scale of novelty, and a comprehensive behavioral history for any putative imitator. (See Whiten & Custance, this volume, for a discussion of related concerns.)

Given these problems, all that can be said is that, like any other putative demonstration of imitation, the bidirectional control effect *may* meet Thorpe's criteria. The joystick-pushing response is novel in that, prior to the experiments in which the basic effect was found, the rats had not been given the opportunity to displace a rigid, vertical object, either to the right or to the left. In addition, as noted above, it is implausible that rats are not only equipped with discrete genetic programs for left and right joystick pushing, but that the sight of each is a releaser for the same in the observer.

In their formal definitions of imitation, considered above, Thorndike and Thorpe did not refer to psychological processes. However, many authors, including Thorpe, have suggested that any behavior which conforms to these definitions must be generated by a particular, complex psychological process, including, most prominently, "ideation," "self-consciousness" and "intentionality," or "metarepresentation."

Ideation

Of these, the bidirectional control effect seems most likely to involve ideation. Following Stout (1899), Thorpe defined ideation as "the occurrence of perceptions, in the absence of the corresponding external stimulation, in the form of images which are in some degree abstract or generalised and which can be the subject of further comparison and reorganisation by learning processes" (Thorpe, 1963). Investigation of the mechanisms underlying the bidirectional control effect is only just beginning, but a hypothesis consistent with the current data is: To push the joystick in the same direction as its demonstrator, the observer forms a psychological representation of what it saw the demonstrator doing (since this is based on visual information, it might be described as an image), and then either "reorganizes" this representation, adding kinesthetic and proprioceptive information, such that it is capable of driving motor output—execution of the same behavior by the observer—or "compares" the image of the demonstrator's action with existing motor programs, thus selecting among and modifying the latter so that one of them will effect execution of the same behavior.

Self-Consciousness

In contrast, it is not at all clear that the bidirectional control effect involves something we would want to call self-consciousness. While it is likely to involve representations of bodies and body movements, those of the observer and of the demonstrator, there is no reason to suppose that either the content or the quality of these representations would justify use of the term self-consciousness. As the content or referent of a representation, "my body" and "myself" are rather different, and there is surely no more or less reason to assume that body representations are conscious, than to assume that any other psychological representations are objects of phenomenal awareness (Heyes, 1994b).

Metarepresentation/Intentionality

Similarly, there is no reason to suppose that the bidirectional control effect is "intentional" (Tomasello, this volume) or mediated by metarepresentational processes (Whiten & Ham, 1992), by an appreciation, on the part of the observer, of the beliefs, desires, or goals of the demonstrator. It would seem that the observer must represent the demonstrator's action, retain a mental image of what it saw the demonstrator doing, and it may be necessary for the observer to detect that this action is followed by positive events (e.g., the sound of the food-delivery mechanism, the appearance of a food pellet). However, it does not appear to be necessary for the observer explicitly to represent the demonstrator's knowledge of the relationship between joystick pushing and reward, or the demonstrator's desire for food. I can imitate a windmill, for example, without attributing to it a desire to grind corn.

Cultural Transmission

A final positive definition of imitation refers not to psychological processes that may mediate the phenomenon, but to a population-level process that imitation may itself support: cultural transmission. A number of authors (e.g., Piaget, 1962; Huxley, 1963; Dawkins, 1976; Boyd & Richerson, 1988; Galef, 1992; Tomasello, Kruger, & Ratner, 1993) have suggested that imitation is distinct from other forms of social learning in having the potential to support the nongenetic transmission of behavior in a way that would allow human-like traditions to develop. The bidirectional control effect is a dubious example of imitation when measured against this standard. In free-living animals, the learning processes involved in the effect, whatever they may be, could presumably allow one individual to acquire a behavior from another. However, what is important with respect to culture is the degree to which socially acquired behavior is retained in the absence of explicit reward for that behavior, and, perhaps, when alternative behavioral variants have positive consequences (Heyes, 1993a), and the preliminary indications are that, under these circumstances, the bidirectional control effect is swiftly eliminated. For example, when observers are rewarded on test for pushing the joystick in one direction, say left, those that have observed right responding make more right responses than left observers only at the very beginning of the test session. When 10 responses have been made, the concordance between observers and demonstrators disappears (Heyes, Nokes, & Ray, unpublished).

GENUINE IMITATION

So, what can we conclude about the bidirectionalcontrol effect? To summarize, the foregoing discussion suggests that the bidirectional control effect does not belong in any established category of nonimitative social learning, and conforms to both

Thorndike's definition of imitation and Thorpe's definition in terms of ideation. However, there is no evidence that the effect is mediated by self-consciousness or metarepresentational processes, and it is, inevitably, unclear whether it meets Thorpe's specifications that imitated behavior must be novel and without instinctive tendency.

This mixed bag of results might lead us to conclude that the bidirectional control effect is not genuine imitation if the survey had not also illustrated some serious weaknesses and inconsistencies in our definitions of imitation. Thorpe's formal definition depends on a superceded conception of the innate-acquired distinction and an underspecified dimension of novelty, while the definition by exclusion strategy, and both Thorpe's and Thorndike's "operational" definitions, circumscribe a different class of phenomena from those that characterize imitation with reference to a particular unobservable process. Thus, as the bidirectional control example illustrates, a behavioral phenomenon can both fail to conform to any established category of nonimitative social learning, and meet operational definitions of imitation, and yet fail to imply self-consciousness, metarepresentation/intentionality, or the potential to support cultural transmission.

Evidently, the established, historical definitions cannot be simply combined; for a usable conception of imitation, they must be assigned differential weight or otherwise modified and refined. But how should this be done? Three potential solutions to this problem, outlined in caricature below, have some support in the current literature on social learning.

The Essentialist Solution

The first potential solution is to assign primary significance to the idea that imitation is a variety of social learning involving metarepresentation or second-order intentionality—this is the "essence" of imitation—and attempt to make the other, definition-by-exclusion strategy isolate phenomena that imply intentionality by elaborating new categories of nonimitative social learning. Behavioral phenomena that we are tempted to call imitation, that do not conform to any established type of nonimitative social learning, and that do not imply intentionality, can thus be recovered from limbo and assigned to one of the new categories.

"Response facilitation" (Byrne, 1994), "mimicry" (Tomasello, this volume), and "emulation" (Tomasello, Davis-Dasilva, Camak, & Bard, 1987; Tomasello, this volume) or "goal emulation" (Whiten & Ham, 1992), are among the new categories of social learning apparently coined with this solution in mind. For example, Galef, Manzig, and Field's (1986) evidence of imitation in budgerigars has been ascribed to "response facilitation" (Byrne & Tomasello, 1995; Heyes, 1995), several potential examples of imitation in nonenculturated chimpanzees have been put down to mimicry (Tomasello, this volume), and it has been suggested that the bidirectional control effect is an example of emulation (Whiten & Ham, 1992; Byrne & Tomasello, 1995).

In general terms, emulation occurs when "the learner observes and understands a change of state in the world produced by the manipulations of another" (Tomasello, this volume), and in the case of the bidirectional control effect, Byrne and Tomasello suggest that "the joystick itself, and its position relative to a wall (any wall) is used as a landmark for orientation. Then, on the emulation explanation, the observer notes the position of the stick and how it moves relative to the wire grid wall and then transfers that orientation to the joystick in its new position relative to the new wall it is up against."

Postponing, briefly, consideration of whether this is a satisfactory account of the bidirectional control effect, it should be noted that the "essentialist" solution has several substantial virtues. First, it is decisive; it recognizes that our conception of imitation is confused and tries to do something about it. Second, this solution focuses attention on one of the most interesting questions about imitation: what psychological processes does it involve? Third, the policy of seeking new types of nonimitative social learning could stimulate empirical and theoretical developments.

The problem with the essentialist solution is that it is liable to stifle empirical research, and isolate theoretical developments from empirical input. This is a risk because it identifies as the essence of imitation a process—metarepresentation or second-order intentionality—which, in practice and possibly in principle, cannot be identified empirically in nonlinguistic animals (Heyes, 1993b, 1994c); a metaphysical rather than a theoretical process. Consequently, if imitation is defined by this process, the decision to treat a behavioral phenomenon as imitation rather than, for example, mimicry or emulation, depends on guesswork rather than empirical enquiry. One can only size up the animals in question, ponder the act they are performing, and see whether, intuitively, it seems likely that the observer attributed a goal or mental state to the demonstrator. Most intuitions dictate that this *is* likely in the case of apes, especially enculturated ones, and *not* in the case of budgies, rats, and other species that are distantly related to humans.

The problem is, to some extent, exemplified by Byrne and Tomasello's emulation account of the bidirectional control effect. It is unclear not only how the emulation hypothesis could account for the effect (joystick movement was parallel to the nearest wall for observers of both left and right pushing, yet the test performance of these groups was different), but also how the hypothesis could be tested against our own, imitation interpretation. However, the problem is deeper than this example suggests. Even if it were possible to show empirically that the bidirectional control effect, or any other prima facie example of imitation, is *not* emulation, mimicry, response facilitation, or of an established type of nonimitative social learning, this eliminative achievement would not be sufficient to imply that the phenomenon in question is mediated by metarepresentation or intentionality. There could be an as-yet-unidentified category or categories of nonimitative, non-intentional, nonmetarepresentational social learning to which it belongs.

The Positivist (No-Nonsense) Solution

In contrast with the essentialist solution, the positivist solution is to assign primary significance to operational definitions of imitation, especially Thorpe's (1963), and attempt to overcome any opacity in these definitions by assuming that, in this context, "novel" behavior means "topographically novel" behavior. Whether the phenomena thus circumscribed as imitation imply ideation, self-consciousness, metarepresentation, or any other psychological process is of lesser importance. Researchers who seem to favor something of this no-nonsense kind include Galef (1988), Zentall (this volume), and Moore (this volume).

The great strength of the positivist solution is that it emphasizes the importance of high-quality empirical work. If imitation is distinguished from other forms of social learning in terms of observable conditions, then it is clear that, to find out which animals can imitate (if any), and under what circumstances, it is necessary to conduct carefully designed experiments. However, the positivist solution is weak on reasons; it is not clear why imitation should be understood to involve the reproduction of topographically novel behavior, and why, if it is not understood to signify complex psychological processing or the potential for culture, imitation is of scientific interest.

The bidirectional control experiments can be used, once more, to illustrate these weaknesses. Imagine that further studies had revealed that, in addition to pushing the joystick in the same direction as their demonstrators, observer rats reproduce the topography of the demonstrator's behavior: use the same forelimb, bend it at the same angle, crouch if the demonstrator crouched, and rear if he reared. What would these further studies have added to our understanding of the psychology of animals? They would no more suggest self-consciousness than the current effect because it would still be "bodies," rather than more inclusive "selves," that needed to be represented, and there is no reason to suppose that a highly specific body representation is more likely to be represented consciously than a gross body representation. Similarly, a more precise topographic match between observers and demonstrators would not bring us any closer to knowing whether rats are capable of metarepresentation. Mental state attribution is no more necessary for precise than for gross behavioral matching. Furthermore, data of this kind apparently would not secure a role for this type of social learning in cultural transmission because even a behavior that is acquired through precise and faithful copying may be lost through relearning before there has been an opportunity for retransmission.

Of course, an advocate of the positivist solution need not argue that topographic matching would imply self-consciousness, metarepresentation, or a connection with culture. However, surely there should be *some* rationale for defining imitation in terms of topographically novel behavior, and an argument or evidence linking it with underlying psychological mechanisms would be one such reason.

It may be claimed, in the form of an alternative rationale, that imitation should be understood to refer to copying of topographically novel behavior because it is only when the behavior is topographically novel that one can be sure that the animal has learned to do the act from seeing it done (Thorndike, 1898). This rationale seems to be faithful to Thorpe's purpose in emphasizing novelty, but it begs two questions: Why does topographic matching of a novel response necessarily provide more decisive evidence that Thorndike's requirement has been met than, for example, directional matching of a novel behavior or topographic matching or a familiar response? Why is behavior that meets Thorndike's criteria of scientific interest?

The Realist Solution

The realist solution, like the essentialist solution, acknowledges that, for most contemporary psychologists, the purpose of studying observable behavior is to find out about unobservable psychological processes. However, in sympathy with the no-nonsense or positivist approach, it emphasizes that the unobservable entities and processes we postulate to explain behavior should be theoretical rather than metaphysical; it should be possible to test hypotheses about them. (See Brody, 1972; Boyd, 1983; Hull, 1984, for more general discussion of the validity of essentialist, positivist, and realist perspectives in science.)

More specifically, the realist approach, which I favor, consists of adhering to Thorndike's definition of imitation, the oldest scientific definition, and acknowledging that the social-learning phenomena it circumscribes are of interest because they are likely to be mediated by complex, as-yet-underspecified psychological processes. Theorizing about these, about the not-directly-observable mechanisms underlying imitation, is important, but the hypotheses must be testable. Of the theoretical discussions in the existing literature, Thorpe's (1963) characterization of imitation as a form of ideation seems most likely to yield testable hypotheses. The tests themselves can be conducted using any species and procedure that has provided evidence of imitation according to Thorndike's definition.

Ultimately, however, if the realist solution is successful in stimulating empirical work and enabling us to find out more about the psychology of social learning, it is likely that Thorndike's definition will be superceded by one that specifies a process, and some behavioral phenomena that conformed to Thorndike's definition will be reassigned to a nonimitative category. If it has not already been found to fall short of Thorndike's definition, the bidirectional control effect might well be one of these. However, in the meantime, bidirectional control procedures may be valuable instruments in the investigation of imitation.

REFERENCES

Boyd, R. (1983). On the current status of scientific realism. *Erkenntnis, 19,* 45–90.

Boyd, R. and Richerson, P. J. (1988). An evolutionary model of social learning: The effects of spatial and temporal variation. In T. R. Zentall & B. G. Galef (eds.), *Social learning: Psychological and biological perspectives* (pp. 29–48). Hillsdale, NJ: Erlbaum.

Brody, B. A. (1972). Towards an Aristotelian theory of scientific explanation. *Philosophy of Science, 39,* 20–31.

Byrne, R. W. (1994). The evolution of intelligence. In P. J. B. Slater & T. R. Halliday (Eds.), *Behaviour and evolution* (pp. 223–265). Cambridge: Cambridge University Press.

Byrne, R. W., & Tomasello, M. (1995). Do rats ape? *Animal Behaviour, 50,* 1417–1420.

Cook, M., Mineka, S., Wolkenstein, B., & Laitsch, K. (1985). Observational conditioning of snake fear in unrelated rhesus monkeys. *Journal of Abnormal Psychology, 93,* 355–372.

Dawkins, R. (1976). *The Selfish Gene.* Oxford: Oxford University Press.

Galef, B. G. (1988). Imitation in animals: History, definition and interpretation of data from the psychological laboratory. In T. R. Zentall & B. G. Galef (Eds.), *Social learning: Psychological and biological perspectives* (pp. 3–28). Hillsdale, NJ: Erlbaum.

Galef, B. G. (1992). The question of animal culture. *Human Nature, 3,* 157–178.

Galef, B. G., Manzig, L. A., & Field, R. M. (1986). Imitation learning in budgerigars: Dawson and Foss (1965) revisited. *Behavioral Processes, 13,* 191–202.

Grindley, G. C. (1932). The formation of a simple habit in guinea pigs. *British Journal of Psychology, 23,* 127–147.

Heyes, C. M. (1993a). Imitation, culture and cognition. *Animal Behavior, 46,* 999–1010.

Heyes, C. M. (1993b). Anecdotes, training, trapping and triangulating: Can animals attribute mental states. *Animal Behavior, 46,* 177–188.

Heyes, C. M. (1994a). Social learning in animals: Categories and mechanisms. *Biological Reviews, 69,* 207–231.

Heyes, C. M. (1994b). Reflections on self-recognition in primates. *Animal Behavior, 47,* 909–919.

Heyes, C. M. (1994c). Social cognition in primates. In N. J. Mackintosh (Ed.), *Animal Learning and Cognition* (pp. 281–305). New York: Academic Press.

Heyes, C. M. (1995). Imitation and flattery: a reply to Byrne and Tomasello. *Animal Behavior, 50,* 1421–1424.

Heyes, C. M., & Dawson, G. R. (1990). A demonstration of observational learning using a bidirectional control. *Quarterly Journal of Experimental Psychology, 42B,* 59–71.

Heyes, C. M., Dawson, G. R., & Nokes, T. (1992). Imitation in rats: Initial responding and transfer evidence. *Quarterly Journal of Experimental Psychology, 45B,* 81–92.

Heyes, C. M., Jaldow, E., & Dawson, G. R. (1993). Observational extinction: Observation of nonreinforced responding reduces resistance to extinction in rats. *Animal Learning and Behavior, 21,* 221–225.

Heyes, C. M., Jaldow, E., & Dawson, G. R. (1994). Imitation in rats: Conditions of occurrence in a bidirectional control procedure. *Learning and Motivation, 25,* 276–287.

Heyes, C. M., Jaldow, E., Nokes, T., & Dawson, G. R. (1994). Imitation in rats: The role of demonstrator action. *Behavioral Processes, 32,* 173–182.

Heyes, C. M., Nokes, T., and Ray, E. (unpub.). Imitation in rats: Observational learning can both facilitate and retard subsequent individual learning.

Hogan, D. E. (1988). Learned imitation by pigeons. In T. R. Zentall & B. G. Galef (Eds.), *Social learning: Psychological and biological perspectives* (pp. 3–28). Hillsdale, NJ: Erlbaum.

Hull, D. L. (1984). Contemporary systematic philosophies. In E. Sober (Ed.), *Conceptual issues in evolutionary biology* (pp. 567–602). Cambridge, MA: Bradford Books, MIT Press.

Huxley, J. (1963). *Evolution: The modern synthesis.* New York: Hafner.

Mayr, E. (1974). Behavioral programs and evolutionary strategies. *American Scientist, 62,* 650–659.

Miller, N. E., & Dollard, J. (1941). *Social learning and imitation.* New Haven CT: Yale University Press.

Morgan, C. Lloyd (1900). *Animal behaviour.* London: Edward Arnold.

Piaget, J. (1962). *Play, dreams and imitation in childhood.* New York: Norton.

Plotkin, H. C., & Odling-Smee, F. J. (1981). A multiple-level model of evolution and its implications for sociobiology. *Behavioral and Brain Sciences, 4,* 225–268.

Stout, G. F. (1899). *A manual of psychology.* London.

Thorndike, E. L. (1898). Animal intelligence. *Psychological Review Monographs, 2,* No. 8.

Thorpe, W. H. (1963). *Learning and instinct in animals.* London: Methuen.

Tomasello, M., Davis-Dasilva, Camak, L., & Bard, K. (1987). Observational learning of tool-use by young chimpanzees. *Human Evolution, 2,* 175–183.

Tomasello, M., Kruger, A. C., & Ratner, H. H. (1993). Cultural learning. *Behavioral and Brain Sciences,* **16,** 495–592.

Visalberghi, E., & Fragaszy, D. M. (1992). Do monkeys ape? In S. Parker & K. Gibson (Eds.), *Language and intelligence in monkeys and apes* (pp. 247–273). New York: Cambridge University Press.

Whiten, A., & Ham, R. (1992). On the nature and evolution of imitation in the animal kingdom: reappraisal of a century of research. *Advances in the Study of Behavior,* **21,** 239–283.

Zajonc, R. B. (1969). Coaction. In R. B. Zajonc (Ed.), *Animal social psychology* New York: Wiley.

Author Index

Numbers in italics refer to the pages on which the complete references are listed.

Subject Index